AF873128

JOHN ASHBURNER

NOTES AND STUDIES IN THE PHILOSOPHY OF ANIMAL MAGNETISM AND SPIRITUALISM

Elibron Classics
www.elibron.com

Elibron Classics series.

© 2004 Adamant Media Corporation.

ISBN 1-4021-9413-7 (paperback)
ISBN 1-4021-0076-0 (hardcover)

This Elibron Classics Replica Edition is an unabridged facsimile
of the edition published in 1867 by H. Baillière,
London.

Elibron and Elibron Classics are trademarks of
Adamant Media Corporation. All rights reserved.

This book is an accurate reproduction of the original. Any marks, names, colophons, imprints, logos or other symbols or identifiers that appear on or in this book, except for those of Adamant Media Corporation and BookSurge, LLC, are used only for historical reference and accuracy and are not meant to designate origin or imply any sponsorship by or license from any third party.

NOTES AND STUDIES

IN THE

PHILOSOPHY

OF

ANIMAL MAGNETISM

AND

SPIRITUALISM.

WITH OBSERVATIONS UPON

CATARRH, BRONCHITIS, RHEUMATISM, GOUT, SCROFULA, AND COGNATE DISEASES.

BY

JOHN ASHBURNER, M.D.

Member of the Royal Irish Academy.

LONDON:
H. BAILLIÈRE, 219 REGENT STREET,

NEW YORK: BAILLIÈRE BROTHERS, BROADWAY. | MELBOURNE: F. F. BAILLIÈRE, COLLINS STREET EAST.

PARIS: J. B. BAILLIÈRE, RUE HAUTEFEUILLE,

MADRID: C. BAILLY BAILLIÈRE, PLAZA DEL PRINCIPE D'ALFONSO.

1867.

" Jesus saith unto him, I am the way, the truth, and the life. No man cometh unto the Father but by me.

" Even the Spirit of truth, whom the world cannot receive, because it seeth him not, neither knoweth him : but ye know him : for he dwelleth with you, and shall be in you."—ST. JOHN, Ch. 14, v. 6, 17.

" But when the Comforter is come, whom I will send unto you from the Father: even the Spirit of truth, which proceedeth from the Father, he shall testify of me."—ST. JOHN, Ch. 15. v. 26.

" Magna est veritas et prevalebit."

DEDICATION.

This Work is, with profound respect, dedicated to an exalted scientific character; who amidst the absurdities of material speculation, dares to conform himself to the rules and exercises of a holy life. He has his reward: for in his habits, in his manners, in his countenance, beams forth the spirit of wisdom and content. He has his reward: for God has blessed his scientific labours by the discovery of the Magnetoscope. All hail!

JOHN OBADIAH NEWELL RUTTER.

Thy name will shed a proud lustre around the Memory of the Civil Engineer at the Black Rock, Brighton.

London, 1867.

INTRODUCTION.

If Animal Magnetism had in our own times, been received by the scientific world with the dignity and patient love of investigation becoming the character of men of extensive knowledge, and philosophic power of thought, or even of high conscientious professional feeling; we might have hoped for some decent attention to the series of truths which we find to be leading us on step by step to the highest pursuits that can engage the studious attention of man.

Reflecting on the extensive view opened of the religious aspects of our science, and its intimate connection with the philosophy of that code of morals which constitutes our rule of life, one could not but be impelled to make this a chief consideration in the treatment of it. But when it was seen how apparently heterogeneous were the materials with which we should have to deal, it must be confessed that it was evident that a very long time was requisite for the composition and elaboration of our work. This has operated as a strong motive for throwing the materials into the crude form of essays or chapters, and has allowed liberty to the author, to be more discursive than he might otherwise have been.

Some of these chapters would appear to have a very slight alliance with others, but being, in truth, portions of the whole subject, they must be regarded only as disjointed links of a chain that require time and a few more links for their perfect catenation.

The chapters begin with some of the religious relations of Animal Magnetism. We proceed naturally to our views of Forces, and to point out the trains in which Natural, or Physical Forces proceed; we may simply remark that Magnetism must be held to embrace attraction and repulsion, and that these form the elementary ideas of all the numerous forces in nature. We hardly need, then, to say that all change is necessarily dependent on these forces—that no chemical compositions, or decompositions can take place without them—that they regulate the great orbs in space, as well as the form of the minutest of the primitive crystalline globules, of which every crystal in existence is built up. Being convinced that matter is completely inert, it follows that no form or shape can possibly exist without the interposition of some force or forces. The simplest manifestation of force in the configuration of shape is that in which attraction and repulsion are simultaneous agents in the production of minute globules, obeying a law of cohesion emanating from Magnetism, a law of attraction that embraces each globule and obliges it to lie in apposition to its neighbour, according to a law of axial and equatorial polarity. In vegetable existence it determines a law of evolution, when it decrees the folding up of embryonic forces in those minute spherules or germ cells, which develope vegetable crystals. It must be borne in mind that we are still in presence of a magnetic law of crystallization which determines the existence of magnetic polarities. Proceeding with these laws, we observe the law of evolution regulating more complicated germ cells in animal existence, but still obedient to magnetic laws of polarity. Thus, are the evolutions of forces to be traced from the lowest grades of mineral crystal force by successive gradations up to that magnetic vital force culminating finally in the magnet, or grand crystal, Man; who remains subject even in the operations of his mind, to an inexorable magnetic law, acting through his phrenological organs, either from the force of his own soul, or from the force of the will of another. The forces here indicated can be proved to exist by not only legitimate logical inferences, but by demonstrable facts founded on experiments repeatedly performed by the investigators of the science of Animal Magnetism—who are witnesses to an incessant pouring out of new truths of the most captivating nature.

We must in the proper place occupy ourselves with important considerations on the magnetic influences exerted upon the human being from extraneous sources, and devote especial attention to the transmission of the influences from one human being through the frame of another. These may be found to relate to the operation of forces exerted by agencies in the atmosphere upon our brain and nervous system; constituting much more completely, than philosophers have yet been able to see, the motives to human action. Without this power of acting by and through forces external to ourselves, we should be at a loss to account for the marvellous phenomena that are revealed to us by the magnetoscope of Mr. Rutter. I know that I do not over-estimate the importance of this philosophical instrument, when I declare my conviction, that in the hands of some man, who shall be so far capable of the virtue of self-abnegation as to ensure the perfect government of his corporeal fabric, it is destined to point out the path by which the accomplishment of the higher moral and intellectual duties of mankind may be attained.

It may be considered to be a startling proposition, but we have no doubt it may be safely predicated that one of the results of Mr. Rutter's great discovery will be to link Psychology with Mathematics.

The connection between the force of Animal Magnetism, and the phenomena of pain, the varieties of sleep and wakefulness, health and disease, will, it is hoped, be made clear in the ensuing chapters, together with the connection between sleep and somnambulism, the true philosophy of the power of the human will, the explanation and illustration of dreams, and then will glide naturally into the phenomena of a most extensive philosophy, which has obtained the name of Spiritualism; but which rests on Animal Magnetism as its foundation. We shall devote one chapter to the history of the Introduction of this vast subject into England. It is a history which redounds to the honour of the names of John Elliotson, Richard Chevenix, and of my old colleague at the Middlesex Hospital, honest and good John Wilson; for without doubt, it is to them that its introduction is mainly to be attributed. We shall have opportunities of devoting chapters to the startling facts connected with the phenomena of Animal Magnetism; and the narration of remarkable cases may occupy space enough to make our work long, although it may be hoped not dull and dreary.

To classify and reduce to a system the whole of the series of facts preserved in the literature of our subject, would be the labour of a long life. We must be content to show its connection with several of the diseases which afflict humanity.

We propose to show that Animal Magnetism is a force which can alleviate the most agonising suffering—which can cure painful cancers, and other malignant diseases—which without endangering life, can render the patient insensible to pain under the surgeon's knife, and in the throes of a painful labour, which can raise up from the deep insensibility of the last stage of typhus fever, a dying patient—and which can do far more than all this, viz.: cure by its vital and healing efficacy the frightful suffering arising from an attack of gout in the peritoneal membrane of the abdomen, for which, I was told, that fourteen of the most eminent physicians and surgeons had failed to suggest any means of relief. May we not assert then, that through the agency of this force, God has enabled man to wield a power almost divine?

We may safely aver that, notwithstanding all the obstructive opposition offered to Animal Magnetism, no train of facts, properly catenated, and logically generalised, can legitimately lead to conclusions more important or more philosophical.

The highest grade of the force to which we affix the name or title of Animal Magnetism, is the magnetic force that resides in the magnet Man. Magnetism being the great trunk force of nature, or Sir Isaac Newton's force of Universal Gravitation, embracing Centripetal and Centrifugal Forces, or the antagonisms of attraction and repulsion; we may assume as a general proposition, that all the forces in nature, being reducible to these two antagonistic powers, are in fact but emanations in various series of modifications from the great primal force:—We attribute then all the forms and shapes in nature to the numerous series of gradations of forces, which it has pleased the Almighty Planner of the Univere to create as his agents, and by which His Will is manifested.

When we reflect on the fact that all matter is quite inert, or passive, one cannot help inferring the necessity of the operations of forces, in order that the forms and shapes of matter may exist. We must admit that the first operation of antagonistic forces is to whirl matter into the globular or spheroidal form. The simple example of this shape, is the spheroidal globule, or drop of water, and which, in fact, is the primitive form of mineral crystal.

This elementary form of mineral crystalline aggregation, is the primitive form of the vegetable crystal, and also of the animal crystal. Thus, we have an illimitable series of forms or shapes of every variety, fashioned each by its own peculiar governing, or regulating force. Nor is this a fancy of the imagination. In 1819, it was simply an hypothesis, based on strong analogies. Since the discovery of the Magnetoscope, which has established the fact that crystals, vegetables, and animals are each and all regulated by a law analogous to Magnetism, each being gifted with undeniable polarities, and each, therefore, being subject to magnetic law; we have a right to infer that the law of crystallization, under various modifications, is one of the general laws of nature. In other words, that the governing, regulating, all-pervading force of nature is magnetic, which endows all matter with shape, and which controls all forms of matter to become obedient in their aggregations to a mathematical law of geometrical crystalline series of forms. If we be capable of reflecting, even a little, on what is known of the laws of magnetism; we may feel assured that no denizen of our earth, which is acknowledged to be a magnet, can escape from the influence of the laws of magnetism.

When we place the question of forces on the simple, but very broad basis of all physical laws being regulated, under countless modifications, by the antagonistic principles of attraction and repulsion, we do not the less admire the infinite complications and variations which these active regulating principles engender, by and through the comprehensive force of God's will throughout his *φυσις*, or domain of nature.

We need not go further back for a refutation of the prevailing materialistic doctrines, than to the first half of the 18th century. Hume and the men of his school ruled the realms of mental science, and works were produced which magnetised men's minds to the implicit belief in the supremacy of matter. Amidst the controversies, rife at that period, there arose a man of the name of Andrew Baxter, one of whose works, "An Inquiry into the Nature of the Soul," I recommend to the study of those who find it difficult to reconcile the facts of Animal Magnetism with the materialistic notions of the present day. Men are now as obstinate in their adherence to the dogma that brain thinks, as they can be to that of the existence of a God. Yet the two propositions are diametrically opposed to each other. It is for want of perceiving this antagonism that many blunders have been committed in reasoning on the phenomena of Animal Magnetism.

Atheism and Animal Magnetism are quite incompatible and I propose to show that without the facts of Animal Magnetism, it would be impossible to sustain the doctrine of the existence of a Deity.

Matter is essentially inert, or passive. The brain is matter, and cannot possibly act without force urging it to act; for like all other matter, when denuded of magnetic or vital force, it is dead and unable of itself to move, much less to think. The advocates of the materialistic doctrine of the nervous system, must therefore ultimately abandon their dogma, that brain matter is the basis of the thinking power of man.

If the power exists, the brain thinks in virtue of that power. The process of life is renewed at each inspiration, and the man acquires continually new force, which is propagated to the brain, and this constitutes the power superadded to that organ. How is this proved? By the fact that the brain is enabled to think, and that thinking, it is enabled to will.

Before we can enter fully on our theme, we must endeavour to explain the vast amount of importance we attach to that which is perhaps the greatest discovery in electricity or magnetism, since the brilliant period when Sir Humphrey Davy shed such lustre on the annals of English Science.

Mr. Rutter's discovery of the Magnetoscope is, essentially, a corroboration of all the labours of Von Reichenbach, for it establishes the fact that all crystals are magnets, and accounts for Von Reichenbach's phenomena arising from the opposite poles of a principal axis. He goes further, and tells why the delighted Endlicher saw magnetic light issuing from vegetable structures; simply because they were living magnets. Mr. Rutter's crowning discovery was that man is a magnet, made up of many magnets; and not only can all this be demonstrated, but we are led to know that we are each of us surrounded by a magnetic sphere of force. This last fact is perhaps the most important of all; for it is our medium of communication with the spirits of our former friends. The phenomena developed by the Magnetoscope will be eagerly investigated, when men become acquainted with its facts, and learn its history. Simple as was its origin, it required the mind of a man of genius to perceive the analogies its phenomena bore to those of magnetism. Mr. Rutter was first interested in the little toy made by attaching a ring or a piece of sealing wax to a filament of silk. He observed that when the silk was held between the fore-finger and thumb, and the ring or wax, suspended over a piece of glass, or about half-an-inch from the bottom of a large tumbler, it assumed, under different circumstances, different movements, some of which were rotations, and others oscillations. A law was soon traced which established, in his mind, that the phenomena were purely magnetic. The great difficulty which occurred to him, was the arrangement of the pendulum, so as to produce an unerring instrument; for it was natural to suppose that while the silk filament was held between the finger and thumb, the movements were liable to many sources of fallacy.

After a multitude of trials, he succeeded in perfecting his Magnetoscope. It became, in his hands, a most marvellous philosophical instrument. Placing a good bar magnet on a table, so as to be in the magnetic meridian, he stimulated the pendulum to move, by embracing, with the forefinger and thumb of the right hand, a brass ball at the summit of a pillar, from whence proceeded, at a right angle, a rod at the end of which the pendulum was suspended over a circular piece of plate glass; the movement generated was the normal rotation. Then placing his open left hand a short distance from the north pole of the magnet, within the range of the magnetic curves, the pendulum, in a few seconds, ceased to rotate, and came to a dead stop. Then moving his left hand two or three inches beyond the south pole, the pendulum recommenced its movement of direct rotation. At the same distance on the east side, there was direct rotation. When on the west side, inverse rotation. And when the hand was brought back to the curves at the north pole, the pendulum again came to a dead stop. Dead animal matter, a piece of ivory, for instance, or a dead fly, had precisely the same effect in arresting the movement. The description of what I have witnessed, on very numerous occasions, of the phenomena attendent on manipulation with this highly important instrument, would alone fill a volume; I must therefore refer my readers to Mr. Rutter's admirable little work, entitled "Animal Electricity," published by John W. Parker and Son, West Strand—it is full of interesting matter. When the world becomes more enlightened on the subject of Animal Magnetism, and more acquainted with the laws which govern our being, Mr. Rutter's discoveries will rank among the most important of the age in which he lived.

No one who has studied music, can be ignorant that the gamut is a scale adapted to every form of musical composition, and consequently that seven is the number selected for the chords of the diapason. Seven also is the number of the movements of Mr. Rutter's Magnetoscope. He at first observed only six movements. The direct rotation; the inverse rotation; the axial, and equatorial oscillations; and the two diagonal oscillations. But Dr. Leger, who was an accomplished musician, discovered the seventh movement. It is an elipsoid, and under normal law, always describes its course in a direction from north to south; but there are cases in which the normal law does not obtain, and in these, the elipsoid described, is from east to west. These considerations lead to reflections on the harmonies of nature, and prove to us, that Animal Magnetism, whatever certain members of the scientific world may have been pleased to think of it, is a science, not only well worthy of our most assiduous study, but that it explains phenomena which have puzzled the greatest thinkers of the world.

With all our admiration for the genius and exalted character of Mr. Rutter, it may be questioned whether his discoveries would have yet attained their high importance, had it not been for an interposition of Providence in directing the steps of an ardent but a modest and humble man of great scientific acquirement to seek his fortune in London. This was a Parisian physician, named Theodore Leger. He had passed through years of strange adventure in Mexico, as well as in the United States. He had profoundly studied Animal Magnetism, and had published at New York an interesting work on the subject, under the quaint title of Psycho-Dunamy. He was an assiduous student, and no shallow philosopher, and when the great facts discovered by Rutter were explained to him, he plunged at once into the same train of inquiry; with an admiration for the genius of his predecessor which evinced a complete appreciation of the immense value of the discovery. Leger was not long in perceiving that the Magnetoscope offered vast developments of truths that would shed their lustre over the age in which he and his master lived; over the obtuse materialistic nineteenth century, which had repudiated the truth acknowledged by the great Van

Helmont, and since re-indicated by Mesmer, and De Puysegur!

How much has the world lost by the death of poor Dr. Leger!

It may safely be predicated, that he who will pursue the same course of investigation will attain a celebrity second to none of his time—and a due consideration of the facts which were poured out upon the world by Dr. Leger, will show, that in this statement, there is not the slightest exaggeration.

We have gone over the ground which has led us to the conclusion that is adopted on the faith of the correctness of our theory of the infinite gradations of force, that all force is not only conservative, but creative, that is to say, that a part of the Almighty Will accompanies every manifestation of force.

We have seen that all matter is arranged in series; that it is subject to forces; and that these are as completely subject to the law of series, as the inert or passive matter they are destined to control. They do not only control it, but in the higher conditions of organization, they are called upon to control more than the arrangements of form they assist in fashioning. What is man's will? How is it often exerted? We may be told man's will is not a force. We contend that it is not only a force but a mechanical motive power. We may refer to a note at page 32 of the English translation of Reichenbach's Researches, published by Baillière, to show that this point is not for the first time insisted on here. It is the result of experiments as irrefragable as any ever made in physics; and the argument is as firmly based in logic as any train of facts warranting any conclusion in Natural Philosophy. The human will is a motive power or force, and the corollaries flowing from this established proposition, are very numerous.

In our general introductory review of the topics which will occupy us in the course of the succeeding pages, if we have not quite specifically enough alluded to the grand magnetic subject, which ties all the forces in nature together, we may be excused on the plea, that we as yet only know that the Almighty Being has directed His angels to give us the conviction, that they communicate with us by agencies external to our souls, and as clearly independent of ourselves as are our fellow-beings who communicate with us upon the earth; we need add no more than that spiritualism is launched into our midst, as completely as the means of intercommunication by railroads, and as the intercourse between different parts of the world by direct and inverse electric currents. If in this work we introduce to our fellow-beings the laws which really regulate health and disease, we shall have accomplished our self-imposed task, in an effort to spread the knowledge of Animal Magnetism.

There remains for me the grateful task of expressing my acknowledgment of the great kindness of those friends who have so willingly rendered me their assistance in preparing this work for the press, and in watching it while passing through it; labour which loss of sight has precluded me from undertaking for myself. I have only to add that if I have in any case drawn a wrong inference from the words or conduct of any of those on whom I have been impelled to make some, perhaps too severe, remarks; I beg to express my profound regret, as my sole object has been to serve the cause of truth.

TABLE OF CONTENTS.

CHAPTER V.

DR. LEGER, HIS FACTS.

CHAPTER VI.

PHENOMENA OF PAIN, AND THEIR CAUSES.

CHAPTER VII.

SLEEP, WAKEFULNESS, TONE, CLONE.

CHAPTER VIII.

A REITERATION ON SLEEP.

CHAPTER IX.

CHENEVIX, ELLIOTSON, AND WILSON.

CHAPTER X.

VAN HELMONT.

CHAPTER XI.

CASES TREATED BY ANIMAL MAGNETISM.

CHAPTER XII.

PHENOMENA OF ANIMAL MAGNETISM.

CHAPTER XIII.

PHENOMENA OF ANIMAL MAGNETISM—JANE MURREL.

CHAPTER XIV.

PHENOMENA OF ANIMAL MAGNETISM, SPIRITUALISM, CLAIRVOYANCE, PREVISION, HUMAN WILL, &c.

CHAPTER XV.

PHENOMENA OF ANIMAL MAGNETISM, SOMNAMBULISM, PREVISION AND CLAIRVOYANCE.

CHAPTER XVI.

PHILOSOPHY OF RHEUMATISM.

ON THE PHILOSOPHY

OF

ANIMAL MAGNETISM.

CHAPTER I.

PRELIMINARY OBSERVATIONS.

We are at a loss to explain the difficulties which men have experienced in accounting for Gout, Rheumatism, Scrofula, and other cognate diseases, so simple as the remote causes of all these maladies have for a long series of years appeared to us. We date our acquaintance with the causes of these evils since the days when our dear old friend and master, Professor Macartney, first taught us the difference between venous fullness and arterial plethora. Macartney, whose mind was naturally logical, would never allow his pupils to use the word *congestion* as applied to an artery, or *plethora* as applied to a vein. This is the origin of all the severe distinctions I have felt myself bound to insist upon in this book.

We enter upon the elements of our enquiries. What are the distinctions between rigors and tremblings? We know that rigors are shiverings; but shiverings are different from tremblings. In what respects do they differ? Shiverings are accompanied by a sense of cold, while tremblings may occur when a sensation of oppressive heat is present. This may appear a very unimportant distinction. In truth, however, it is vastly important. For rigors occur when the arteries are almost empty and consequently when the veins are very full;

but, on the other hand, tremblings may and do occur, when we feel oppressed from a state of heat, in which the arteries are loaded with blood. This distinction can well be illustrated by numerous facts. Our purpose, at present, is to point out the relations of these two states of the sanguiferous system to the production of disease. No man can have a clear view of his duties as a physician who does not always keep the distinctions we have laid down, in his mind's eye. We proceed to ask, how far this distinction affects inflammation? We must first know what we mean by the term.

Inflammation consists of five principal stages. The first is the stage wherein we witness redness of the inflamed part. This we cannot always see. The next is that wherein we trace the phenomenon of either pain or throbbing. The third, is that of a certain amount of swelling, or of effusion of lymph into the cellular or areolar tissue of the neighbouring parts. The fourth is a change of structure, denominated solution of continuity, or formation of pus; for these terms are implied when either abscess or ulcer takes place. The fifth is a stage leading to a death of the tissues involved, beginning with gangrene, and proceeding to slough. We must be very chary of talking of inflammation, without having all the conditions fulfilled, or in course of being fulfilled. We must not be content with pain, or with redness alone. The one accompanies toothache, when no inflammation is present. The other is seen when a sensitive person is made to blush, or when the fire in winter has caught the cheeks of a delicate female. We must not say a part is inflamed when we have cut our finger, and no *pain* is present, for simply the blood coagulates, and by uniting the sides of the cut, heals it. These are examples of the fact that inflammation is not a necessary part of the process of uniting severed portions of the living body. But we have to make further observations on inflammation. First, our attention is directed to our own body in a state of health.

We feel absolutely no uneasiness. If we do, we cannot be quite well. What is the meaning of perfect health? We are told there is no such state of existence, because we can never live in a garden of Eden. Perfect health implies the normal performance of the functions of the whole body. But it must be remembered that we live in a state of society in which we cannot accustom ourselves to the habit of *self-control.* We may be assured that without self-control we should be ill, even in that delightful garden in which our first parents wandered with so much happiness. How did they manage to lose the privilege of self-control? Precisely as we daily manage to swerve from the paths which lead to perfect health. If we were asked to define what we mean by the words we use, we should reply, a sound mind in a healthy body.

We are proceeding to take the calibre of our own meaning. It embraces a large area. Our readers will be shocked at our introducing into a medical work the idea of religion and of piety. We cannot help it. Without both of these considerations we cannot hope to attain to perfect health. We are about to explain the circumstances under which people fall a prey to Gout, Rheumatism, Scrofula, or other cognate diseases. We must introduce the considerations which bear upon the causes of the production of these terrible evils. We are obliged, then, to enumerate among others the vicious habits into which we are apt to fall, from a want of attention to a proper education of our faculties in the habit of self-restraint. We are bound to display our regard for the Being who has placed us in circumstances of health. We ought not to swerve from those rules which make us abandon our tendencies to evil habits. These rules are the rules known as divine precepts. Where are they to be found? Some will obstinately adhere to the Vedas, and the Shastras. We are not Hindoos. We prefer progress. Some of our friends in London, who are most uncommonly eager to find a loop-hole in order to allow them to escape from the strict rules of Jesus Christ, are apt to talk to us about the immense learning displayed by a certain Godfrey Higgins, in two very thick quarto volumes, entitled "*Anacalepsis*," just as if we were not familiar with all that either he, or our old acquaintance M. Dupuis wrote. We were once proud of our acquaintance with twelve volumes entitled "*Origine de tous les cultes*." We now regard the tendency of all such reading to be retrogressive, because, having devoted much time and study to such subjects, we came to the conclusion that the great law of Nature was one of progression. Every change in the world's history leads us onwards. We are aware that in health and in morals we should neither remain stationary nor go backwards. So in the history of creeds and of religions, we ask ourselves why should our friends delight so much in worshipping the labours of those men, who desire to shew them that the earliest revelations of God to heathen nations, were vastly superior to those He deigned to offer to men more advanced in civilization. We do not now propose to recur to the numerous arguments spread through the works of numerous learned divines, who have incontestibly proved this point. We only demur to the excessive vanity, want of logic, and absence of all knowledge of the history of argument on such subjects, which characterize the speeches and writings of those who are so proud to enter into the lists, as opponents, both to the incontestible evidences of christianity, and to its scope and tendencies. We do not propose to prolong this part of our subject, because we are anxious only to show, that whatever may be thought of Christ and His mission, the precepts laid down in the New Testament form the best rules for the attainment of that desirable condition of human existence, we have designated as perfect health.

We now demur to the doctrine that we are amenable to our own evil propensities, and are at the same time justified in expecting to obtain the advantages of good health. To many, this introduction may be very tedious;

but our work is on a set of diseases which defy the resources of the medical art, and we must boldly grapple with all the causes productive of them. At the bottom of the whole lies the absence of the habit of *self-control.* That is the great evil of social life. That is *the* social evil. We may seek far and wide for the causes of misery, and its attendant poverty, and we shall invariably find that want of self-control can be traced at the root of all their causes.

We are sufficiently aware of the difficulties which lie in the way of those who try to make the world wiser than it is. We are not ourselves so wise as we should be. We are so much given to follies, that when we reflect upon them, we feel quite ashamed of ourselves. But, we argue, is there any one amongst us who could not honestly say the same of himself? There are very few of us so perfect as we should like to be. This, however, is no reason why we should not examine into the causes of our follies. Having arrived at the origin of the evil, we have only to wade through the shallow waters of repentance, and we are again washed clean. Do we hold the doctrine, that being clean, we have the right to plunge again into the foul waters of sin? We are far from advocating so very wild an idea. We know that no man can return to bad habits without degradation. We know that degradation implies a dereliction from right principle. Where is the eve of the day of regeneration? Certainly not in the relapse into the follies and errors of our former vices. We are not aware of our own proneness to slide into the causes which deteriorate our mental as well as our bodily condition.

We feel that we should like to become wise in the knowledge of curing the evils which beset us, but we have not the courage to probe ourselves in order to obviate the causes of those evils. If we ask our gouty friend to reflect on his past life, the chances are, that he will jump down our throats for that which he considers as our impertinence. How few are the men who acquire gouty habits, who do not lose the power of calm reasoning. They are notoriously an irritable race. Their irritability often leads them to conelude that every one is wrong, except themselves. No matter if you can bring abundance of evidence to prove the insanity of their conduct, it is of no avail. You are wrong, not because you are really so, but because you have told the truth, which is disagreeable to a man, who is wilfully blind. You are not obliged to succumb in argument; but you feel that discussion is out of the question; for the man is no longer a reasoning animal. What is the inference? We either are or are not accountable beings. We are not accountable to our neighbours, for if we have wealth, we may forget we have neighbours or even moral obligations. We may hug onrselves with the consciousness that our friends may ridicule us, but we do not care for ridicule, unless we are disagreeably reminded that it may come from a quarter we have endeavoured to cherish as a pet *protégé*. Then, indeed, it is somewhat annoying. But the gouty man is wilful. He wills to believe that the objects he cherishes are specimens of perfection. We have no inclination to quarrel with his choice of associates. We only know that so long as he persists in his follies, so long will he continue to derive the blessings flowing from those follies. Gout is not a mental disorder, but mental disorders are engendered in a gouty diathesis, and the man who will not in time reflect upon the consequences of his follies, inevitably slides into insanity. We have seen this so often exemplified in life, that we have learned the necessity of studying our profession as moralists, while we are bound to study it as physiologists. The profession of physic, then, is a compound of several callings. We regard the duties of the man of science to be more varied than those of the mere physician. But the mere physician is a wrong term. No one can be a physician, who is not prepared to study his profession as a physiological psychologist. It is not to be understood that the physiological psychologist is a metaphysical dreamer, who has read books on physiology. He is a man far removed from imaginative distinctions. He bases his knowledge on the elements of physics. He must be aware of so much astronomical science as is necessary to his comprehension of Sir Isaac Newton's problems on the force of gravitation. They are more easily acquired than bother-headed people imagine. They are based upon very simple laws, not very difficult to understand. The sense of all this is, that the medical man, who aspires to understand the subject of gout, must be a person versed in the elements of physic, while he is not deficient in what he considers the essential part of his knowledge, anatomy and physiology.

The time is not arrived then for our endeavour to make the whole medical profession understand the doctrines we advocate of the nature of gouty diseases. It is quite clear that the medical world are not physicists. The most advanced of our profession are apt to get angry, if they learn that you hold them cheap, when you test their knowledge of electricity and magnetism. Many occasions have occurred when it has happened that the common instrument known as the electro-magnetic coil machine has been proposed as a therapeutic agent, in cases where the medical attendant has been quite ignorant of the principles on which he proposed it should act. So many men of distinguished eminence have proposed the apparatus to their patients, while they have been perfectly ignorant of its *modus operandi.*

They have thought one thing, while the instrument was an agent to effect the thing quite contrary. We are not exaggerating. The cases of ignorance and awful blunder have been very numerous. Can we wonder then, if men will refuse to master the facts of so simple an affair as the electro-magnetic coil, we should find them so stolid on matters of a much higher philosophy. We are not yet accustomed to regard Fellows of the

Royal College of Physicians, as familiar with the little book of lectures which was read to them at the College by one of themselves, the late Dr. Golding Bird. If they could imbue themselves with the elementary facts which are there published, they would be better fitted to receive our doctrines. But the misfortune is that the besetting sin of vanity is so overwhelming, that they will not consent to go to school again. Often have the great celebrities of the day replied, when urged to look into facts relating to the highest studies that can occupy the attention of men, "*Do you intend to force us to go to school again?*" This is one of the great stumbling blocks in the way of progress towards knowledge. Men glory in their own vanity, instead of being humble philosophers. One might almost say the whole College of Physicians is tinctured with the silly sin of a love of self-glorification.

Numerous are the occasions on which men fly from the point, when pressed to give a reason for their desire to obstruct the increase of knowledge. That establishments like the Royal College of Physicians and of Surgeons, that congregations of associated learning, like the Royal Society should be found throwing obstacles in the way of an advancement of science, is a very melancholy fact to be accounted for, only, by the influence acquired over the large body of members, by the intrigues of the petty minded men, whom they, in their indolent love of ease, permit to lead them. We shall very soon have occasion to denounce this shocking indolence, this tendency to succumb to the stolid ignorance of real charlatans. What in fact is a charlatan, but a man full of conceit, and distinguished for ignorance? We are sorry to be obliged to lead our readers through a quagmire of diatribes. There are, however, occasions on which we may use the old adage—*Judex damnatur, cum sclerer osus absolvitur.*

Many are the opportunities which, in a course of years, have been afforded to our opponents to retrace their steps. They have a pigheaded desire to continue in a course of ignorance and folly. We have now no alternative but to show the strong grounds on which we rely for public confidence, and to denounce, in unqualified terms, the stupid obstinacy of those opposed to us. We are sorry not to be able to find excuses for those who have been led on to their own destruction, by their own sordid folly, by their own overweening conceit, or by their pernicious habits of intrigue. Time hastens on, but these men appear to believe they have power to crush the events, which time pours forth. We are sorry for them, and sorry, moreover, for the destiny which urges us to the higher mission we have chosen. We cannot compromise. We have but to comply with the inevitable mandates of our fate. Truth requires of us the severest sense of duty, and that sense of duty is quite imperative.

Our object now is to make it clear to our readers that we are compelled to walk in a groove the direction of which is not agreeable to our sense of kind feeling:—that a sense of duty obliterates all other sentiments, and that if we be found denouncing those we have known in other days, as men of urbanity, we are obliged to apologize to our readers for our apparent harshness. All we can say will have no effect upon certain cliques, but we do not write for cliques. Our office is that of teacher, that of promulgator of truth; and we cannot condescend to care for the opinions of cliques. Time will settle them. For the most part they are wanting in moral courage, and when they read the steady and cogent arguments, advanced against their pet notions, there is a great probability they will change their cowardice for the safe valour which delights in associating with the majority. At all events, fear does not become us. We proceed towards our goal, fearless of all consequences.

CHAPTER II.

PRELIMINARY OBSERVATIONS, CONTINUED.

Convenient as are the dogmas of the learned, there is something far more important in the world. The truths and fallacies of religion are the sources of endless dispute. We are not now to discuss these matters. We merely assert the general proposition, that beaten as the whole subject has been by the hammers of the disputants, no one has yet forged the matter into a good shape. The world continues to be plagued by the varieties of opinions, not on facts, but on quibbles. Few men can have the boldness to suffer themselves to be divested of a clothing of religious superstition. All adhere to the dogmas of sects. If we could bravely discuss the facts relating to the existence of a God, and to the probabilities of a revelation, we should soon be convinced that the arguments in favour of each were of a nature not to be doubted. Again, if we could summon courage to seek out the fallacies that overlay our creed, we should very soon cast off the slough of the absurd nonsense, many of us profess to believe. We are strange beings. We quarrel about trifles, and neglect the most essential subjects. We dispute about that which cannot be proved, and we neglect to ratify the common sense we could easily clear of the surrounding rubbish.

We are positive of our ground, when we assert the iudisputable nature of the truths of Christianity. They have been assailed on all sides, but the assailants have never been able to make a formidable attack. The book so much insisted on of late years, Strauss' "Life of Jesus," is no more than a repetition of the absurd objections which have emanated from the sciolists of the last century. To call it a profound book is to cater for the vanity of superficial thinkers. We are not inclined to question the right of mankind to puzzle themselves with the quagmires of obsolete objections to truths that lie in a very small compass, and that can be demonstrated with the greatest facility. Dr. Smyth, the learned Professor of History at Cambridge, has put the evidences of Christianity in so clear and condensed a point of view, that any one who reads the second edition of that admirable little work, and rises from its careful perusal, unconvinced by the arguments in favour of the Resurrection of Christ, must be a man imbued with the evil spirit of the mule. There is no accounting for the structure of men's minds. Some are led away by vanity. Some are puffed up with pride. Both are equally entitled to be accounted stupid, when such causes obfuscate their clearness of intellect. Truths strike men differently. This is an undeniable fact. We are to seek for its causes. They will be found in the shape of our heads. It is to phrenology we must look, if we would learn why men are unable to receive the holy truths laid before them. What does phrenology teach us? We are aware of the general acceptation of the doctrines of phrenology. We know that many talk on the subject who know nothing of the matter they discourse upon. We even know, that among those pretending to a deep knowledge of psychology, there are some who are not only profoundly ignorant of the facts of the great science, but who are most discreditably engaged in battling against its truths. We should desire no better proof of a man's superficial mind than that of his having pretended to master the subject, and then publishing his decided opinion against the truth of the science. No proof of the absolute folly of mental superciliousness could be adduced more striking than this. The man who could so commit himself in the face of the facts, now so widely spread, must be a most infatuated fool.

The time has come to draw a severe line, our book is calculated to become the stronghold of the cause of philosophic truth. We propose, in treating of gout and other cognate diseases, to limit ourselves to the truth on all subjects that come within the range of our discussions. It will be no fault of ours, if our readers do not find us outspoken. We have found the world too long trammelled by the fetters of false philosophy. On all sides we find ourselves surrounded by the most silly theories; the most absurd hypotheses; the most wild hallucinations. Men have so lost themselves in the maze of conceit, and in the vagaries of imagination, that they would appear to be regardless of the taunts of those who have exerted their common sense. We do not propose to spare any. We trust we have no foes. At all events, our feelings tell us that we are ourselves not only no foes to any being alive, but verily, on the contrary, most desirous of being good, and most anxious to do good, even to our enemies, persecutors and slanderers. We are not, however, disposed to forfeit our sacred right to make clear the pathway leading to truth. We are going to assail all those who oppose the progress of our holy mission. Truth must not be left in the hands of those who would do battle against it, rather than abandon their love of intrigue. We shall in time expose our meaning more clearly. Now, we propose only to introduce the subject of phrenology, in order to show its importance in explaining the causes of the existence in the world, of so many men incapable of reasoning as they should do. We are surrounded by a world of spirits. These are beings of so subtle a nature, that it requires more than ordinary power of abstraction to realize their existence. But without the postulate of their actual existence being granted to us, it is to be feared that even the truth of Sir

Isaac Newton's grand discoveries will stand but a poor chance of being clearly understood. In the course of this work, we propose to show that Mr. Faraday does not comprehend the philosophy of Sir Isaac Newton, and that in the egregious vanity he has displayed in comparing his own mind to that of England's glorious astronomer, he has been acted upon through the agency of the phrenological organs of his brain, by the unseen spirits of the air surrounding him. How far he is a believer in the poetry of Shakespeare, we will not stop to enquire. We only know that he might, if his humility would allow of it, read much in the works of our great dramatic bard, that would be calculated to pull his vanity down a peg or two. It is not pleasant to know, that engaged, as we have been, in the ardent pursuit of truth for many years; having arrived at conclusions in accordance with the convictions of some very profound philosophic minds, we are still at variance with the dictators of science in England. We have, however, this advantage, that our opponents have had the want of discretion to commit themselves most egregiously to acts from which they cannot escape, and which must enable us to come down upon them with a force of reasoning that must quell all their railing. We have asserted our belief in a fact, which is a legitimate corollary to the astounding truths that have been developed to the world in the course of the last eighty-three years. These truths are acknowledged by numbers of persons in Europe, America and Asia. They are susceptible of very easy investigation. They relate to the philosophy of the mind, to that of vegetable and animal existence; to the laws regulating the creation and maintenance of these kingdoms of nature, to the laws regulating sleep and wakefulness, to those regulating health and disease, to the philosophy of life, to the existence of angels, and to the best proofs of the existence of a Deity. One should conclude, that any set of facts which embraced so important a range of subjects would meet with a welcome reception from any unbiassed set of philosophic thinkers. It has not yet been my lot to have them so favourably regarded. Can it be wondered at if I become anxious to show my superiority to those who have opposed me? It is only a superiority arising from a fearless disregard to the authority of names. With every respect for facts, I must now boldly assert my right to vindicate the truth; and though I may possibly hurt the feelings of some I respect, and of some whom I much love, the stern dictates of conscience will not allow me to pass unheeded the cry that calls aloud for justice. Either the cause for which I battle, must perish, or I must remain a martyr to the truths I have advocated.

I cannot allow myself to doubt the result. I know my present appeal is an eloquent one. It is the appeal for the protection of the most holy truth that calls for the respect; nay, for the veneration of man. It is not to be smothered by the tame and cooly calculated pressure of numbers, however powerful. The ground on which I stand is firm, and the arguments I use are cogent. The science I adduce, is without objection. No reasoning can subvert it; no ridicule need assail it; no efforts to undermine it can prevail. It stands on an impregnable rock; and should it happen that after all the calculations of years, as to the soundness of its bulwarks, they could by some wretched subterfuge be blown into the air, it must inevitably recombine its elements and reappear as God's holy truth. We do not fear the result. We only quail when we think of the consequence to our friends, who may be rash enough to allow their want of self-control to master them, and to force them blindly to be the victims of their vanity.

We must pass from this to another part of our theme. Many considerations point out the necessity of proceeding with caution, when we insist upon the promulgation of perfectly new views. However new to the world, they are not in the sense of time so very new. They have been developed for years; pondered upon in every shape;

allowed to rest on shelves till they could be re-digested; made the subjects of thought over and over again. They are new to the world; they are old to the author. Yet they will not be accepted as they should be, simply because the world is not yet wise enough to reason fearlessly. People think they must bristle up at any idea that appears new about religion. In fact, take the subject of religion calmly, nothing now proposed can do aught but strengthen the bonds which tie it to the soul of man. Take religion as a matter to be strengthened and supported; these views will point out the means by which its foundations may be well buttressed. Many fear for their dogma of the Trinity. Spiritually considered, the material doctrines which assail that dogma are here demolished, and we may fearlessly assert that no subject has ever been broached which avails itself of so many resources for the great object of fusing into one whole or really catholic system all the discordant plans of religious faith. This, however, is like parliamentary reform, for no party likes to compromise; all insist on orthodoxy, and phrenology alone can tell us why religionists, and politicians, and men of deep science are such fools.

We cannot afford to yield to the persuasions of bigots. Unhappily the world is made up of these silly and obstinate folks. We are averse to class our friends in this category. Some are so liberal, that we admire them for their freedom from all prejudice. But this is a very rare case.

You will, no doubt, class those who are wedded to atheistical notions with the bigots. There are proportionably more bigots among the atheists than among any other class, having always regard to their numbers. Everything is relative. In relation to their numbers, out of all question, the atheists have more fierce bigots among them than any other class of persons. Their very creed is one of exquisite contempt for the opinions of others, and based, as their convictions are, upon a solid foundation of frivolous vanity and selfish pride, they cannot conceive of any error so gross as differing in opinion with themselves. It is not now the place, nor is it the time to discuss a subject of this nature. We were upon the causes, derived from facts in phrenology, of the aberrations from strict logical conclusions, in the train of reasoning, in which men indulge. None are so apposite in illustration of our point as the causes which obfuscate the intellects of the atheist. He is always a conceited man, and often a dull, stupid man; he is most frequently very vain and silly.

In proceeding to demonstrate the causes why we have so long withstood the knowledge God pours out into the world, we must take into the account, the interests of men, who have, by intrigue, or some other route, arrived at a factitious eminence. Their positions are valuable to them; and not only to them, but to a host of others, who fancy that their interests are bound up with theirs. Thus, in the instance of a man who has been many years about a great hospital, we find that his master was in a similar position before him. His master might have formed a reputation upon the unpublished works of a predecessor. The pupil, who succeeds to his post and to his practice, is the successor to no unpublished works, for may be they are all destroyed for the purpose of concealing the thefts of which his predecessor had been guilty. The successor is obliged to work, but he works very much as if he could not help it. He works to establish a reputation in a particular line. His facts are published, but though they are a huge mass of useful knowledge, they are ill put together, illogical in all their bearings; and wanting in the power that should distinguish a real thinker. We are drawing the picture of not one man, but of a class. We may be said to furnish a sample of the London pure hospital surgeon. Few men of the class have trodden a new path. Few of them would thank us to attack one of their number, for they believe that they are all equally implicated in the blow we might level at one. It matters not, our duty must be performed. Those who have cautiously abstained from meddling with Animal Magnetism, in the hope that the public will pardon them, when this book appears, are much mistaken. They will get their share of blame. What is their position? The world is advancing in knowledge. It walks on, with giant strides, on all subjects connected with physical science. The civil engineer is the leading man of the time. He studies all that can possibly relate to the pursuits in which he is engaged. He taboos no branch of new science. He is ready to listen to facts, come whence they will. Even the old fashioned clergyman thinks it time to walk into the fields of physical science, and cull here and there a bunch of sweet smelling knowledge, to heighten the aroma of his odour of sanctity. But the medical man, call him by the title best suited to gratify his empty vanity, and never was there a time when this frivolous weakness was so rife, the medical man must turn up his nose at Magnetism; unless, indeed, it is brought to him by some humble artisan, who has invented a new form of the electro-magnetic coil machine, and is humbly desirous of obtaining a certificate of the valuable opinion of his patron, who proud to advertise his own fame, grants a few lines to certify the paramount excellence of that of which he may be profoundly ignorant. The mischief accruing to society from this intermixture of ignorance and servility is melancholy.

There are few occasions for our intermeddling with the motives which prompt our neighbours to render themselves notorious. Having for a long course of years felt that our pursuits were of a high philosophical character, we could not be led to descend into the arena to compete with men who were anxious to exhibit themselves to the public before they had realized the amount of knowledge which could distinguish them as teachers. Our life has been spent in learning. We are not desirous even now to leave off the pursuit of acquiring knowledge; but the herd of practitioners fear the advance of knowledge. They would not for the world be questioned on Animal Magnetism. Their views on the subject are vague. They have no idea of being edified by the few, who have looked into the matter. They think the best way is to pronounce a decided judgment against it. First, there is really very little in it. The effects are so uncertain. Then, whatever there may be in it, is so mixed up with immoral tendencies, that their advice is not to look into the matter at all. Then, it has ruined the reputation of every one who has taken it up, and deservedly so, for it absorbs the whole attention from practical pursuits. These, and other equally veracious statements, are hazarded by those whose profound ignorance seeks a refuge in the generalities of base detraction, or more cowardly slander.

Animal Magnetism needs no defence from such men. It is on its last legs when such men favour it with their notice. It is not necessary to have recourse to their assistance to enable the world to emerge from ignorance.

Animal Magnetism is a very large science. It begins with its analogies to the forces regulating the universe. It proceeds to the facts belonging to all the various forms of matter. It takes cognizance of each force tending to regulate each form of matter. It is itself a creative force, for it creates the animal structure. It is a solar force, for it acts by the creative agency of light. It is an animal force, for it builds up the structure of every organic being. Of course, it must be understood to embrace the considerations of all the creative forces in universal nature. It is not only a creative force, it is more, for it produces the phenomena of mind, and mind is obedient to laws which have hitherto been very imperfectly studied. Mind is itself a phenomenon dependant on the laws of Animal Magnetism. This exposition is only a very general view of the classification of our subject. We might enlarge to a great extent. Our object is not to bewilder, but to state with calmness the vast importance of a science which has been shamefully, and one might say wickedly, opposed by the narrow bigotry of professional cliques. The time is come to vindicate the claims of this science, and in doing so, to vindicate our own claims to be considered not only a philosophic thinker, but a practical physician.

CHAPTER III.

PRELIMINARY OBSERVATIONS, CONTINUED.

Faulty as are all the systems which man has constructed for the explanation of the phenomena of the human mind, we are nevertheless bound, in deference, not to the stolid prejudices of narrow minds, but to the industry of past labourers in the field of psychology, to yield respect to the facts which have been developed. If we look back to the progress of knowledge on the phenomena of mind, we shall find that philosophers have worked hard, have pondered much, and have not reaped a great and abundant harvest. The time could not then have arrived for a clear comprehension of the causes of mental phenomena. Hugh Farmer, a man who wrote an admirable book on Miracles, was a reasoner of close and vigorous power of intellect. He saw things in a clear point of view; but the knowledge of Animal Magnetism was not extant in his time. The facts of this science explain much that to him was obscure; and, consequently, seeing facts that could not be utilised, he dogmatically rejected the natural consequences flowing from them. He wrote a work on the worship of human spirits by the ancients, in which he attempted to prove that the Heathen Mythology took its origin from the facts of what we now call Modern Spiritualism. Names are a great stumbling-block to the progress of knowledge. We do not contend that Farmer knew anything of Spiritual Manifestations, but it is evident he believed in the existence of a world of spirits. It is wonderful that any one who has studied the Bible can doubt of this fact. We are to take advantage of Hugh Farmer's work, in order to show that he was gifted with the power of making out one of the most important theories that can refer to the human mind, that of the prevalence of hero worship in the earlier ages, and its consequent corollary, the worship of the spirits of the heroes, who were supposed to become immortal, and to dwell in a new sphere of existence. The immortality of the soul, and the power of deification was then an article of current belief. Christianity came to confirm the creed of the world. The mission of Christ was essentially a superaddition to the faith already in existence. This is a doctrine, startling to many who have not looked into the causes which produced convictions in the minds of men, in what is called Heathen Mythology. We are not now to enter upon a dissertation of the Gods and Goddesses of the ancient world. We advert to the subject only for the purpose of shewing that Animal Magnetism and its phenomena must have given rise to the religions of the world. In this point of view, our subject will not be considered by any means trifling.

The consideration we now offer on the early influence of spiritualism on the moral government of the world, is not a new one. We are proceeding to shew that, if we can believe in the views put forth by the learned author of the work on the worship of human spirits, we may well defer the consideration of Animal Magnetism to a later period of the history of the world.

We take the liberty of stating that the facts of modern spiritualism are only a repetition of facts, that must have been familiar to mankind in the earlier history of the world. The bearing of this idea may not at once strike the reader. We are proposing to show that Animal Magnetism is neither more nor less than the origin of all religion. We are warranted in this idea by all we read in the works of the ancients; in all we know of their games; in all we have learned of their arts and sciences; in all we can gather of their penetralia. There are few subjects which can illustrate so completely the manners and customs we have been taught as belonging to the ancients, so well as our theme. It would be difficult to account for the Delphic and other oracles, and for the numerous facts relating to ancient mysteries without reference to the abhorred phenomena of clairvoyance. We are struck, whenever we recur to this subject, by the ignorance of the London physicians, who, a few years ago, combined to represent Dr. Elliotson and myself as madmen, for believing in facts which have since been as common as blackberries. As physicians, we are obliged to stand alone, to be the targets for the cowardice of the multitude who link their faith to the creeds of the doctors. Many suppose such a position is a very unenviable one. Time will shew.

We are not one of those who cry out at being selected for a noble mission. It is felt that we cannot but be distinguished for a love of truth. Independant of the history of an atrocious conspiracy to promote our ruin, we are clear that time must develop the conviction in the minds of the public that the work we are now producing will carry its weight. We are engaged in proving that the history of the human mind, the progress of knowledge of religion, the efforts of science, all combine to elucidate the truth of the propositions we have undertaken to lay before the world. We may not be able to go so far as we would wish, but we can, at all events, make it clear that man is surrounded by influences which at once keep him in life, and are the sources of all his mental, moral and religious phenomena. If we can shew that all these are linked with Animal Magnetism, we need be under no apprehension that the learned body of physicians and surgeons will be able to crush our efforts to enlighten mankind.

Few, if any, can boast that they have done much to elucidate the history of disease. We propose to show that no subject can possibly offer so many channels of thought in the direction of great progress as this. Some have written on theories of physic; some have encumbered the shelves of medical libraries with heaps of learning; none have ventured to shew that the whole of medical science was involved in the forces which are engaged in the electric telegraph. None have dared to say that the late President of the Royal Society was ignorant of the principle of life, although he missed so fine an opportunity of displaying the fact, that Animal Magnetism was the source of animal heat, and the cause of the power of breathing, of thinking, and of acting. When a man can stand forward and challenge the world of science to deny the truth of these assertions, he may not be considered to be in an unenviable position. Madman though I be, for believing in the best proofs which God has pleased to afford to man, on the existence of a state of future life, I am not sorry to be scouted for that which must soon convince the whole world. It was the fate of the harbingers of truth, in all ages, to be persecuted. It is not the part of a brave man to repine that he meets the fate of those who have preceded him in a holy mission.

Many are the vicissitudes of scientific progress. If we can give good proofs that the arguments we use are based on facts, quite incontrovertible, we need not fear the assaults of those, upon whom we throw the onus of disproving our facts. In the year 1848, appeared a pamphlet written by myself, which begins with these memorable sentences. They are memorable, because though published seventeen years ago, they are equally applicable to the present day, and I can as fearlessly refer to them as I can undauntedly declare my conviction, that the physicians and others, who then opposed the philosophy of Animal Magnetism, are, at this day, equally answerable to the public for their want of that principle which I then so vigorously denounced. I leave out the names, simply because I pity the individuals, and should be sorry again to hold them up to the contempt of mankind. It would never be my desire to be personal; a sense of duty alone prompts the necessity of indicating names. "Physicians who know little of metaphysics, almost nothing of the physical facts relating to the philosophy of mind; who are grossly ignorant of the splendid truths developed by the brilliant discoveries of Gall; many of them, perhaps, unable to appreciate the knowledge of these from the habits of illogical reflection in which they indulge, or may be from a deficiency in the organization of their brains, continue to chatter before courts of justice, and before committees of the legislature, on subjects they do not understand, and which should be studied and comprehended in some elementary fashion, at least, before they are ventured upon in public. . . . . . . As a physician, no man should be ignorant that many things, important to the best interests of humanity, are to be found in a periodical, called the 'Zoist,' which for several years was published by Mr. Baillière, in Regent Street. No physician should consider himself fit to practise his profession, who is ignorant of the facts contained in Petetin's book on Animal Electricity. The different works of Mesmer, de Puysegur, Tardy de Montravel, Foissac, Deleuze, Mialle, Townshend, Colquhoun, Esdaile, Reichenbach, Rutter and Leger, not to speak of the Harveian Oration of Elliotson, should have had attentive consideration, before a man ventures to commit himself solemnly in a court of justice, or before an assembly of Members of the House of Commons, on the subjects of which these authors have recorded most important truths."

Such were the words which flowed from my pen, when I wrote my paper on the facts in clairvoyance. Those who smart under the lash of such words, may well think themselves small in public estimation. They richly deserve much more than they have got. When time passes, and one reflects on the indignant feelings which prompted the few quoted sentences, it is not difficult to conceive that men will not lay aside their ignorant prejudices, from sheer pride. It must, however, strike our readers that unless some spirits, more bold than others, are found to continue the stimulus of flagellation, the physicians and surgeons of the age will relapse into their sordid and stolid indolence, and the science of which they are so ignorant, will be suffered to meet with the neglect which has been its lot heretofore. We are not disposed to allow this pertinacity of obliquity of moral principle. Medical ethics are much talked of. Let us see those who strive to stimulate their brethren on this subject, becoming alive to their duties as students of physiology. Let no man call himself a physiologist who is ignorant of the principle of life, as every one must be, who is ignorant of Animal Magnetism.

This introduction relates to the science of mind which is intimately tied up with the consideration of life, and the history of the religions of mankind. It has often been said that a sense of religion was a very universal feeling. We not only reassert the proposition as true, but undertake to shew that the physical laws of the universe are linked with it. Religion is a part and parcel of every man's nature. There may be instances in which habits of thought have blunted the edge of the religious element, but in all men, unless malformation or disease have injured the brain, the sentiment of religion is a part of the constitution of the human mind. We do not now discuss the causes of the boastful ejaculations of those weak reasoners who take the negative side of our question. Our business is to display the connection of the history of the religious phenomenæ of the human mind, with the subject of Animal Magnetism.

More could be said than we are at present disposed to write on the religions of the world. It will suffice to indicate their dependence on the forces regulating certain portions of the human brain. Religion is an attribute strictly of humanity. It dissolves sympathy, benevolence, and the dependance on a superior being to which we give the name of veneration or religious sentiment, into one. It matures all the finer feelings of our nature, and becomes the regulating force superintending all our moral sentiment. Religion, then, is more than a simple faculty of the mind. It is a superior magnetic force, operating on several cerebral organs. Gall's idea, perhaps not published, was that no single organ of the brain could act independantly. His view of the complication of several organs, obviated one great objection to the supposed material tendency of his doctrines. No organ is supposed to be self-acting, consequently it must lean for support on other organs. What, then, becomes of the generation of magnetic force from the matter of the magnet? the matter being dependant for its force or power upon its neighbour, which cannot originate force by itself. We may be told that the neighbours are in opposite conditions of plus and minus, or positive and negative. If this be so, why should not the forces be sufficient to account for the phenomena, without the necessity of a recourse to the hypothesis of a power of action in matter, which is known to be quite passive, and incapable of action. People confound the idea of motion and that of action. A thing moves, because it is either attracted or repelled; but motion, in this sense, is not action. Action implies a will to act, and the consequent power to produce motion in the thing acted on. This is a very different thing from simple attractive, or repulsive force. It would be easy to account for the fancies that have led men to confound these ideas. They are in the nature of things. Men do not acquire clear ideas without some education of the reasoning faculties. When these have been trained to become dormant, it is curious how soon feeling becomes the law of action. A man has been educated, for instance, to imagine that his friend was vastly clever. His affections influence his admiration. His friend held that matter was the source of life. Consequently, though he had reflected much on the various arguments which had a tendency to shake his conviction in the infallibility of his idol's doctrine, he could not divest himself of the essential fallacy. We shall give the example of his definition of life, for the purpose of shewing the influence of habit on the faculty of thought.

"Life is that attribute, by virtue of which, tissues manifest their appropriate functions, through the waste of their own substance.

"At its origin, the function of *assimilation* exceeds that of *waste*, constituting a *period of growth*, but the proportion between the two gradually changes, and becomes reversed, constituting a *period of decay*, ending in death, a law, the necessity of which is not apparent.

"Vital functions are *tissue forces*, generated by chemical changes in the elements of the tissues manifesting them."

Here is what is called a definition. We shall not endeavour to class the several propositions, which enter into the fabric of these sentences. Suffice it to indicate the fallacy which runs through them, and our friend is too liberal to be offended at our criticism. It is the misfortune of those who are wedded to the idea that matter can think and act, to be thoroughly confused on the subject of forces. They must believe that matter owes its origin, not to the will of an existing being, but that it has itself existed eternally. The inevitable consequence of such a conviction is, that we cannot think when we are awake. We can think only when we are acted upon by forces which are attractive. This is a startling mode of posing our question. We know that sleep results from attractive forces, and that wakefulness is the consequence of repulsive agencies. What do we mean by attractive and repulsive? We mean the forces regulating the orbs in space, which would be annihilated, if they were not so regulated. But those orbs are each of them composed of particles of matter, held together by attraction. How can that attraction be at the same moment, a force regulating matter, and holding its particles together, and a force subsidiary to matter, and being an attribute of matter. Either the one or the other must have precedence. "Oh!" my friend would say, "they are coincident." If so, there never was a beginning. All that we see around us is the result of stationary law. We need not point out the absurdity of such a conclusion, which implies that no creative mind is necessary for the construction of houses, for the cultivation of crops, for the processes of manufacturing industry, for the thousand occupations that engage the attention of man. If matter were eternal, what need of change? What need of the law regulating unceasing new beginnings and new endings. If life be an attribute of material tissues, and death simply a puzzle, what need of the changes attendant upon assimilation, and upon waste? Why should not life be eternal as matter has been? Why should change of any kind be a necessary law of nature? When we own to a puzzle in not being able to account for death, we may as well cease to speculate on the phenomena of life. To say that it is a property or attribute of tissue, is to acknowledge that a pair of horses are the attributes of a carriage, or the wind which moves a balloon or a ship is the attribute or property of that balloon or ship. No fallacy is so glaring as that which prefers to worship the effect and to ignore the cause; or which is perpetually insisting on the consequent being the antecedent in any train of causation. To endow matter with intelligence, and to make intelligence the servant of inert and passive substance is a gross fallacy; but it is one so common and so engrained in the minds of the present generation that one is at a loss to conceive of the power of habit in fixing such a fallacy in the mind.

Much as those who do not choose to adopt the evidences of Christianity, despise the learning displayed by the powerful thinkers who wrote the Gospels, there is a passage which is a monument of the deep thought of the apostle, who was the author of the fourth of these marvellous productions. The four first verses of St. John's gospel form an epitome of the reasoning on the subject which has been occupying our attention. It is unnecessary to advert to the learning of those commentators who have written so much to elucidate the doctrine of the Logos. We only wish to shew the beautiful train of thought that characterises the words, "In the beginning was wisdom, and wisdom was with God, and wisdom was God. The same was in the beginning with God. All things were made by Him; and without Him was not anything made that was made. In Him was life; and the life was the light of men."

We shall not endeavour to trespass on our readers, beyond an earnest request that they will remember the uses to which we shall have occasion to apply not only the words but the ideas contained in these magnificent verses, which comprehend the idea of the wisdom, power and goodness of the Creator; which embrace the only rational and intelligible view of the dogma of the trinity.

We shall be impugned for our introduction of these topics into a medical work. The reader may be assured that nothing irrelevant is admitted into this treatise, and it will be seen how easily we glide into the fundamental principles of medicine, connected as they are with religion and animal magnetism. We have traced the connection of forces with matter; we have endeavoured to shew that matter was subservient to force, and consequently to mind; that mind was a force appertaining to intelligent beings; that these emanated from Almighty will. We shall proceed to shew that no part of intelligent forces is lost in the vast expanse of space; but that we are warranted in concluding, however subtle may be the intelligent agencies actually existing in space, we are not only safe in believing them to be vital, but in believing that their vital forces are constantly operating on the brains of mankind, and in forming what we call the motives of human action. It is not long since men were shut up in lunatic asylums, who could be found capable of writing and of publishing such doctrines as we now put forth. We are aware of the risk we run in advocating any portion of the science of Animal Magnetism. When we made up our mind to put forth our views, we were sure of one thing, that of an earnest desire freely to investigate the truth, and having found it, not to be alarmed at any of the consequences which might attend upon its publication. Truth is arrived at by a severe examination of facts. Facts are here stated boldly. If they can be impugned, let those who have the capacity as well as the courage to deny their existence, come forward and confront our statements. We know it is easy to deny, but we are ready to prove, and we trust that our opponents will be equally ready to disprove our veracity. This is not an easy task.

CHAPTER IV.

FORCES.

MAN is not accountable for the phenomena he may exhibit as an integral portion of the material world. He is in this point of view subject to the laws which regulate matter. All matter is subject to law, and it is the province of the natural philosopher to study, not only the varieties of matter, but to endeavour, as far as the limits of his capacity will permit him, to make himself master of the laws regulating those varieties. The physiologist and the physician are included, always, when their ignorance and mulishness do not rebel against the compliment in the term now made use of. The Natural Philosopher takes into his scope a wide range of knowledge, and though, by some sciolists, he may be forbidden to enquire into the philosophy of the mind, he is bound, by his researches into the laws of matter, to enquire into those of mind.

The really philosophic man is not to be trammelled by the traditions of the elders. He inhabits this globe; and he learns that it is a sphere. His attention is directed to other spheres in space; and he is led to enquire. His enquiries result in the knowledge that the most powerful minds have turned their attention to the same objects; and have developed a multitude of facts which have warranted the belief that one grand and simple basis is the foundation for all the laws of the Universe, and that however multiform and various may appear the works of all the human minds on our planet, and however wondrous are the results of human mental exertion and ingenuity; however deep may be their researches into science; however abstruse may be their mathematical formulæ; there is one creative mind that plans according to its own grand formula, and the scope of that formula is so vast and incomprehensible as to leave the largest powers of contemplation in a maze of wonder, the natural effect of which, in a capacious intellect, must inevitably be a feeling of awe and veneration at the most sublime grandeur. Yet the contemplation descends, and if the humility of the enquiring child is the characteristic of the thinker, the feeling of awe engendered by the power of that comprehensive creative mind produces a sentiment of homage. Is it the wild Indian alone who is disposed to entertain a veneration for sublime power? It has been asserted that the untutored mind is the proper receptacle for the simplicity of humility; and that the wisely proud philosopher rejects with disdain the abasement implied by such a sentiment. What is there in pity, that we so freely bestow it upon the exhibition of arrogance? or in charity, that we would fain extend it to the victim of pride? "Suffer little children to come unto me, and forbid them not!" Theirs is the beautiful innocent simplicity that seeks for new knowledge, and is the real humility that asks for new truths. Theirs is the delight of sympathy which thanks with fondness the benefactor who gives it the welcome facts. To a small degree, according to its capacity, the child evinces the philosophic tendency. The tendency to expand its affections, and to feel gratitude for benefits. Here is the germ of politeness; the gentleness of the gentleman; the humility of the Christian; the self-knowledge of the philosopher; and when the philosopher knows himself, he is naturally led to be humble, and in just humility seeks for the gratification of his desires in those occupations which lead through Nature up to Nature's God! The philosopher contemplating the ideas which have occupied the minds of Newton and of La Place, arrives at those which have become the subjects of discussion among a host of modern thinkers, and must note that the progress towards truth has been among them all in a ratio depending upon the degree of humility with which equal intellectual capacities have approached the investigation of a subject.

There is no topic more hacknied than the one to which we have turned our attention, as that which has occupied the thoughts of the greatest philosophers: the Sphere. If we abstract our minds in reflection upon matter, can we possibly imagine a shape more simple? and yet we know that it involves every possible modification of form. This needs no illustration. But that which is very remarkable in relation to it is, that we can conceive of no manner of existence for matter which could allow the inevitable dominion of change besides the globular condition. No freedom of motion could exist if the ultimate particles of matter could assume a form that was not spheroidal, and hence what Whewell has justly called the conservative law of Nature, and which Edmund Burke, with the striking terseness of genius called the great law of Nature, Change, could have had no place in the operations of the Universe. Change necessitates motion, which is its parent; and no motion could exist without the presence of force. Now what is force? We have thought of matter without explicitly defining it, otherwise than by an hypothetical assertion of its existence in globular forms, but if we seek to have any more clear view of matter, and its relations, we must indulge in considerations on ideas which have puzzled the greatest philosophers, and have led them into error on the nature of matter.

In the present state of our knowledge, one of the most difficult questions to answer is, what is Force? It is as difficult to define the word force, as the word *soul*. If you define the one you accomplish the definition of the other. Men, who think themselves great philosophers, find nothing more easy. But a very humble individual, who desires to have a clear idea, and a power of crisp definition, does not find it so easy. He abstracts his mind, and endeavours to imagine the idea. You must not *imagine*, interrupts the man of strict science. Imagination is not admissible into the halls of science. We discard all *hypothesis*. A theory is adopted only, at the suggestion of some man of superior repute. So the inquirer is to be rebuffed by a man who is clever in showing the force of his petty temper, which most often means only his envy, malice, hatred and all uncharitableness. These are forces, and they are more retarding to progress than is generally imagined. Why so often use the word *imagined?* Simply to express the originating cause of all ideas, which exists in our faulty imagination. Let us try, notwithstanding the obstructors who eagerly bother us, to form a clear idea of the word Force.

In its widest acceptation, we should say the idea includes all motive power. We know when we think of matter that it is only *inert*, or *passive substance*. Force, on the contrary, is not substance. It may be *brute*, or *sensitive*, or *intelligent*. These terms are not only strictly applicable, but are often applied by us to indicate the nature of the force we desire to distinguish. In its simplest acceptation, we should say that force is the unseen agent in the production of change. If it be true that matter be quite inert, then it must follow that no change can be produced without the interference of force. But force is attractive or repulsive. It attracts particles, or masses of passive matter towards each other, or it repels these passive substances from each other. We must *imagine* that before force acted upon them, they were static, or stationary. But, in fact, no matter can be quite in a static condition. God, having said, Let there be Light, there was Light, and from that instant the great trunk force of gravitation, with its centripetal and centrifugal powers were ushered into existence, to rule the passive inert matter of the Universe. Centripetal forces mean forces which attract to the centre of a spheral magnet; and centrifugal forces mean those which repel from the centre of that magnet. They are, then, magnetic forces. But all magnetic forces permeate the magnets in which they are destined to take up their abode. They permeate every particle of the magnets, and not only this, but they shed their influences spherally around those magnets, forming in the language of the men of science, a magnetic aura, or sphere of force around them. We, human beings, as well as all animals, vegetables, and minerals, within the magnetic sphere of this magnetic earth, must necessarily partake of the magnetic influences emanating from the grand trunk force of Universal Gravitation. We must inevitably be subject to all the conditions of our residence on this earth. If all under the operation of magnetic influences, or laws, become magnets, it is very difficult to conceive how we can escape from the powers of those forces which permeate us, and make us really and truly magnets. The proof of this simple fact, is irrefragably established by Mr. Rutter, by the aid of his ingenious philosophical instrument, the magnetoscope, which is not accepted, forsooth, by the world of science.

Let us ask, if the two opposite poles of the magnetic force be so decided, and if we be so completely subjected to their influences, can we not fairly infer that many objects around us are imbued with a polar force that is attractive to us, and many others with a force that is repulsive to us? We may not only infer this, but we may proceed further, and generalising on the numerous facts within the scope of our knowledge, we may conclude that all the forces in nature are referable, ultimately to the elements of all force; attraction and repulsion. All change in every kind or form of matter, whether the change be mechanical or chemical in its nature, must be dependant on attractive or repulsive forces, or an alternate combination of both. If these views be correct, then the all pervading force of nature, to be modified so infinitely for the creation and for the destruction of organic arrangements which constantly proceed around us, must admit of an infinite variety of emanations in all directions, from the grand trunk force of universal gravitation. My friend, the Baron von Reichenbach, no common-minded philosopher, and a man experienced in delicate scientific manipulation, has arrived at many and very interesting new facts in the philosophy of the all pervading force of nature, to which he has given the name of *Od*, a convenient word, derived from the old German root which attaches to the Scandinavian deity, supposed to typify the all pervading spirit of Nature. Of Odin, the admirers of our English poet, Gray, will not be ignorant. When the English public become possessed of the Baron's work on Man considered as a sensitive, they will find a great mass of deeply interesting facts, especially on Odic light; a subject well worthy of philosophic investigation. How far my hypothesis on the gradations of forces may dovetail with the facts established by the Baron, in his "Researches,"* published in an English form by Baillière in 1851, is not now the question. Certain I am that the facts on the existence of chemical and of coexistent forces cannot be so well explained as by the hypothesis I have put forth, of gradations of creative forces emanating from the great trunk force.

These new ideas may lead to others, based upon the observation of the gradations of crystalline forms leading up to higher geometric arrangements in vegetables and animals, the creative forces in each and all being found by the magnetescope to obey magnetic laws. These laws, like the forms they regulate, being the graduated links of the great chain of causation, that obeys, necessarily, the mathematical rule of series.

If we contemplate the stupendous fabric of the universe of worlds held together in complete unison, each

* One Vol. 8vo., 608 pages, 1851, 15s.

globe performing its revolutions in space, obedient to the laws that oblige it to move, with wondrous and unerring velocity, according to the rules dictated by the original creative force of the will of God, we little think of the source of that will, the *wisdom* of that all beneficent Being, who has decreed the harmony that attunes all the creation around us. We little think that the *will* and the *wisdom* are Forces. We little dream that such forces are creative of all the magneto-electric phenomena in nature. We do not reflect that this earth we inhabit is a vast magnet; that all the animals which surround us—that all the vegetables of whatever kind—that all the minerals in nature are consequently and necessarily magnets.

"In the beginning was Wisdom, and Wisdom was with God, and Wisdom was God. The same was in the beginning with God. All things were made by it, and without it was nothing made. What was made had life in it; and this life was the light of men; and this light shineth in darkness, and the darkness hindered it not."

The same law ordained that life was the ordinance creative of the grand trunk force which the will of God, in his own good time, permitted Sir Isaac Newton to discover to mankind. What is force, but the life and soul of the matter, which it controls and regulates? The life and soul of the universe is the force of universal gravitation. But for gravitation, all matter would be a chaotic mass. It is an axiomatic principle of matter, that it is necessarily inert, quite passive. Without the agency of force it could not possibly move. Death could not be more still. Life is animation and motion. Death is that inert condition which would be the equivalent to annihilation, but for that Change which, as before mentioned, has been well designated as the conservative principle, or the great law of Nature.

Change is the result of motive forces. From all the phenomena around us, it is easily perceived that the motive forces in Nature obey strict rules. The atoms of inert matter are placed by them in exactly their proper relations. We speak of abnormal, and of amorphous conditions of matter for the sake of convenience; but the extent of our knowledge is so limited, that we may not be able to perceive the exquisitely beautiful symmetrical arrangements, regulating forms, that may appear to us irregular in shape, and without any definite or normal form. If, as La Place suggests, that the Almighty planner of the Universe defined his work according to a grand mathematical formula, there can be no doubt that, in the classifications we weak mortals might adopt of any arrangement of his extensive kingdom, we should, from imperfect knowledge, allow of categories of things abnormal and amorphous, which might, nevertheless, be susceptible of fitting into their constituted places of harmony, grace and beauty.

The wondrous perfections established in all mechanical and physiological as well as psychal laws explain and proclaim the existence of a perfect plan, embracing the immaterial or psychal, as well as all material considerations, is so vast as to subordinate even the will of the Almighty to his own stupendous conceptions. We may, in deep worship and veneration, attribute to him illimitable power and might, but he is himself quite capable of that control over his own will, and over the laws ordained by that will, that, perhaps, we can conceive of no grander attribute of divine virtue, than his perfect Self-Control, exercised in fulfilling his laws of justice and mercy, and his unceasing industry and vigilance, in concentrating his attention upon the laws regulating his most minute arrangements.

It has often been said that the union of the most marvellous complication of detail, combined with the most sublime simplicity, was observable in the plans of the all-wise Creator. In all the unlimited variety existing in the details of Nature, there is not a regulating force, nor an obedient atom, which is not amenable to a law of series. All Nature is full of life, light and joy, and in its gladsome enjoyment, it revels in the elasticity of its mathematical precisions. It dances to the measured melody attuned by unerring magnetic forces, the harmonious vibrations of which accord to the music of the angels, who sing in unceasing praise to the glory of God.

The universal law of Nature may be regarded in another point of view. Truly, one of change, but not only of that. It is a law relating to the forces productive of the changes constantly occurring among the particles of matter. Now, a force to expand them to become liquid, then to become aerial; then it elicits an attractive agent to combine or dissolve actinised light from the sun to form oxygen into ozone; or to magnetise that oxygen and light in order to render the compound a vitalizing agent. Oxygen, in itself, is inert gas. When polarised by a force, as in becoming magnetised, it is the mesmeric vehicle of the vital force, and thus, at each inspiration, gives us a new dose of life. The magnetic force of vitality, then, is simply a detached parcel of force, shed from the trunk force. It has become modified, and still a positive or direct current, it is an attractive, or health-inspiring, force; a modified force, now destined to forms of higher arrangement; the force, the Baron von Reichenbach designates, the od force. Thus, in detail, necessarily, numerous complications arise, constantly modifying the simple change producing attraction and repulsion.

If the conservation of the Universe be dependant upon the unceasing changes, everywhere in progress, and these changes cannot possibly proceed without the agency of forces in illimitable variety, for each change is modified by the law of its production, although obedient to a law of series, it must follow that all changes are the results of motive forces, under the dominion of one grand universal law. This law is the grand trunk force of universal gravitation,—a force every where present, shedding its influence to regulate the minutest as well as the grandest operations of nature. It requires deep abstraction of mind to grasp these ideas of an unseen and impalpable agent, every where coercing and regulating large masses, as well as minute particles of material substance, and obliging them to yield their static conditions for changes that tend to fulfil the creative behests of the all-wise Supreme Being.

One cannot doubt that the universal force of gravitation contains, in itself, the germs of the forces in nature. It originates all attraction and repulsion; and, therefore, it must be that which Burke chose to designate as the great Law of Nature. There is no change which is not dependant upon it.

We have never been in the habit of reflecting sufficiently upon all the phenomena which depend upon the universal law of gravitation. We talk of the forces of heat, light and electricity, and we know very little of that of which we are talking. If we reflected as deeply as we ought on these subjects, we should find that all the phenomena involved in the so called forces, are dependant on attraction and repulsion. No matter could change its static condition, unless it were operated on by each, either, or both the agencies into which universal gravitation could separate itself. These are, manifestly centripetal and centrifugal forces, or attractive and repulsive forces. We need not mince the matter, the whole of the arrangements of nature, under Almighty guidance, has been framed upon the simple antagonistic forces of attraction and repulsion.

In considering the influence of these forces on the several conditions of matter, we need not long detain ourselves to illustrate the solid, the liquid, and the fluid or aeriform conditions of matter. It would be trite to observe that solid and crystalline ice was water when it was in a liquid condition; or that water, as vapour, was steam. The gradations of solid, liquid and aeriform are here effected by gradations in the repulsive energies of heat. But what is heat itself? We know it only as the phenomenal illustration of the meeting of opposite polarities. All those frictions, all those mixtures, solutions, or combustions which generate heat, or which favour the evolution of heat, are simply phenomenal manifestations of the forces of attraction and repulsion, under modified conditions. We have alluded to matter appearing under three forms or conditions: the solid, the liquid, and the aeriform.

Now, under the present laws regulating matter, agreeing in the proposition that matter is quite inert, we cannot, under the present laws regulating the universe, conceive of matter appearing in any other forms than those we have alluded to, and this is a natural consequence of the inertia of matter. We have heard philosophers exercising their ingenuity in supporting the hypothesis, that the Almighty could, if He willed it, change, at once, all the existing laws regulating his grand universe; but the idea, though one acknowledging, as we are bound to do, His immeasurable omnipotence, is at variance with the basis upon which He has fixed the rule of right and wrong. He has given us a wondrous range of faculties, and He has destined that our law shall be one of progress. How should we progress if we had not the grand principle of self-control. Without self-control, how can we be honest and virtuous? and if the Great Being, whose will rules the universe, have established honesty and virtue as the principles which are to guide our conduct, How could he consistently abandon his present laws of Nature?

We may be assured that however philosophical our imaginations may be, we never shall have a warrant for believing in the infallibility of men who boast of their being positive philosophers, men who would abrogate the great law of series, and this law is bound up with all the phenomena that regulate the changes in what we may regard as the forms of matter.

All changes, necessarily obeying physical laws, are dependant on series; and this is a mathematical expression. It is difficult for any human being to evade the law of pure mathematics.

We may be sure that we are not accountable for the decrees of fate, and grand would be the aspirations of that being who would venture to controvert the laws, regulating the inevitable decrees of fate.

We are now ready to resume the train of our investigations. The law of Series is our starting point. We consider it no shame to acknowledge that our first ideas, upon this subject, were due to the reflections forced upon our mind by the study, and repeated use of the elaborate tables in the work on Logarithms, by the late Dr. Hutton. No one can ponder on the figures in that book, who is not impressed by the consecutive series of numbers, which tell of the consecutive events throughout creation. What is all our philosophy, if it be not based on series of numbers?

The beacon lights of the age in which they live have a reliance of this consecutive system. We are not obliged to individualize the thoughts of persons on this subject. If, however, we were to point out for the admiration of our contemporaries, the name of any individual who has wrought for himself the fame of a discoverer, we should be inclined to indicate and honour the name of the inventor of the Calculating Machine, Charles Babbage, whose philosophic 'Ninth Bridgewater Treatise,' and whose deeply interesting 'Passages from the Life of a Philosopher,' should be attentively read by all who can respect a deep and original thinker.

We may proceed to consider some of the changes undergone by matter in the various forms we find it assuming. We know this expression is inaccurate, while we are yielding ourselves to the powerful force of habit. We should speak of matter being forced to assume, for matter is inert, and of itself cannot act or assume.

We think we know what water is. We ought to

know that it is an aggregate of drops. What is a drop of water? Is it simply a small liquid spheroid—a little globule, separated from the mass of which it was a part, held to the mass in virtue of a cohesive force? Cohesion is only a magnetic force of attraction, forcing the particles of water to cohere in the form of drops. But each drop, or spherule, of water enjoyed the first, or primitive, form of crystalline aggregation. It was a crystalline globule, and may have been the elementary primitive crystal;—the first crystal in existence. But crystals require electrical agencies for their production. God said, "Let there be Light, and there was Light." We know that actinised light produces electrical phenomena. Who shall say that the magnetism of the great orbs, in space, did not once owe all its varied forces to the Light which God called into existence? Who shall say that this was not the primitive, or creative, force which set the universe of worlds in motion, and prompted their dancing to the harmonious melodies of Nature; the chaotic mass of hydrogen and oxygen to unite and form crystalline water, the grand solvent of all other chaotic substances. The grand force of gravitation developing all its multitudinous powers, or creative agencies, at once flashed the active obedient light, which conveyed the will of God,—the wondrous fiat for the universal creation.

Wherever a current of electricity exists, light accompanies that current, and hence is the messenger and guide in all the grand, as well as in all the minute operations of change, which thus is the great formative, creative, as well as the conservative principle of Nature.

A certain low degree of atmospheric temperature affords us the specimen of the solid crystal of water, each of which, though varied, is a geometric form of crystalline aggregation, and is a promotion to a higher grade, or rank, of geometric structure, evincing a tendency to a higher condition of being. Water may thus be regarded as a most frequent essential in the composition of regulated arrangements of natural structure. It has been said to belong to all crystals; but this is not a fact, without numerous exceptions. A very remarkable fact relating to the crystalline being, or magnet Man is, that seven eighths of his weight are due to the water in his system.*

We would that the past could serve as an introduction to the theory of the universal gradations of forces. It would be vastly agreeable to prove every step of our geometrical staircase of propositions before we proceeded to the next grade; we should be able to illustrate every link in our social system. The gradation of ranks so unpalatable to the ardent advocates of liberty, equality, and fraternity, would be proved to be an unbending law of nature, and how often, soever, in the revolutions engendered by the cyclopean excitements of wild passions fostered by the pscychal and phrenological epidemics, by which Destiny clears the political atmosphere, the unbending fabric breaks by the force of circumstances, and the elements of the constitution are remodelled by attractive and repulsive forces into a crystalline fabric of a higher grade. Who that reflects at all on the events which pass around him, and knows that magnetic influences are every where present, that he cannot break a stick from a hedge and have it examined by a magnetoscope, but he must inevitably have it proved to be a vegetable magnetic crystal. He plucks a flower, or an apple, and he obtains irrefragible evidence of the facts. He buys an orange, or a pear and he takes them to Mr. Rutter's truth-loving pendulum, and no judge can reject the evidence sworn to by the clear affidavit of the magnetoscope. It is God's truth vouchsafed for the benefit of mankind, and revealed to us through the modest, humble but expanded intellect of Mr. Rutter.

How far then I am warranted in the theory of the universal gradations of forces, I may humbly leave

* See Prout, 5th Edition, page 454.

my readers to decide. About the year 1819, being then a lecturer on chemistry and mineralogy at the Middlesex Hospital, I engaged myself in experiments on various oily and resinous substances. On one occasion, having procured a quantity of that solid oil, which is in this climate the ice of that beautiful limpid oil known as cocoa-nut oil, which we were wont to burn in our wallshades in Bengal, I cleared a parcel of it and digested it in hot alcohol. Carefully decanting it into tall phials, such as are used by spirit merchants, I put by a couple of these in a dry warm cupboard in my laboratory. Two days afterwards, on bringing these bottles to the light, I found a number of apparent clusters of clear white spherules suspended in the liquid. They were not unlike small ripe white currants; and I concluded that I had found the primitive form of the crystal of cocoa-nut oil. My friend Prout, who was at that time a member of our club of Licentiates, meeting at the Crown and Anchor Tavern, would not be persuaded into a possibility of a wide range in crystallography than that belonging to the families of mineral crystals; a great chemical physiologist he had not looked deeply into botanical dissections, nor was he aware of the original views promulgated to the Royal Society by the late Dr. Allman, the modest professor of botany in my younger days in the School of Physic at Trinity College, Dublin. Professor Allman strove to establish the law of botanical structure in alliance with a law of mathematical forms. We have related the having observed a clear spherule of what, after a study of Haüy's captivating work on Crystallography, we conclude to be the primitive form of vegetable crystal. We were led by subsequent observations and reflections to conclude that Allman had arrived at a great truth, when he was able to ally the spherules, the cells, the tubes and the various shapes of vegetable structure to a geometric law. Without reflecting upon the fact, he thus established the continuation of the law of geometric rule into the vegetable kingdom. Here was vegetable crystalline arrangement. We have now only to follow the forces which have insisted on the obedience of mineral and vegetable matter to the strict military rules of unpurchased promotion, and we arrive at the magnetic laws of gradation in the forces regulating animal structures, in fact to the laws of animal magnetism, or mesmerism, or Van Helmontism. What may the profound philosophers, physicians, surgeons, and apothecaries of Great Britain and the shades of Brodie and Wakley have to object to this?

That I may not be supposed to derogate from the merits of a sagacious medical philosopher, for whom I always entertained a high esteem, and a profound respect, I shall take the liberty to quote his own words. Dr. Prout (Stomach and Renal Disease 5th Edition, page 450) "divides the objects of nature into the inorganised and the organised. Under the head of inorganised bodies are included all elementary principles and their mineral compounds. . . . . Organised bodies include vegetables and animals; . . . yet vegetables and dead animals so gradually approximate, that their characters appear to coalesce, and at, and near, this point of coalescense, it often becomes difficult to discriminate between the two divisions of organised beings, and to pronounce with certainty which is vegetable and which is animal. Nor does the *chemical* composition of vegetables and animals assist us in overcoming this difficulty. It is true, indeed, that vegetable substances in general contain essentially no more than three elements, *hydrogen*, *carbon*, and *oxygen*, while animal substances, usually involve a fourth, *azote*. Yet there are many vegetable matters of whose composition azote forms a considerable part; while certain animal substances are entirely wanting in that principle."

Prout proceeds* to describe that "besides the four constituent elements mentioned, of which all organised

* Pages 451, 453.

substances are essentially compounds; other principles generally enter into their composition," and we are told that these, "which are in very minute quantity, are not so essential to the existence of organised substances, as the four constituent elements, yet, however minute the quantity, the influence of these other principles seems to be most important; they are *sulphur, phosphorus, chlorine, fluorine, iron, potassium, sodium, calcium, magnesium,* and probably more besides." These have by chemists been deemed extraneous, but Prout admitted them into the range of the ultimate elements of organised bodies. Hydrogen, carbon, oxygen and azote, regarded as the *essential* elements, he viewed *sulphur, phosphorus,* &c., as the *incidental* elements of such bodies.

He proceeds to state that though the *structure* of organised bodies is not within the scope of his inquiry, yet, "that of the many opinions advanced by physiologists on the subject, the prevailing opinion seems to be that the ultimate structure of organised bodies is vesicular or cellular; in other words, that organised bodies consist of vesicles or (nucleated) cells, each one of which is, in a certain sense, a complete and independent system or *organised* molecule." In the next page, Prout proceeds in these words, "It is sufficient for our present purpose to know, that in the ultimate and more elementary condition in which organised molecules are recognised by us, their composition not only differs altogether from mineral bodies, but that they never assume a crystallised or mineral form."*

I take the philosophy of my old friend as the best specimen of the philosophy of his age. He was a re-

* It is quite evident that in the book of the age on chemical physiology, we do not find the wide generalization warranted by the knowledge of our time. The great mathematical thinkers of the day have worked at subjects far beyond the understanding of those who have pottered at urinary inquiries, and yet these latter are physicians who push themselves into notice as men of science. Perhaps Dr. Garrod ought to be excepted, as he has legitimately and not empirically wrought himself into a just reputation.

markable man, and certainly, in England, his mantle has not fallen on the shoulders of any of those who are his successors in the speciality to which he devoted himself.

One may fearlessly say, that in Prout's book, there are more passages having the aroma of the deep thoughts belonging to Animal Magnetism than in all the works of his successor put together. This may be a sad suggestion, for it implies a vast ignorance of the highest physiology, and a sordid disregard of those truths which ought to distinguish the tastes and aspirations of the real philosopher.

Having clearly established that the law of crystallization has, by the best of our chemical physiologists, been limited and confined in its influence to the mineral kingdom, we may feel ourselves at full liberty to shew that all nature obeys the laws of Magnetism, which is equivalent to the subversion of the idea that crystals are simply mineral aggregations of molecular matter. Prout has fully admitted the existence of organised molecules; and the building up of these, according to geometrical forms, and under the influence of mathematical laws to produce crystals of a higher grade of physical configuration is no great stretch of scientific hypothesis.

It is proposed, however, to prove that there exist laws which will carry us much further on our journey towards the capital of the science we have been cultivating. It is not a very bold assertion that all crystals owe their existence to currents of electricity. The facts at which my good friend, the late Andrew Cross of Tyne Court, had arrived, were sufficient to establish this truth. The corollary flowing from it explains the influence exerted by all crystals upon sensitive human beings, for it has been proved that all crystals are composed of duly apposited magnetic particles of matter in such relations as to enable the forces regulating them to send their combined influence through the principal axis of the crystal of which they form the minute particles. These particles are magnetic. Each spheroid constituting a crystallic molecule is a magnetic particle. It has not the property of attracting iron filings, but with the exception of this, and of the tendency, when suspended, to point to the north, it has all the properties of the magnet. It has certainly not only the property of attracting the human hand of a sensitive person, but the tendency of influencing that person to sleep very deeply, as I have witnessed very frequently. The apex, or attractive pole, of a minute crystal of sulphate of magnesia has, many times, when directed to the forehead of a sensitive woman, caused her to become profoundly asleep; and the opposite pole of the same crystal has awakened her in a quarter of a minute. If the facts now recounted were singular, we might relate them as really curious, but in an experience of, at least, twenty-five years, they have so frequently recurred, that all circumstances and conditions being equal, I may state them as quite established facts. It is meant here, that with persons of equal degrees of sensitiveness, the same striking facts will be invariably developed. As far as my experience goes, I should say, that, with great patience, and steady energetic perseverance, persons, at first, not remarkably sensitive, may be made to become sensitive by continued mesmeric passes, by pointing to the eyes with two fingers, or by gazing into the eyes, with strong will, for half an hour, for an hour, or for a longer period, once or twice every day, according to the susceptibility of the patient, or the vigour, perseverance, or energy of the mesmeriser.

If these be facts, sufficiently established by numerous similar results, we must conclude that Animal Magnetism is an established science. In tracing the universal gradations of forces, from those appertaining to mineral crystallic structures, to the forces connected with organic arrangements of more elevated ranks in the scale of being, all odic or magnetic influences, exerting more or less beneficial agencies on the human sensitive, we may well believe that, on this earth, the highest grade of magnetic force culminates in the magnet, Man. If this view be correct, then we must be prepared to yield great glory on earth to the name of the Baron von Reichenbach, who has so assiduously traced the striking differences that exist between the mineral magnetic force, and the higher grade he has discovered as the odic force; and the distinctions which he has protected by so much assiduous labour, and by such logical precision.

If I be right in my generalization, I may well be proud that my trains of reasoning, which resulted in the conclusion of all oils and resins being part of an extensive natural family of vegetable crystals, and again, afterwards, from extended observations, to the fact that all Nature was subject to crystalline laws. It may be imagined, with what delight, under such convictions, the Baron von Reichenbach's great discoveries were hailed. He not only corroborated the general principle which had taken possession of my mind, but he went further, and gave additional scope to a vague idea which had more or less wandered in my head, after reading the charming discussions pervading the four volumes of Murray's System of Chemistry: this was, that a force, universal in its extent and relations, was the actuating principle of each and every change of form in all nature's varied types of configuration. It is not pretended that this is a new idea. We are here tracing the steps of the mind towards what may be well claimed as *clear*, if not as new ideas; and this becomes important, because Mr. Faraday, in claiming merit for a most confused statement of that which he is pleased to call conservation of force, has retained what it would have been far more graceful in him to omit, after his egregious blunder in his dictation to the world, on the forces engaged in Table Turning, and perhaps after his attempt at definition in these words, "I believe I represent the received idea of the gravitating force aright, in saying, that it is *a simple attractive force exerted between any two, or all the particles, or masses, of matter, at every sensible distance, but with a strength varying inversely as the square of the distance.* The usual idea of the force implies *direct* action at a distance; and such a view appears to present little difficulty except to Newton, and a few, *including myself,* who in that respect, *may be of like mind with him.*"

Mr. Faraday's idea that Newton ever confined his view of the universal force of gravitation to *direct* attraction is of a piece with the concluding limb of his sentence, that the great astronomer might " be of like mind with him." Many men would hesitate before they put forth to the world such vanity as appears in the sentence just quoted. Sir Isaac Newton was no shallow thinker. He distinguished between science and a limited smattering confined to one branch of knowledge. Newton's mind was eminently humble. He never gave room for frivolous vanity. He thought of gravitation as a force not inherent in matter, and had, very evidently, far different views from his would be compeer. If we were to state, as we are now bound to do, what Mr. Faraday calls " the received idea of the gravitating force," we should declare that it is *not* that of " *a simple attractive force,*" but, on the contrary, of a decidedly compound force,—compounded of a force attractive,—and a force repulsive; in Sir Isaac Newton's own language, of a force *centripetal,* and a force *centrifugal.* How the confusion amounting to absolute ignorance of some of the most important propositions in the 'Principia,' of the world's great astronomer, could have obfuscated Mr. Faraday's mind, one is at a loss to guess. It is clear that a *direct* simple force was no part of Newton's doctrine on gravitation. Let us try to imagine what Newton intended to express. Bryan Robinson wrote an octavo volume to illustrate some of Newton's propositions, which, if Mr. Faraday has attacked them in the 'Principia,' may have been too much for him. The calculations submitted by the illustrator, tend to shew that such an idea as a "*direct*" force, without modifying influences, could never have been Newton's meaning. Mr. Faraday himself, who has some pet ideas on Magnetism, and Diamagnetism, ought to have known better, than to have limited his ideas to a "direct" force. He must have known how far he was warranted in assuming the existence "*of a simple attractive force*" in gravitation, when he himself claimed a discovery, which I have shewn to belong to a French philosopher, named Thouvenel, who had discovered the facts on which Mr. Faraday rests his claim to discovery, as long back as 1783. Many are the evils which arise from too eager a desire to be thought a far superior man to one's neighbour. This is exemplified in the case of Mr. Faraday. He was educated to be a chemical manipulator, and he wisely took up Electricity as a subject of investigation. His mind was well adapted, by habits of ingenuity, for making out new facts dependant upon combinations of instruments. Instrumental manipulation is his forte. His previous acquirements had not fitted him for a deep thinker. He was surrounded by men of inferior calibre of mind, and was soon puffed up with an idea of his own importance. It was not long before such a character was called upon to witness facts far beyond his capacity of understanding. He was brought to my house, when I resided in Grosvenor Street, by my friend, the late Andrew Crosse, who entered into my views with great ardour; but was not able, from deficient knowledge in physiology, to comprehend the important bearings of the facts I shewed to him and to his friend. I was much struck with the want of philosophical acumen in Mr. Faraday. I found in him a vain man, with somewhat of a supercilious temper. His language, and the tone of his general bearing did not convey the idea of a man of enlarged mind. His objections were small and petulant, and of a nature by no means calculated to convey the idea that he deserved a hundreth part of the reputation he has acquired. Famous as he is, in Europe, he will not fail to fade before many years are over. His blunder in the attempt to explain Table Turning, and his lecture on the conservation of Force, must for ever settle his reputation.

All that I desired to shew to the Royal Institution Professor, related to a set of facts dependant upon his own subject of investigation—Electricity; not, however, simply to the lower grades of electricity, that of the mineral class of phenomena; or, I claim to expose to the world, facts appertaining to a higher electricity—that belonging to the more complicated arrangements of organic molecules. I assert that, in the humble and modest spirit of a real physiological investigator, I offered to Mr. Faraday, in the presence of a man of genius, the late Andrew Crosse, a full opportunity of investigating the facts of the highest electricity of which we can be cognisant while we inhabit this earth. I mean the Magnetism of the Human Being. I now humbly submit the facts. When my friend Crosse came, accompanied by the Professor, I was prepared to exhibit to him, that which I had exhibited to numerous other persons. They were astonishing phenomena. They related to the influence of crystals, or, if I may be more strictly philosophical, to the forces temporarily taking up their abode in crystals, upon the human body, in the production of sleep and wakefulness. I had a very interesting and sensitive young woman present, whom I had cured of epilepsy in the Middlesex Hospital; and I was prepared to shew the curious facts relating to this case to the singularly obtuse professor. She was so susceptible to the higher influences of organic magnetism, that I had only to will that she should sleep, and she slept at once—instantly. If, in her ordinary state of wakefulness, I made a single pass of my hand, one inch from her person, beginning at the forehead, and proceeding to the pit of the stomach, she slept instantly. The fact I wished particularly to shew to Mr. Faraday, was illustrative of the Baron von Reichenbach's discoveries on the influence of crystals on the living human subject. Many and many a time had I shewn to Mr. Andrew Crosse, and I may say to hundreds of others, the fact, that when Mary Ann Douglas was placed *awake*, and standing, in my hall, with her face towards my dining-room door, and absolutely ignorant of the fact, that any one, in the dining-room, was holding the pointed end of a rock crystal to the door, in the direction of her person, she was obliged to sleep instantly.

Standing asleep, with her face touching the door, for the attractive force of the crystal obliged her to approach it, she was speedily awakened by the opposite end of the crystal being held towards her. She was attracted by the pointed end of the crystal sending its influence through the wood of the dining-room door, and approached the door asleep, touching it with her forehead, and soon afterwards, she was repelled from the door and was wide awake, because the opposite, or repulsive pole of the crystal was held, so as to shed its repulsive force, and oblige her to become awake.

Is there any one of my readers so stupid as not to estimate the importance of these experiments. The facts disclose attraction and repulsion in their naked truth. The fact just stated, has been demonstrated twenty times successively in half an hour, while the door was held ajar, persons in the room being able to see the crystal, and to witness the opposite influences on the subject, according as the pole, positive or negative, was held in the direction of the subject; and the demonstration of the young woman being absolutely unable to know how the experimenter was manipulating the crystal, being quite perfect. I have no hesitation in declaring my conviction that Mr. Faraday, brought to my house, by the kind, single-hearted, and ardent Andrew Crosse, behaved quite unworthy of a man, pretending to possess philosophic tastes. He exhibited petty temper, and evinced the possession of a small mind. He could not soar above the poor and miserable minds of those who endeavoured to obstruct Dr. Elliotson, while he was promulgating new and important facts, leading to great discoveries, at the London University College, when that devoted martyr to the cause of a holy truth, exhibited the marvellous phenomena, in the cases of the two sisters Okey, which, in many important respects, were quite analogous to the case of Mary Ann Douglas, and of several others in my cognizance. But Mr. Faraday and all his followers may be assured that, strive as they will, they cannot overwhelm the truth. They have done their utmost, not to say their best, to obstruct it. It will prevail, notwithstanding their mulish and obstinate desire to crush it. It is strange, but very melancholy, that men who have wrought for themselves a certain reputation, cannot perceive that there are higher aims in true philosophy, than any which have occupied their limited minds. Well might Andrew Crosse, that amiable and poetic man of genius, for real genius can well appreciate a genuine truth, exclaim, when estimating one night, in my library the claims put forth by the so-called men of science, "I would rather have one grain of the milk of human kindness, one grain of kindness of heart, than waggon-loads, aye, ship-loads of learning and science." He had had some experience, as I have had full knowledge of the extent of those dominions of detraction, envy, malice, hatred, and all uncharitableness, that spread their sway where the regal rulers of so-called science and professional intrigue hold their sordid courts.

When we turn from cliques and their base and narrow schemes of self-glory, to higher characters, we may well remark quite another man in Mr. Rutter, of Brighton, who as the discoverer of the means of testing the most delicate Magnetic forces, must exert an influence on the future progress of science which it is quite impossible to estimate. No man can predicate the extent of range in the discoveries to be hereafter made through the agency of Mr.

Rutter's Magnetoscope. Already has it been shewn that all the objects in nature are subject to Magnetic laws. We will not say that Mr. Rutter would himself admit the full conclusions to be drawn from the premises he has established, that man is both a crystal and a magnet. Happily, we are not called upon to agree unanimously upon all points. We shall soon know many facts which will make the vast importance of the Magnetoscope more apparent. Phrenology will, in time, lean upon this instrument for the astounding discoveries for which the world will have to thank the late philosopher Gall, and Mr. Rutter's disciple, Dr. Leger, a man of transcendent genius, who did not live long enough to leave the statistics of the most important facts relating to the human mind in a train for publication. These will, in time, force themselves on public attention, and when they do, men will conclude that the present race of scientific men are a set of pigmies. What can be more absurdly in contrast for its frivolity than the volume of Pyschological Enquiries, by the late President of the Royal Society, and the labours of Mr. Rutter and Dr. Leger?

There are few subjects that require less originality than the facts of Psychology, as they were treated by Sir Benjamin Brodie. A youth in his teens might write just as profound a book as that produced by him. In the time of Professor Jardine, I have known boys in the logic class of the University of Glasgow produce works really far superior, in close reasoning, to any thing the late President of the Royal Society has ever issued from his pen.

It is time to shew that the pursuits of Magnetism, electricity, physiology, and psychology have not only received a heavy blow by the choice of that president, but that he was worse than ignorant of the great interests of science. Luckily, this is a very easy task. We shall commence by shewing how we propose to ameliorate the condition of Mr. Faraday's mind on his crotchet of conservation of force. He is little aware that his own facts, made out by convenient adaptations of Grove's battery, and other apparatus of coiled wire and magnets form the basis of our explanations of the impossibility of Sir Isaac Newton's views, being those of a "simple, direct, attractive force." Mr. Faraday has made out that when, with his electro-magnetic coil, he passes a direct current from the battery, the course of that current becomes reversed, the moment it meets a magnet. This is a reversal of polarity. It is very extraordinary that, with common ingenuity, Mr. Faraday could not apply this fact. The source of the magneto-electricity which reaches the earth is a direct current from the great magnet, the sun. When it strikes the lesser magnet, the earth, its polarity is reversed, and it becomes an inverse current. We learn from the facts of Animal Magnetism, and Dr. Noad might have taught Mr. Faraday this, that the inverse current is a repulsive force, or what Sir Isaac Newton called a centrifugal agency. This explains the difficulty which has given rise to the mistakes in the lecture on Conservation of Force. But we have more in store, in the way of explanation. Mr. Faraday claims great credit for his doctrine on Magnetism and Diamagnetism. Let us ask, what is the difference between a hazel-rod or twig, and the apparatus by which the great discovery that in the magnetic field, between two powerful poles of a horse-shoe magnet, certain metals are obliged to assume an axial, and certain other metals are forced to lie in the equatorial direction. M. Thouvenel, making experiments with a hazel-rod, and sometimes a small twig, placed on the hand of a sensitive man, named Bléton, found, as the man walked upon ground under which iron or other of Mr. Faraday's magnetic metals was found, the hazel indicated the axial direction; and if the metals were of Mr. Faraday's Diamegnetic list, the hazel turned on the hand, and placed itself, or rather was placed by an unseen force, so as to indicate the equatorial direction. Mr. Faraday would simply observe that important facts arrived at by the aid of the hazel-rod, in the hand of a sensitive person, could not be by any possibility, *science;* whereas the same facts, determined by the aid of Grove's battery, were deep science. If facts relating to mineral Magnetism are to be discovered at the Royal Institution, they are all right. If far more important facts relating to the higher class of Magnetism, that belonging to the highest forms of organic structure, human Magnetism or vitality, then, forsooth, it becomes quackery.

We say Animal Magnetism is quite true; that it is fraught with benefits and blessings to mankind beyond most other truths; that it is leading us on to the highest inquiries which can occupy the human mind. If the investigators into the truths of Animal Magnetism have hitherto suffered dire obloquy, slander from the inner nests of certain cliques of medical colleges, abuse, vituperation, and outrageous falsehood from the wicked stupidities of their day, we have only to pity such poor vituperators and the cliques to which they belong, for their ignorance, and to pray that God may forgive them, for they know not what they do.

* There is yet another force to be mentioned, namely, *the force of inertia;* or that force which is in opposition to any force productive of motion, as that of death, is to that of life. One of the most beautiful examples of the force of inertia, is the dead stop of Mr. Rutter's pendulum, which takes place when his left hand comes into contact with some animal substance that has once had life in it, but is now quite dead. Here we may fairly say is an example of the action of a lethal current opposed to that of a normal current—in other words, the force of death opposed to that of life; a striking example of the force of inertia, and even in this vibrating magnetic sphere on which we exist, an example of a static condition.

CHAPTER V.

DR. LEGER, HIS FACTS.

ARRIVING at the consideration of facts that have been lost sight of by the world, because, in all probability, the small envy of the competitors in science with minds far more expanded than their own, was too much for the time for a man ill with a diseased heart and struggling against poverty. We must endeavour to do some justice to the memory of Dr. Leger, and if our readers can appreciate the importance of his discoveries, we shall not find it a difficult task to narrate, as simply as we can, the story of a set of facts which will be found to belong not only to the science repudiated by our so called *savans,* but really to be a part of the great subject of Psychology. We are not bound to respect the opinions of the wretched intriguer, who was the author of the "Psychological inquiries," whose absurd pretensions, when placed in contrast with the statement we shall now make, will be as completely smashed and demolished as if they were made of the most flimsy material.

Dr. Leger was a physician who had been educated in Paris. I had no means of becoming acquainted with his early history. When I first saw him he was in great poverty. Understanding that I took an interest in the facts of Animal Magnetism, he called upon me, and introduced himself as the author of a book then on my shelves, with the quaint title of Psycho-Dunamy. It was published in New York, where he had resided for a time. We may not scrutinize this work too minutely.

It was remarkable chiefly for a short but careful memoir, perhaps one of the best extant, of the celebrated Mesmer. Dr. Leger was glad to act as a mesmerizer, and I had the good fortune to be able to recommend him occasionally for employment. He was a painstaking modest man, and no one would have predicated of him that he was a philosopher, destined to become one of the world's celebrities. His manners were unassuming, and he was not able to conceal the simplicity of his heart. He was a man without guile. A Frenchman by birth, he was of too expanded a mind to be trammelled by national prejudices. I never knew a man more easily influenced to lay aside preconceived notions. Finding him thus a man after my own heart, I became very intimate with him. These particulars are stated only to elucidate some remarkable facts in his character. When he first applied to me in the year 1850, I was in the habit of having patients at my house in Grosvenor Street, who were daily mesmerised, and who offered phenomena of an intensely interesting nature. At the period of our first acquaintance, Dr. Leger and I conversed freely on many subjects of science. He was an excellent anatomist, and I found him *au fait* in all the French philosophy of anatomical development. He had been a pupil of Broussais, and I was grieved to find him a decided enemy to all Gall's philosophy; for the powerful and clear mind of Broussais had yielded its convictions to the truths of Gall's doctrines. Leger chose to think for himself, and the time of his conversion had not arrived. I quietly drew his attention to the facts of Phreno-mesmeric physiology. I had some fine cases very illustrative of these. Leger's attention was rivetted. He expressed a desire to look more deeply into the subject, and he was soon induced to lay aside his prejudices and to read the six volumes of Gall's works. I was too much delighted at this change to twit him with his former expression, that all that Gall wrote was heavy stuff, full of error, and not worth the trouble of reading. Dr. Leger truly was enchanted by the genius of the great discoverer of Phrenology. He was anxious that I would lend him the volume of the lectures on Phrenology by Broussais. This book he devoured with great avidity. With the true simplicity of a man of genius, Dr. Leger commenced his studies on his new subject. I watched his assiduity, and the untiring industry with which he laboured. He was not a man apt to follow blindly the dictum of anyone, however exalted his scientific position. Dr. Leger was not only an original thinker, but he was a man of great acquirement. Perhaps he was to be considered as characterised by singular habits. A physician of a high order, because he had regarded his profession with the singleness of mind that belongs to the real philosopher. Having based his knowledge on the sure foundation of anatomy and physiology, he was not apt to swerve into frivolities.

Nevertheless, he had lost himself by a fondness for sensual pursuits, a failing too common in his day, among a certain class of ardent and scientific medical men who formed his *entourage*. He lived in the midst of eminently literary and scientific society. He had formed a close friendship with a great celebrity of his time, Benjamin Constant. Who that has heard of that man could doubt the deep interest he took in his friends. The misfortune that attended the society of Paris at that period, was one that is fraught with melancholy ideas to the reflective mind. Paris was, *par excellence*, gastronomic. Nothing is so much calculated to undermine health as the habit of eating more than the stomach can easily digest. Even in London, poor Dr. Leger's love of eating and of drinking wine of a *recherché* character, undoubtedly accelerated the fatal termination of his serious disease. I learned that Dr. Leger had been improvident, but I did not too minutely inquire into the causes that had induced him to proceed to Mexico, where I understood that he had a brilliant career of prosperity. He had realised two handsome fortunes, which he had lost improvidently, partly through good-nature, but chiefly by a careless disregard of attending to the principle of insuring remittances by sea, and thus losing a large amount of property by shipwreck. It was the steady friendship of Benjamin Constant that secured him warm recommendations in America; but during his absence, death had removed his friend.

It was strange that a man so gifted with mental faculties, far above those of the herd of medical practitioners, for these are ordinarily bigots with some small power of progress, should have so far have forgotten the duties of a pupil of Broussais, as to return to Europe with prejudices against Phrenology, as strong as those he had carried abroad with him. This was the subject upon which we totally disagreed, animated discussions were of no avail. The numerous opportunities I gave him of witnessing and studying the facts of Phreno-Mesmerism captivated him at last: and he not only declared his regret at having so long entertained his old prejudices, but he expressed, in warm terms, the same admiration I have ever felt for the original genius, the capacious intellect of the discoverer of the truths of Phrenology. As for the great work of Gall, I know of none, which, for scope of thought, or for argumentative power in controversy, so forcibly reminds me of the intellectual grasp displayed in some of the prose works of our great John Milton.

It has been stated that Leger was a man of great simplicity of character. He had neither pride, vanity, nor envy. He was thus prepared to enlist himself as the humblest disciple of Mr. Rutter. Having once apprehended the importance of the Magnetoscope, he worked at it very steadily, and according to my ideas, improved it so far as to obtain a very simple, as well as correct measure, of the magnetic forces, which formed the subject of his studies. In order to shew the delicacy of those forces, I may diverge a little while I refer my reader to Mr. Rutter's experiments in the presence of Dr. Quin, a gentleman whom I introduced to my friend, at the request of the late Earl of Belfast. These are detailed in a little periodical which was supported by the believers in Homœopathy. Dr. Quin having placed in my hands a dozen tiny tubes or bottles containing globules, which, he said, had been made with extreme care at Paris by a very expert chemical manipulator, and each of which contained the salt of a metal, prepared by submitting it to successive solutions, so as to ensure a complete subdivision into parts, having each a relation to a grain of its respective salt, which required for its expression in relation to that unit grain, sixty places of decimals; these were separately, at proper intervals, dropped by me into the hollow of Mr. Rutter's left hand held at his back; and he successively announced to me, from the signs communicated to him by the pendulum of his Magnetoscope, the name of each metal, the diminutive quantity of the salts of which had fallen into his hand. We cannot conceive of delicacy surpassing this, unless we are prepared for what may be the real truth, that the forces of the class of magnetic phenomena, appertaining to the human mind, be more rare and more delicate still. We are sure that time is required for the influence of these, in operating changes in the form of the hard and compact vault of bone, we call the skull. Formerly, people talked of the influence of mind upon matter, but they had no direct facts to appeal to. We now know that the phrenological organs of the brain become much modified aud enlarged by the exercise of the faculties of the mind. They not only wear the bony skull sometimes so much, as to make it diaphanous, but they alter the shape of the head. One of the remarkable facts, shewing the influence of education in altering the shape of the head, was fully established by Mr. Deville, of the Strand, who with great zeal cultivated phrenology, by securing casts of the heads of boys whose career he could watch; and he found, that when at several intervals of time, he repeated the process of casting the same head, he had evidence of an increase of size, most decided in those organs which had been most cultivated and exercised. This fact admits of no cavil. The late Mr. Holme, of Highgate, took a cast of Mr. Samuel Taylor Coleridge's head, which became much diminished in size from his pernicious habit of eating opium. Coleridge sneered at Phrenology, and had no idea of the purpose for which his friend Holme required him to submit his head, a second time, to a disagreeable process. Sir E. Bulwer Lytton was with me when Holme took the second cast of my head, and he then witnessed Mr. Holme's application of his calipers to Coleridge's two casts. The contrast between them was very striking. In my own head there has been a great change. The interval between the two casts of my head was seven years. I had abandoned the use of wine; I had acquired improved power of self-control, and I had much exercised the intellectual faculties. These created a magnetic accession of mental forces, which completely altered the shape of the head. My head must now be again much improved, for under the direction of high spiritual aspirations, I have worked harder, and have eaten less than I ever did. I am in better health; I have become more calm, more contented, more intellectually powerful. These facts improve the argument so often insisted upon, that mind is the creative force, that matter is subservient to it, and of this we may rest assured, when we trace up the gradations of forces, independent of conservations of force, and of correlations of force, simply because the ideas, now reproduced, were published in the "Zoist," years before Mr. Faraday or Mr. Grove appeared to be dreaming of their confused ideas.

Many considerations have to be entertained on Phrenology before we quit this part of our subject which cannot be here lucidly given to the world as our views on forces. They are of a more vast importance than people may imagine. It is not to be assumed that time alone can give them the importance they merit. They must be now made as clear as we can make them, for independant of the clumsy filching they may undergo, we are anxious for the world's sake, that our ideas may be well understood. We are not writing a common-place book. We are endeavouring to establish principles that shall guide men, when we have quitted their sphere, to work at the real and substantial improvement of the facts which constitute medical science. I stand forth to declare that our science is really a holy one, and my work shall remain as a monument of the efforts I have made to prove it so.

Dr. Leger made out a great deal more than we can now prove, simply because our time and space will not allow us to go into a statistical detail that would occupy a volume four times the size of our present book, but we may try to make our readers understand the general propositions involved in Dr. Leger's discoveries. We may at once say that he was the discoverer of the Magnetism of the phrenological organs of the brain, and of the vastly important facts which establish Psychology as a mathematical science. Mr. Rutter had already led the way to this very important series of conclusions, by proving Man to be a magnet. His disciple earned a share of his master's claim to have his memory handed down to posterity as a benefactor of his race, by his ingenuity, by his modest assiduity, and by his marked humility in declaring that Mr. Rutter's mind had been the cause of his discoveries in phrenological science. He used to head all his papers of phrenological examinations by making a similar acknowledgment, and it was charming to witness the expressions of respect with which he adverted to the discoverer of the Magnetoscope. Regard him apart from his sensual love of the pleasures of the table, he might well be considered as an extraordinary man. We are apt to remark on what would have been the feelings of a man, could he have anticipated the magnificent discovery which it would soon fall to his own lot to make. But, with the humility, that characterised Leger, it was his privilege to feel that he was but an instrument in the hands of the Almighty Being.

The studies in which Dr. Leger had now engaged himself, prepared him for the use which he subsequently made of the Magnetoscope. On one occasion of my return to London from Brighton, where I had repeatedly witnessed the experiments of Mr. Rutter, I communicated the facts to Dr. Leger. Having himself verified all the results, he reflected much on the subject, and then modified the form of the instrument by suspending the pointed sealing-wax from the upper part of a brass tube, which was about an inch and a half in diameter, surmounting a bell-glass. Instead of the ball which Mr. Rutter embraces between his thumb and fore finger, Dr. Leger conceived the idea of a flat disc, upon which he rested the middle finger of his right hand. There were other details introduced by him, chiefly with a view to obviate the cavils of sceptical objectors, some of which need not be detailed, others, more interesting, may be noticed. The bell-glass curiously corroborates Mr. Rutter's original idea on the magnetic relation of this extremely delicate instrument. It is a remarkable fact that the vibratory magnetic force of the human subject is communicated by the operator to the glass of the bell; for, if the most minute star, or other lesion of the glass, should exist, the instrument would become useless and could not possibly be made to act. Dr. Leger found it necessary to contrive some accurate measure of the forces indicated by the pendulum; and this he effected by the simple expedient of a circular card placed upon the plate glass used by Mr. Rutter. This card had described on it, at one time, twenty-five concentric circles, the centre of which was exactly under the pointed extremity of the pendulum at less than the third of an inch distant from it. At a subsequent period, finding that he required a wider range for the pendulum, he increased the number of his concentric circles to thirty. The circles were about a fifteenth of an inch apart, and a contrivance was resorted to in order to enable him to read, with facility, the number of the circle to which the point of the pendulum reached; he divided his thirty circles into six series, each of which was distinguished by a separate colour. Dr. Leger found occasion, as he gained increased experience in the use of the instrument, to vary at different times the construction of the pendulum, altering its shape, and even the material of which it was formed. He changed the sealing-wax for lignum vitæ, and finally found that ebony, turned into a shape somewhat resembling an inverted pear, with a fine point, was the most desirable to use; this wood enabled him to form a hollow cavity, which he loaded with mercury, and thus obtained the means of ensuring greater accuracy of definition in the range of his pendulum, while it described its rotations and oscillations, on the area of the circular card, with the most precise regularity. For all this, it was requisite to have the instrument level, and this object was secured by adjusting screws.

When the instrument was ready for use, I witnessed Leger's incipient experiments, which will be found detailed in his book on the Magnetoscope, dedicated to Sir David Brewster, and published in 1852 by Baillière.

Conversing with him on the prejudices formerly indulged in by Broussais on the science of Phrenology, and the subsequent conversion of that man's large mind to the truth of Gall's doctrine, Dr. Leger proposed to test the magnetoid forces of the various phrenological organs of the head by his Magnetoscope. To his astonishment and delight he found that each organ had its own influence on the pendulum; and that the characters and the forces of the movements, whether rotation, oscillation, or elliptic course, varied in energy and intensity, according to the organ tested, and that the extent of movement communicated to the pendulum by each organ varied in different individuals. This was a great and an important discovery. With unwearied zeal and patience, he set to work to improve his apparatus, and by dint of great assiduity, he became a very expert, practical phrenological manipulator. I was often with him, and can testify to the ardour, zeal and steady industry with which he energetically pursued his new subject. Amidst vexations and serious difficulties, for with a wife and two children to support, the *res angusta domi* pressed upon him, he struggled on, and tried to draw public attention to his pursuits, by delivering Lectures at Hungerford Hall. The scheme was not successful. He then printed cards, announcing *séances* for the examination of heads, at a small fee. This idea rapidly brought him into notice, and such numbers of persons flocked to his humble apartments in Gerrard Street, that he began to realize a good deal of money. Some days he received as much as ten pounds, but he had, with broken health, been working too hard. Just as prosperity began to dawn, a disease of the heart entailed pulmonary apoplexy. An illness of several months succeeded, and notwithstanding the close attention and fostering care of his friends, Dr. Chapman and Dr. Quin, who aided him with their purses, as well as their medical knowledge, and notwithstanding the unceasing surgical attendance, rendered with a brother's friendship, and administered with the well-known skill and untiring benevolence of Mr. Kiernan, he died on the 6th October, 1853, aged 54.

If we ask the question, where are the results of his labours? Or the question, did not the principal scientific men of England care to inquire about the matter? We may rest assured we shall get no satisfactory reply. The statistics of his experience, now in the possession of his kind benefactor, Mr. Floris, of 89, Jermyn Street, would, if published, be almost astounding. Dr. Leger was enabled to deduce certain laws from the various combinations of the numbers representing the magnetical forces of the different phrenological organs, which allowed him to predicate with curious accuracy the offences for which any given set of prisoners, in a jail, were committed. In the House of Correction, at Cold-Bath Fields, Colonel Chesterton, the governor, was kind enough to allow Dr. Leger to examine the heads of 126 prisoners, all unknown to him, whom he had never before seen, whose names he did not know, and whom he distinguished simply by the numbers attached to their cells. Colonel Chesterton drew the attention of two of the visiting magistrates, Mr. Pownall and the late Mr. Serjeant Adams, to the Magnetoscope. These gentlemen were much interested, and selected the numbers attached to twelve individual prisoners, and, having assured themselves that Dr. Leger had no other clue to the offences of these men than that of the numbers by which they were distinguished, and the indications of the magnetic energies of the phrenological organs of their brains, afforded to him by his instrument, they took their seats in the board-room, and asked him to give them an account of the facts he had elicited. To their astonishment, Dr. Leger, from his notes, deduced *minutely* the offence for which it was probable that each man had been committed to that prison, and this was done with so much accuracy, that especially Mr. Serjeant Adams was struck with the truths of phrenology. He subsequently frequently visited Dr. Leger, and brought several persons to witness the phenomena of the Magnetoscope. Mr. Serjeant Adams had two sons whose heads were examined by the Doctor. I did not know them, but I can testify to the Serjeant's conviction that Dr. Leger had fully satisfied him on the powers afforded by the instrument of distinguishing minutely the opposite characters of these young gentlemen.

It is not necessary to go into a detail of the cases which formed the subject of inquiry in the House of Correction. It is sufficient to state that the offences of the prisoners were so unmistakeably indicated, that, were it proper to mention all the facts to the public, unbounded would be the astonishment at the accuracy with which Dr. Leger distinguished, not only crime, but abnormal tendencies to aberration, from healthy degrees of propensity.

It may be observed that much detail is here avoided, but it becomes a natural and important question: how could all this be done with so much accuracy? The answer is very simple. The concentric circles have been explained. While Dr. Leger's right hand middle-finger rested lightly upon the immoveable disc of his Magnetoscope, which was connected with the tube whence the pendulum proceeded into the bell-glass, forming a part of the instrument receiving the magnetic influence of the human crystallic or vital force, the thumb and middle, or index finger of the left hand, were applied to the organ to be tested. The pendulum, by a wider, or less extensive range, indicated the magnetic force of the phrenological organ. Each organ had its own movement, indicated by the pendulum, as well as its amount of force, measured by the number of circles to which the pendulum reached. When the sums representing the organs of greatest activity were added together, the inference as to the character of the individual was easy, and seldom failed in being perfectly accurate. It was a numerical process, and was the germ of what must be, at no very distant period, the application of mathematical laws to the functions of the human mind.

Dr. Leger found means, by patient study, and great acuteness, of discovering tendencies to aberration from natural or normal manifestation of propensity. These, by the aid of his instrument, were certain of detection. In the asylum at Colney Hatch, he examined epileptic cases, and some of the insane. The failure of his health prevented his pursuing his investigations, but could his statistics be published, all scientific Europe would be set agog at his discoveries. Our stolid cavillers in England are not worth a moment's thought. They *will* to travel in their own grooves. They obstinately persist in hugging their antique and ancient thoughts. Their wills are not regulated by humility, and the self-control necessary for the purpose of eliciting the delicate truths, their stiff and gnarled minds would fain love to grasp, but cannot, is wanting. That one law of Dr. Leger's relating to the absence of the power of self-control, in the epileptic and in the insane, is so illustrative of the doctrine of possession, that in future times his name will be regarded as that of a most acute observer. Standing alone, unsupported by the scientific spirits of his age, he made out a fact that must exert hereafter an immense influence in the appreciation of the causes of insanity. That fact was that, in most of the insane, and in all cases of incurable epilepsy, there existed a very striking disproportion between the magnetic force of concentrativeness, and that of some other organ in the moral or intellectual group, generally ideality. Whenever this disproportion was observed, ideality was represented by twenty or twenty-five, and concentrativeness by a very low figure, perhaps two, or by *nil*, which was indicated by the curious phenomenon of the pendulum coming to a dead stop.

An epileptic gentleman, who was a patient of my own, had his head examined for me by Dr. Leger, and the accuracy of the law he had reached at Colney Hatch was here strikingly verified. Ideality was represented by twenty-four. All the other organs of the head were under twelve, most of them ranging about five. Concentrativeness did not allow the pendulum to exceed the limit of two. I had carefully abstained from giving the Doctor the remotest hint on the malady afflicting my patient, but, taking me into another room, he communicated to me his fears that the gentleman was an incurable subject of epilepsy. This has proved to be quite true, although, from the pendulum acting, and taking even a small range, I was led to hope that there might exist a chance for his recovery. I should have had no hope had the organ of concentrativeness brought the pendulum to a dead stop. This curious and remarkable phenomenon, be it observed when it may, cannot fail to arrest the attention of the reflective mind. Mr. Rutter finds that all dead animal matter, held in his left hand, while operating with the Magnetoscope, arrests the movement of the pendulum in a few seconds. Arsenic, held similarly, produces the same phenomenon. Several other deadly poisons would appear to have the same effect. The left hand held within the range of the curves of the magnet, near its north pole, while the right is occupied with the instrument, stops the pendulum. A body suspended in the air, rotating, gyrating, oscillating, with various degrees of force, appearing as if it were endowed with vital energies, to be at once arrested in its antics, as it were, and to be shot dead as if by a rifle bullet, is a marvellous phenomenon. Who can regard the fact of a sudden death, before one's very eyes, without emotion? This phenomenon is calculated to inspire one with some degree of analogous awe.

The various causes of all the steps of Dr. Leger's progress towards his great discovery of the application of mathematical laws to human motive are very interesting. One looks back to the early observation he made that, relatively to himself, five was the average number representing each organ. Relatively is the word used, and naturally. When I could manipulate and work the Magnetoscope, the scope of my average in examining the organs of the head was three. Why have I lost my power completely? My right-hand wrist and elbow have suffered from gout, and thus I am rendered unable to carry out Dr. Leger's investigations. As a general rule, all persons who adopt the pernicious habit of smoking lose the inestimable faculty of perceiving the refinements of high mental and moral perception, and none of them can manipulate the Magnetoscope. Narcotics of all kinds are not only injurious to the brain, the size of which they tend to reduce, but they so decidedly influence our sentiments and feelings, that their tendency may truly be said to extinguish virtue and to encourage vice. Hence, all tendencies to evil are the pernicious sources of vice, as those to good are the healthful sources of virtue.

But it is to be hoped that, keeping in view this fact of the relative difference of different individuals, labourers will be found in time to continue the investigations, arrested in their progress by the removal of Dr. Leger. It has been said that five was the number this philosopher adopted as the figure representing the average force of each organ of the brain. So that, if one organ gave a higher result, there were others that gave lesser numbers. In this way, he arrived at the fact that the average number representing the head of the individual we ordinarily meet in our walk through the streets of the metropolis is about 180, or five times thirty-six, the number of organs on the phrenological busts mapped according to the best authorities. This led to the further examination of heads of remarkable persons, and the average number was sometimes exceeded. Thus Lord Ellesmere, who had remarkable mental powers, had a magnetic activity of the whole brain, amounting to 350. A person who was anxious not to be known, and who had somewhat disguised himself, always carrying an umbrella under his arm, in a peculiar manner, for which he was distinguished as the Paul Pry visitor, was found to have a magnetic activity of brain represented by 320. The philanthropist, Robert Owen, had an activity of brain reaching 314. Lord Mahon and Babington Macaulay, afterwards Lord Macaulay, had 308 and 310. A great many distinguished men had numbers varying from 270 to 300. Very few exceeded that aggregate number. On the other hand, where mental activity was small, as in the poor Irish servant of-all-work, a number of individuals gave results varying as low as between 130 and 140.

Dr. Leger's researches extended to upwards of 2,000 cases. These enabled him to collect a series of statistics of the most remarkable nature. But his work was left unfinished. I have said before, that his "illness and death were a great loss to the progress of phrenology, and it may be said to have been a real calamity to the department of the medical sciences to which his researches appertained. Insanity would have had a brilliant light shed upon its philosophy. Psychology has to wait for the observations of some competent successor to Dr. Leger."

We must now advert to certain topics connected with Phrenology, arising from the considerations in which we have been indulging. We are far from asserting that our claims are to be compared to those of the great spirit of his age. Gall was a man of no common calibre of mind. He was born in Vienna, but treated with want of consideration among his own countrymen, he went to reside in Paris. Here he published his great work. We may well call it his great work, for he was not able to make mankind understand it, until he was becoming too ill to care for anything. A few were found to be his disciples, but the scientific world threw every obstacle in his way. They were addicted to their own conceits, and muddled themselves with details of fibres and decussations *ad infinitum*. His Anatomy of the Brain established his character as a man of master-mind. Still the world was not advanced enough to appreciate his philosophy.

We ought not to omit all notice of the disciple of Gall, the almost wondrous Dr. Vimont, whose labours were so incessant, and who published the magnificent book on the Comparative Anatomy of Phrenology, and which made Leger wild with admiration at the patient philosophy of a man who could produce such a work. Much may be said in favour of his doctrines without embarrassing our progress, but we know that we have no occasion, nowadays, to advocate Phrenology. We are prepared to hear that none but scientific men can be found to take up such a subject. It is enough to know that we are not scientific. We are glad to class ourselves among the herd of those who are daily increasing in numbers. They are men of education who have trained their minds to think correctly. We are glad to find that we are not quite alone, numbers now believe in the truths of Phrenology, who at one time disputed the possibility of any truth belonging to the subject; and it will, in good time, become more creditable to a man's intellect to be classed with the phrenologists, than to be classed with the select few who could be found to vote that small-minded Sciolist, Sir Benjamin Brodie, into the chair of the Royal Society.

Time passes, and men change; we feel ourselves becoming old, and we do not hesitate to acknowledge that our age is an advantage. We know that, during a long life, we have spent our time in acquiring knowledge. We have regarded the labours of our brethren as works to be respected, notwithstanding the license we always give ourselves to criticise them very freely. We are not sure whether the habit of criticism has not sometimes led us too far, but this we know that we should be sorry to be told that we had spared a man who had mounted, by the force of cunning pawky intrigue, step by step, into a position to which he was not entitled, and without an atom of pretension to fling abroad the assertions of his gross ignorance, in the place of establishing for himself a high reputation for great genius, and its accompanying characteristics, a high mind, and a pure love of truth. Seldom have we had to denounce so forcibly any man who has attained a high position in society; but we feel assured that our motives are unimpeachable, our object being to shew the world that Animal Magnetism is a science which ought to have been investigated by men having pretensions to high scientific position—and when we find men who aspire to that position, abandoning their imperative duty in a manner so flagrant, we are compelled to shew not only their dereliction from the standard of high moral feeling, but their gross ignorance on the subject they profess to denounce. We are aware that we shall be stoutly assailed for our exertions. We do not choose to care one straw for all our assailants. They may know beforehand that the great Harvey was ruined by his assailants. We would glory in the same ruin. We know that the cause of truth is the cause that God protects, and whether that cause had the accident to be assailed by the small Sir Benjamin Brodie, or the host of pigmies who may be ready to do battle for him, we care not? We shall pursue the even tenour of our way, and endeavour, by cogent facts and clear reasonings, to shew that the holy truths connected with Animal Magnetism are not to be overthrown by ignorance and prejudice.

Enough, then, has been said on this topic, we proceed to inquire how far the world is prepared to accept our explanations of the phenomena of the science in question. The facts are undoubted. They are of great antiquity. But men of science have not turned their attention to the inquiry many years. Mesmer is said to have been the first who was struck by the analogies between what has been named after him, Mesmerism, and the phenomena of Mineral Magnetism, Mesmerism is, in fact, Animal Magnetism.

If we admit that we can measure the magnetic forces of the various phrenological organs with so much accuracy, what is our next question? Do we admit that the forces of the brain influence our conduct? Are we, or are we not subject to moral laws? What are those moral laws? These are questions to be answered by our psychological philosophers. I mean by philosophers, men worthy of the name. We do not want answers from stupidities. We are not now engaged in clearing the road of the rubbish of sciolists. We are earnestly asking those capable of appreciating our questions. Well then, if the forces of the brain do influence our conduct, in what way can these forces reach us? for it is admitted by all sound reasoners, that the brain does not and cannot act of itself. It is inert matter. We know, as mesmerists, that the force of our will can compel the organs of the brain to yield to its influence. We do not say the organ acts. It suffers. We may figuratively say it acts, as we say a muscle acts, or a rope acts in tightening a sail or in tying a spar; but we know very well the brain, muscle and rope are obedient to control. Our language being figurative, we must admit we do not mean what we say. How obstinately we nevertheless adhere to our fallacy.

We proceed to inquire how far we are warranted in the strict use of terms? If we do not acquire the habit of saying what we mean, we slide into a life of hypocrisy. Are there any excuses for such a life? Can we say that it is necessary? Can we plead that we are not satisfied with too severe a habit of definition? Is truth so very irksome that we will not have it at the price we are bound to pay for it? None of these excuses will serve us. We profess to a philosophic love of truth. We are not to swerve from that profession. We have then no choice left. We must go on with our sequences, whithersoever they may lead us. We now come to the reply to the question. The strict use of terms is absolutely imperative. There is no philosophy without it. Many would be glad to escape from the necessity of using proper definitions, simply because it delays their progress. Thinking correctly involves doubts; doubts are perplexing. They must be satisfied before we proceed. Nevertheless, though they correct our rapid progress, they are the forerunners of truth. For he who does not doubt, cannot feel satisfied that he is not mistaken. It is after doubting that we become sure there can be no error in our train of reasoning. This is so important a point that we may be sure that doubts serve as the landmarks of truth. Few, then, can be said to be thoroughly religious, in the philosophical sense, who have not doubted at some period of their lives. But this is more a question of theology than of physics. We hold that all theology is based on physics, as we hold that all psychology is based on Animal Magnetism. He who writes on pyschology makes a sorry hash of his book in these days of facts if he be ignorant of Animal Magnetism. This is indeed the essential science to the Psychologist. All the old bases of metaphysics have been blown to the winds, and he who would make himself master of the science of mind, must first become familiar with the science of magnetism, and then with all that has emanated from it. Animal Magnetism is, in fact, a corollary to Electricity and Magnetism. We deal, from the beginning, with the forces influencing matter. First with those influencing its simpler forms, then with those influencing crystals, which begin with water, and go on to minerals, vegetables and animals. Our science is thus made out to be relative in gradations to all the various natural sciences. No part of the Cosmos is free from its influences, and man would be at a loss to point out his relations with the Great Being who created him, if he could not refer the relations of his own life, of his own mind, of his own soul to the great trunk force, which constitutes the governing principle of the Almighty's rule, both in the moral and physical government of his kingdoms. We are about to shew that these statements are not a whit exaggerated, by entering upon a detail of facts, as deeply interesting and as important as any that can occupy our mental faculties.

More need not be said on the great subject we are now dilating upon, than that we are sure of the interest it must excite in all minds that are rightly constituted. There is no fear of our being submitted to the obloquy of small cliques of dirty governors of petty hospitals. We are not afraid of the large public; but we wish at all times to be delivered from the influence of those who attend public boards for the sole purpose of gratifying their intense vanity, or of cajoling their neighbours into a belief that they are wondrously charitable, while we, who have been behind the scenes, know the dirty motives which actuate many such persons. The picture drawn by Mandeville in his book entitled "Fable of the Bees, or Private Vices, Public Benefits," may well be a lively portraiture of some vain and ostentatious men whose names I could mention.

Forces that impel men to act as they do, then, are our present topic. These have existed long before the Bible was written. These are the forces alluded to in that volume of deep knowledge, as actuating the conduct of men, but not in the very language we now make use of, for the simple reason, that we live in more advanced times. We do not now talk of daimons, or demons, or of the devil, because we are better aware of the nature of the spiritual intercourse between the world of spirits and the world of flesh, obedient to magnetic laws. There are few of us who would not be startled to be told that the devil was at our elbow, prompting our thoughts and controlling our actions. If we clothe the same fact, for it is a fact, in the words of Science, it loses all its alarming effects. Suppose we say that man is a magnet, that his habits have led him to neglect his power of self-control; that the laws which regulate his conduct result from the agency of spiritual forces, forming part of the divine government of the universe. This, though humbling enough to our pride, is not so disagreeable as to feel that we are subject to the suggestions of the gentleman in black—that spirit of evil, whom folks call the devil. What better example do we require, that weak mortals as we are, we are apt to be led away by empty words? Things are facts. Words are but their shadows. Let us ask how it happens that we of the two prefer the shadow to the substance. Mr. Faraday's facts, as far as they go, are very valuable; but when he ventures away from the philosophy of facts, and gets into the world of shadows, he would fain persuade himself he becomes clearer in his intellect. He loses sight of Magnetism and his grasp of forces, and flatters himself that he is a clear-headed philosopher, whose great function is to see that no force is lost in the shadowy haze in which he allows his mind to wander. He need be under no apprehension that the Almighty is so clumsy as to allow of the loss of the smallest function of any of the forces which emanate from that trunk force it was His pleasure to allow Sir Isaac Newton to reveal to the world. Mr. Faraday need not fear that Animal Magnetism will be dragged into his world of shadows. Notwithstanding his contempt of it, there is little fear but that it will prove his world of shadows to be a world of unrealities, and the world of spirits to be one of real facts. Animal Magnetism will shew that Mr. Faraday's scruples about table-turning and the phenomena emanating from this manifestation of what is now a hidden magnetic force, will be dissolved by a conviction that his darling science is at the bottom of all the phenomena which have so awfully puzzled and confused his brain, and those of some of his lieutenants.

Time passes, but events flow on in rapid succession. Who shall say when this conviction shall come to pass? We are no prophets. We may, however, venture to assert that if common sense can influence numbers of minds, the time is not far distant, when the arguments of the cogent advocates of free thought and exalted truth must prevail over those who constantly fail in their experiments, in their blundering contrivances, and in their efforts at bad reasoning.

It is clear that the time has not yet arrived, for it is necessary to carry on this sad battle against the enemies of truth. It would be more agreeable to plunge into the excitations on Gout and Rheumatism; but people are little aware how the prejudices which the enemies of mankind have succeeded in establishing against our science, and even against us personally, have obfuscated the intellects of those, who but for such erroneous feelings would be willing to be led on to perceive at once that health and disease are parts of the science of Animal Magnetism; and that without a knowledge of this science, no physician can clearly understand the elements of that physiological knowledge, which is supposed to guide him in the distinction and treatment of disease.

It is simply absurd to believe that any accession to the store of scientific knowledge can injure the interests of medical practitioners. This is the sordid consideration at the bottom of all the objections to Animal Magnetism. It matters little who are the successful practitioners so that disease is cured. But it is a very serious consideration how far the motives of what ought to be a high moral and learned profession are to be scanned, and found to be debased by the lowest and most profligate love of gold. There is no doubt that the conspiracies against private character and the wretched subornations of perjury are a disgrace to a civilized age. Still we cannot yield to those motives, which we assert to be emanations from evil spiritual forces. The men who have willingly lent themselves to these influences are our enemies, persecutors, and slanderers. We feel for them the pity arising from a conviction that we have reasoned correctly on many physical facts, which it has pleased God to make us the instrument to discover and to lay before the world. If these lead to a more just appreciation of the motives of human conduct, and consequently and necessarily to a more solid recognition of the noble law, developed in the First Epistle of St. Paul to the Corinthians, chap. xiii, it will be clear that neither all our sacrifices in the cause of Animal Magnetism, nor all the contempt bestowed on us by the proud science of England will be thought of the slightest consequence.

CHAPTER VI.

PHENOMENA OF PAIN, THEIR CAUSES.

THE Magnetoscope of Mr. Rutter having irrefragably established the proposition that all the gradations of Mineral, Vegetable, and Animal existence form one vast series of Magnetic or Crystallic arrangements—that in fact, all separate parts of crystallic forms obey a magnetic law, we arrive at the conclusion that all the forces in Nature governing the shapes of natural objects are as completely in steps or ranks as the objects they were created or commissioned to rule; and thus, in all their mysterious complications, are ultimately referable to the one grand trunk force of gravitation.

We might enlarge indefinitely on the sources of Light, Heat, and Electricity, and on their mutual dependencies. We might enlarge upon the phenomena constituting the science of optics, and on its relations not only to the facts so abundantly numerous as they are, and so closely connected with the phenomena of these subjects, but with the causes that are involved in the dioptrics and the catoprics of the earth we inhabit, and in the share which belongs to these departments of optical science in the magnetic attractions and repulsions which the sun generates in our globe. It is not necessary to follow out these considerations further than to shew the mutual connection of the series of forces which bind all nature up in the gradations of crystalline and magnetic structures, and which commencing with the spheroidal, mineral, magnetical molecules, proceeds with geometric law to rule the three kingdoms of nature, until here it ultimates in the magnet—Man.

The Magnetoscope, the instrument that enables us to establish such a series of facts, must immortalize the name of its inventor, and of the discoverer of the remarkable and important facts that have enabled us to catenate so sublime a series of gradations. It would be easy to reiterate the facts and propositions that have culminated in Mr. Rutter's great discovery that Man is a magnet. After having stated some of the evidence to shew that Mr. Rutter has established this great fact, we may proceed to reason on its natural consequences.

The Baron von Reichenbach, arranging a vast number of facts on human sensitiveness in relation to crystalline objects in Nature, with a view to questions on vital forces, arrives at a conclusion most logically deduced, that he had discovered a new force, to which he applied the name of *Od*. This force had close analogies to Magnetism. Mr. Rutter experimenting with his new instrument, the Magnetoscope, established the existence of forces in the living body, to which he gave the name of Magnetoid. Both these philosophers brought to light the existence of new facts on vital forces by the illuminating power of rare genius. It is quite unnecessary to quarrel about the claims to priority between these two enlightened men. They would each modestly disclaim any identity of pursuits or labours, although their discoveries throw brilliant lights on the progress of inquiry into the nature of forces generally, and of vital forces particularly.

We do not propose, at this moment, to enter into minute details of the steps by which the invention of the Magnetoscope was perfected; suffice it to say that these steps were worthy of Mr. Rutter's philosophic acumen. They consisted of a series of experiments made with a toy, which previous to his time had, in vain, for two centuries puzzled the ingenuity of men. It was a simple affair, and had been written upon by several old authors, who had occupied themselves on physical phenomena. It was a toy made by attaching a spheroidal piece of sealing-wax, no longer than a marrow-fat pea, or a small coin, to a fine filament of silk, the free end of which was wound round the end joint of the fore finger. Thus the bob of sealing-wax, or the coin, could be suspended freely about half an inch from the flat bottom of a glass tumbler. In this manner the arm of the operator being made steady by resting the elbow on a table, the pendulum, for such is the toy, finds itself obliged to assume certain movements. Now, in reflecting upon them, Mr. Rutter found that they obey a magnetic law. No one was sagacious enough to trace the phenomena attendant upon that law, until it fell to the lot of this acute philosopher to expose them to the world in his very unassuming work on "Human Electricity," the second and complete edition of which was published in 1854, by Parker, West Strand. These phenomena discovered an entirely new range of functions belonging to the pendulum, and when means were found to perfect the philosophical instrument, which its inventor has called the Magnetoscope, a most extensive field was opened for discoveries that are of an importance which no man in our generation has a sufficiently large scope of mind to embrace. For no man can imagine a state of knowledge in which the barbarous love of words shall be superseded by the just comprehension of things, when metaphysics shall have passed away, and all our science shall be concentrated in a knowledge of physical laws.

It may be perceived that I assign to the Magnetoscope no trifling importance as to its influence on the future progress of science. It has already indicated, in the hands of Rutter, the pathway to a close analogy between the forces regulating the series of vital phenomena and those of the sublime science of Astronomy. Leger, by the aid of the Magnetoscope, has opened the door to the store houses of facts which will link phrenology to the science of numbers, and join mental science in indissoluble union with mathematics. A man who has led the way to such developments as these can be no contemptible philosopher, whatever may be the estimation in which he shall be held by those who constitute the present world of science.

Mr. Rutter has established the fact that man is a magnet. Let us inquire into some of the consequences flowing from this proposition.

When we regard man as a crystallic magnet, or as a magnetic crystal, we have a regard to the series of crystals from the simplest form of mineral obeying the law of geometric shape to the most complicated aggregation constituting the magnet Man. We do not say that it is characteristic of all magnets to be insensitive. The gradations of forces lead us to a vast series of facts relating to the magnetic characteristics of many varieties of vegetable and animal tissues. Macartney was wont to speak of the organic sensibility and of the organic instinct of the separate tissues of the living beings in animated nature, we have many gradations of the varieties of the great fact that vegetables and animals exhibit these magnetic attractions and repulsions. The shrinking or retiring from the approach of an obnoxious offer, is only the same fact as the repulsion exhibited when similar poles of a steel magnet are offered to each other. Man, then, is a far more delicately sensitive magnet than a crystal of flint, the rock crystal, and yet the force residing in this substance, and constantly emanating from it is capable of being tested by the Magnetoscope. So beautifully transparent as some specimens of this crystal are found to be, no crystallization of flint in solution will ever take place where sensible light is present, while the magnetic crystal Man becomes diseased, and dies if deprived completely of sunlight. The reasons for, and explanations of these phenomena are not far to seek. They are dependant upon the attractive and repulsive agencies of light derived from the sun's rays, and modified in the substance of the earth by the laws of dioptrics and catoptrics in regulating the formation of mineral crystals in the one case, and by the necessity of greater freedom in the higher conditions of organisation by the interruption of magnetic or actinised light to unite with oxygen in the other. The conclusion we must arrive at is, that if a man be a magnet, all animals are magnets. We speak of the instinctive forces of insects, and these must be like other forces attractive and repulsive, and hence illustrative of Animal Magnetism.

I may be told that I am not advancing the cause of Animal Magnetism, by affixing so high a value to an instrument which can be manipulated only by a small number of persons, that even among these, in order to obtain results, the necessity of watching the pendulum with the eye of the manipulator is fatal. People will not assent to such a condition. Suppose we were so pig-headed as to refuse assent to the return currents of the telegraph wire, or of any other analogous phenomenon. A few years hence, it will be hardly credited that a splendid discovery, fraught with such tremendous consequences, should be met by the world of science as a flock of sheep would meet a strange dog. Some smelt, some only sniffed, some gazed, some looked as only sheep can look, very silly, some tried to discover whether the dog was mangey, some thought the dog was a bothering phantasm, some agreed that a dog requiring to be looked at with eyes, while it moved about, could be no dog at all. At last the flock, full of sheepish wisdom, turned about, and, true to their accustomed gregarious habits, huddled themselves off in one compact flock, determined to ignore the existence of a foolish dog, who was no dog at all. This is literally the only analogy that the men of science of England deserve. They may not feel themselves complimented, they have themselves to thank.

Mr. Rutter proceeded to demonstrate the laws which regulated his instrument, and it may be stated, with perfect truth, that there is in all philosophy no series of facts more exquisitely delicate, more ingeniously beautiful than those demonstrated by this extraordinary man.

We do not propose to describe the Magnetoscope in minute detail, nor to enter on the numerous disquisitions on the numerous facts that it has been the means of developing. We must refer for some of these to the interesting and excellent work on Human Electricity, to which we have already alluded. We are only now to refer to the fact that this work of Mr. Rutter's fully bears us out in our first assertion that man is a magnet.

Now, it must follow that if man be a magnet, he must necessarily originate currents of magneto-electricity. If he does originate these currents, what is their normal course? They must have a normal direction, and when their polarities are reversed, they must have an inverse direction. This need not be questioned. But whether ignorance chooses or not to question it, there can be no doubt of the truth of the fact. The question, such as—how do you prove it? may be asked. But we need not stop to reply to such obvious frivolities. Folks must be ignorant indeed of the researches which have determined the facts of reversals of polarity, when they indulge in such questions. Suffice it to say that proofs have hundreds, not to speak of thousands of times, been afforded of our experiments. Not to be teazed by frivolous smatterers, we must refer our questioners to the schools of electricity, where they may be imbued with the elements of Magnetism. Let not the reader imagine that we describe scenes which have not often occurred. The two rooms on the ground floor of the house, where I resided for several years, No. 55, Wimpole Street, have, on many occasions, been witness to exhibitions such as we here describe; the actors on the scene having been scientific or medical men, not ashamed to expose an amount of ignorance too melancholy to reflect upon.

Man is a magnet. This is our theme. He has, like all other magnets, poles and equators. But, being a magnetic machine of very complex structure, his magnetic apparatus is divided into many parts. The brain is the chief magnet, and the trunk and extremities are separate magnets, having intimate relations with the chief source of Magnetism. We infer from these facts, what is the truth, that the normal currents take a normal course from the brain to the caudal extremities. Mr. Faraday and Signor Matteucci had already established this fact on the gymnotus and on the torpedo. I have written on this subject in a paper on a theory of sleep, in the Zoist, in 1846. In that essay, I demonstrated the same fact. Dr. Noad was coadjutor in this discovery, and I refer with pleasure to his letter to me, inserted in the fourth volume of the celebrated periodical which sustained the cause of Animal Magnetism, in England, for several years. We need be under no apprehension that our facts will perish. The Zoist, a monument erected by Dr. Elliotson to his own zeal and energy, will live much longer than the enemies of the truth might desire. But the truth has marched, and it will continue to march. Ignorance will oppose it in vain. Pride and envy will not prevail against it. Currents of electricity are demonstrated to exist by too many facts in living bodies. If living bodies be magnets, let us seek the striking illustration of the truth by an examination of the phenomena of a subject not understood. What is the cause of the phenomenon of pain? How does pain occur? Many will say, it is a natural phenomenon not to be accounted for. Sir Benjamin Brodie, or perhaps Dr. Copland, who, equally with the small-minded baronet, was an opponent to the truths of Animal Magnetism, said that pain was a wholesome provision of Nature, and their wise speeches at the Royal Medical and Chirurgical Society led to the inference that they believed it to be almost sinful for a patient in the Anœsthesia of Animal Magnetism to deny that he suffered pain under the surgeon's knife.

Pain is, however, easily to be accounted for. It is a magnetic phenomenon. Polarity belongs to magnets, every part of the living body is a magnet, and every magnet is capable of being submitted to a reversal of polarity. Those who are conversant with the common instrument, called the galvanic battery, may easily satisfy themselves of the fact of reversals of polarity, by uniting the extremities of the positive and negative wires from the two poles of a battery properly charged with the fluid calculated to set in motion the electric current. At the moment of contact, the fluid in the cells effervesces. The polarities of the current have been reversed. This fact of reversal of polarity applied in the case of the living body, explains the phenomenon of pain. But how does it explain it? To answer this question, we must trace some preliminary facts, and it may safely be predicated of these, that they are as consecutive and as convincing as any series of facts ever offered to the attention of the Royal Society.

We well know that sleep and its proximate cause has never been properly explained in any work on Physiology. We will now endeavour to explain it. There are no events more common to mankind than those we designate as diurnal. We go to sleep daily, and we become awake daily. The phenomenon of sleep then is sufficiently common not to require a description. But how would any one define it? It is simply a loss of consciousness. How a loss of consciousness? We lose consciousness if we receive a heavy blow on the forehead, or if we get dead drunk, or if we take a large dose of opium. But if we reflect on the facts connected with the loss of consciousness in each of these instances, we must be struck with the strong analogy existing between the states dependant on the several causes. In every case, the victim suffers from pressure somewhere about the brain. It is proposed to assert that approximation between the particles of brain matter is the cause of sleep. This is no new fact. It has been exemplified very frequently.

Summarily, then, we view the phenomena of sleep. What is sleep? Our best philosophers have described it as what it really is, a close approximation to the idea of death. We cannot call it absolute death, because life is present. But we may say it is so like death, that the only difference is that we can waken the sleeping individual. Is this the only difference? for in truth this is no difference at all. The person who is asleep, and the person who is said to be dead, may each be merely in a state of trance. Many have been thought to be dead, who were only in the sleep of trance. In the year 1818, I was called to visit a patient in the Tottenham Court Road. His name was Leckie, and he was the near relative of the late Sir Richard Birnie, chief magistrate at that time in Bow Street. Mr. Leckie had been ailing for some time, and when I was called upon, he had been rapidly getting worse. I found him feeble, pale, emaciated. He was not otherwise remarkable for any serious disturbance of the principal functions of the body. The case puzzled me, and my friend Dr. Southey was called in. We could detect no lesion of any important organ. Our patient, however, did not rally. His debility increased, and he fell asleep, in death, as we thought. His friends, alarmed at the striking change that was going on, for he appeared to be dying, summoned us to his bedside. The resources of our art did not produce a restoration of the pulse. We remained administering stimulants, until our patient ceased to breathe, while he lay on his back with his glassy eyes wide open. Closing his eyelids, we applied the looking-glass to his face, to ascertain if he breathed. Not the slightest sign of moisture appeared on the glass. We left the house, pitying the disconsolate young widow. The body was laid out, but it did not become cold. Three days afterwards, I met our patient walking on the sunny side of Cavendish Square, looking cadaverous, but strong enough to walk steadily. Here was really a case of death in sleep. The man recovered, but in three months was again attacked, and succumbed to the power of death, as far as his life in this world was concerned.

Now, we may be permitted to ask, how far can we be sure that in such a case of trance, death is only simulated? There was absolute loss of consciousness; there was no pulse; there was no respiration. The man was to all intents dead. How does such a death differ from sleep? Only in one slight particular: the sleeper is alive. The man in trance is dead. The sleeper remains alive. The man in trance is resuscitated when he recovers from death. In both cases, the principle of consciousness has quitted the body. We cannot say what may be the conclusion of the physiologists, who have not gone into the subject of sleep with a view to determine the ultimate nature of the phenomenon; but this we know, that sleep regarded in this point of view is no easy matter to be accounted for.* It is easy to be dogmatical, and this is a common failing with superficial philosophers. The subject now under consideration is, however, too cogent and too serious to be treated as a matter to be determined by authority. What is sleep? is our question, and it must be answered. If our sceptics and adversaries cannot answer it, we will do it for them. Sleep is a magnetic phenomenon. It depends upon the attractive forces which operate between the minute particles of brain or nerve matter.

We proceed to the proof of our proposition. When an animal is deprived of the calvaria or bony upper covering of the brain, its brain is exposed to view. We suppose a cat, or a dog, or a rabbit properly secured on a board, for the purposes of experiment. It is alive and awake. A touch on the end of the nose will soon convince the experimenter of this fact. The brain is

* Unless by attraction between the soul and its former habitation.

exposed; it has no bony protection. The palm of the hand is applied so as to press upon the animal's brain. Instantly, sleep supervenes. If the pressure be steadily kept up, the condition, known by operating surgeons as anœsthesia, is present. The animal no longer starts at the touch upon the nose. It is quite unconscious. Remove the pressure and the animal is awake. Here we have sleep and wakefulness illustrated. We have caused the particles of the brain to approximate, and we have an example of sleep. We remove pressure, the particles have room to expand, and we have wakefulness. Can we desire a clearer illustration of a simple fact? But we must shew how far we are borne out in Nature, by the analogies offered us. We know that men are struck down by apoplexy. We are aware that fractures of the skull will cause pressure of the brain. What are the conditions in such cases? Blood or serum effused on the brain produces the deep unconscious sleep of apoplexy. Pressure of fractured bone, or from injury to other parts of the bony vault oppressing the brain, is quite analogous to the pressure causing apoplexy. These facts only strengthen the assertion that approximation of brain particles is productive of sleep. How, then, do we determine that all sleep is a magnetic phenomenon? We know that there is a great difference between healthy sleep and the sleep of disease. Let us quote a passage from Macnish on the Philosophy of Sleep.* He tells us that "The sleep of Health is full of tranquillity. In such a state we remain for hours at a time in unbroken repose, Nature banqueting on its sweets, renewing its lost energies, and laying in a fresh store for the succeeding day. This accomplished, slumber vanishes like a vapour before the rising sun; languor has been succeeded by strength; and all the faculties, mental or corporeal, are recruited. In this delightful state, man assimilates most with that in which Adam sprang from his Creator's

* "The Philosophy of Sleep," by Robert Macnish, L.L.D. Glasgow, 1845, page 2.

hands, fresh, buoyant and vigorous; rejoicing as a racer to run his course with all his appetencies of enjoyment on edge, and all his feelings and faculties prepared for exertion."

This description is sufficiently accurate for our purpose; but we might add our own view of normal sleep, by saying that it is characterized by the easy semiflection of the limbs, and a free, deep respiration, with the calm countenance and attitude of repose. Even rendered very intense and deep by the healthy attraction of a powerful internal magnetic force, it is not accompanied by snoring.

On the other hand, abnormal sleep may vary in its characters according to many accidents,—from the uneasy restlessness attendant on weak digestion, or on mental annoyance, to the deep stertorous coma of dead drunkenness or of opium, or of apoplexy.

It is evident, then, that in very deep sleep, in *anæsthesia*, we have neither pain nor even consciousness. Our question was, on the essential facts of pain, what is pain? It would not be a very profitable spending of time, if we were to refer our readers to the ordinary dictionaries in vogue for a definition of pain. Consult the quarto edition of Todd's Johnson's Dictionary, and you are told that *pain* is a sensation of uneasiness, and the author has been fain to quote the glorious Milton, who says that "Pain is perfect misery, the worst of evils, and excessive, overturns all patience."

Physiologists should have definitions more definite, and in respect for their habits of logical precision, we naturally revert to more technical works on the question, where we expect that learning and ingenuity, aided by industry, shall collect together numerous facts to produce a marrowy definition. We find ourselves, however, quite disappointed.

Each tissue of the body has its own organic sensibility. We do not affirm that the material substance has an inherent force, but the magnetic force which endows it with sensibility constitutes a peculiar characteristic of that form of structure which is endowed with that particular sensibility. The Sensitive plant and the Dionœa Muscipula in vegetable life, and the Zoophyte Actinea have repeatedly afforded examples of simple instinctive sensibility. The trains of reasoning in which Professor Macartney used to indulge, illustrating the views he took of vitality by these facts, among others, have been very curiously corroborated by the discoveries of the late Dr. Leger, by means of Mr. Rutter's invention of the Magnetoscope. Often has he asserted that that most delicate instrument, the Magnetoscope, has demonstrated to him that every living particle of the vegetable and animal kingdom is magnetic, and that the varieties of instinctive sensibility depend upon laws essentially magnetic. It is well known that philosophers have inferred that each iron particle in a magnet, which attracts iron filings, is itself a magnet, gifted with the same polarities that pertained to the magnet, of which it was a part. If we proceed from a metal to the crystal of an earth, or of a salt, we shall find that we have again instances of polar arrangements.

The Baron von Reichenbach clearly established the facts relating to the attractive and repulsive influences of crystals, and had observed that these effects resulted from a power concentrated in two poles of a principal axis. Mr. Rutter's Magnetoscope has enabled us to proceed further, and to establish the point that crystals are magnets, without the power of attracting iron filings; although they have attractive and repulsive agencies in relation to many other forms of matter.

If the magnetoid or odic force of the crystal cannot attract and repel iron, and can attract and repel the human being, is it not a fair inference that it is a magnetic force which has relation to a higher grade of magnetic arrangement, to a higher grade of organic structure?

Any portion of a branch of a living tree may, by the

aid of Mr. Rutter's Magnetoscope, be demonstrated to exhibit opposite polarities at its opposite extremities. A carrot, a parsnip, a radish, an orange, or an apple may be used to illustrate the fact of the existence of each being endowed with opposite poles.

A walking stick cut from a hazel tree retains most remarkably, for years, the property of exhibiting its two opposite poles. Among the striking facts appertaining to the hazel stick are some very curious phenomena, induced upon certain susceptible somnambulists by the attractive and repulsive agencies of this peculiar kind of wood. If a highly sensitive sleep-waker have a hazel stick held at a certain short distance from the face perpendicularly, in the direction of its growth, the individual will be attracted towards it. If, on the other hand, the stick be held in the direction contrary to its growth, that is to say with its tapering end downwards, the individual will be repelled from the stick.

These attractive and repulsive agencies, and the analogies yielded by all vegetable substances, to the test of the Magnetoscope, lead to the fair inference that the magnetic analogies are complete, varying, perhaps, according to the organic sensibilities of the substances.

It may be concluded, as indeed that which experiment has corroborated, that each living organic particle is magnetic and is gifted with its own peculiar sensibility, that belonging to the generic mathematical intimate structure of the vegetable, of course destroyed by completely destroying its vitality; in other words, by the removal from it of its magnetic vital force.

Perfectly, or completely, dead vegetable or animal matter no longer affects the pendulum of Mr. Rutter's Magnetoscope, unless it be allowed that a powerful negative force, representing absolute death, exists which can speedily arrest all movement in a suspended body. The pendulum stops, and its complete, apparent static quietude, under the influence of death, is a most remarkable and striking phenomenon. The awe engendered by this perfect stillness becomes a sublime idea! To the contemplative mind no imitation of God's power of death can be aught but sublime! Is it not an instinct of the human being to become appalled at the sudden catastrophe of a fellow-being instantly deprived of life?

What is an instinct? All vitality is instinctive. All living beings, by their nature, recoil from injury, and covet, or are attracted by a benefit. Instinct, then, is the tendency of a living vegetable, or animal organism, to attract or repel. Instinct is an endowment for the purposes of self-preservation and progression.

In the more complicated forms of organism, life is constituted by a *harmonious aggregation of instincts.* Let us try to reword our proposition. Several ideas strike us when we reflect upon instinctive forces. We may again ask what is an Instinct? To reply to this question, our first proposition must be, that it is a complete absence of reasoning power. The animal is gifted with instinct alone. Why do we not wish to mix the two ideas. They are incompatible. The instinctive force, residing in an organ, cannot be compared to the reasoning faculty. The reasoning faculty takes cognizance of facts, and no one can say that instinct arranges facts. We proceed step by step to shew that no instinct could exist without an arrangement of matter we call organic. We come to a fact allowed by all men. But we must reason upon the fact. We shall be told that we admit that organization is the parent of instinct. We did not say so. All we assert is, that organic structure is *a necessary condition of instinct.* No instinct can exist in a stone.

Much has been written on the points of discussion relating to instinct and reason. It is not proposed to enter now upon the topic further than to define the meaning of terms. The power of reason must depend upon an exertion of will. No individual can exercise the faculty of will without an implied freedom of the faculty. That freedom must depend upon the power of governing certain organs. If these organs be not present, the individual cannot use them. He must reason without reasoning power. But this is an absurdity. All those animals deprived of reasoning power, depend for their preservation upon the faculty of Instinct. When organs are superadded enabling them to use the additional instinct of reason, they belong to a higher class and become reasoning animals.

Let us, then, who belong to this higher class, reason on the elements of *pain*. While the animated being proceeds in a direct train, his sensibilities are not the sources of conscious feeling. They are to all intents quiescent. The polarities of the magnet are undisturbed. In all directions the ultimate organic molecules, constituting the organism, are undisturbed. This is sleep. The moment the relations of these molecules to each other are altered, the individual becomes conscious of the change by an altered state of sensation. This is not necessarily a pain, or a person could never become awake from a condition of sleep, without suffering pain. The conditions of sleep and wakefulness are striking examples of altered opposite polarities. The whole body being a congeries of magnetic molecules, must necessarily be subject to the laws regulating polarities. Any change in the relations of the poles of living animal molecules, must be productive of a change in the sensibilities of the part. Whether the change be the cause of pleasure, or of pain, must depend upon the faculties of the individual. Endowed with a nervous system, the animal is susceptible of sensations, without which, the idea of pleasure or pain become absurd. The inference, then, remains that pain is an extreme disturbance of the polarities of a part.

CHAPTER VII.

SLEEP, WAKEFULNESS, TONE, CLONE.

HAVING discussed some of the consequences flowing from the proposition established by Mr. Rutter, that Man is a magnet, and especially the influence of electrical or rather odic currents in the production of sleep, and of wakefulness, and in affording us the only true and correct explanation of the phenomena of pain, we may request the indulgence of our readers, while we recur to certain facts and arguments in order to make our thoughts more clear to them. The present stage of our inquiry, naturally involves questions on the phenomena of sleep, and of wakefulness, and these are most intimately connected with health and disease, so that we are reaching the marrow of our subject.

Knowing that Man is a magnet, we cannot doubt that electric currents exist in the human body. Prout, and with him, the deepest thinkers on vitality have inferred the necessity for their presence, in order that the chemical phenomena of the animal economy may proceed in their due course. Faraday and Matteucci have not only acknowledged the presence of electrical currents in the living body, but their labours have verified the important truth, that these currents in the gymnotus, and in the torpedo, take a course from the head to the caudal extremity. This, in fact, is that which Mr. Rutter's Magnetoscope has established as the normal course in the living human being.

The direction of these currents cannot now be considered a matter of doubt. They must, in all animals, be subject to laws. In health, they must be normal, in disease abnormal. From the facts adduced, from the labours of the best experimenters, we naturally conclude that the currents coincident with health must be those pursuing the direct course. These considerations naturally lead to the facts involved in the study of sleep and wakefulness which are, to us, of a deeply interesting nature, and are, in themselves well worthy of the serious attention of any really scientific and philosophical mind. If we reflect on the sublime simplicity that reigns through all the arrangements of the stupendous universe, the marvellous handiwork of the omnipotent mind, we are struck, in tracing the details of the universal law of the great trunk force of gravitation, to find that attraction and repulsion are present everywhere, regulating all around us. For centripetal and centrifugal forces rule, not only the relations of the orbs in the solar system, but even the magnetic phrenological organs of the animal body, and are productive of the phenomena of sleep and wakefulness, and of health and disease, and of all the deeply interesting phenomena belonging to the great truths of somnambulism, clairvoyance and other great mental functions, so much despised by some of the paltry minds that distribute the so-called honors of Science.

Sleep and wakefulness are no trifling matters to occupy the minds of profound thinkers. They are linked with all the considerations bearing upon corporeal strength, vigour of mind, and upon all the questions of health and disease. Their relations and ramifications then fall under the general laws of Animal Magnetism, a subject contemptible only to very contemptible minds. The great trunk force of gravitation thus influences all nature. We proceed to shew that sleep is the result of an attractive force, analogous to the attraction of gravitation, and that wakefulness results from a repulsion, analogous to the centrifugal agency constituting a part of the phenomena attendant on the great trunk force. Attraction and repulsion are opponent phenomena for they belong to Magnetism. Without *Animal* Magnetism, however, Mr. Faraday would be at a loss to classify the belongings of these remarkable facts. We knew that magnets attracted; but this was a fact relating to dissimilar poles. We did not know absolutely or definitely to which current, the direct or the inverse, the phenomena of attraction or repulsion belonged.

It is, however, a very important question, and the world is indebted to Animal Magnetism for its solution. Few would imagine that sleep and wakefulness were to settle an important question in physics. But so it is. In sleep, the force of attraction prevails. In wakefulness, it is the repulsive force which is operative. Called upon to prove this undeniable fact, we must have recourse to Mr. Faraday's observations on the reversal of polarity in the magnetic electric current. He has told us that when a current of electricity is sent through a wire, it pursues a direct course unless a magnet be interposed, and then its course is reversed. It becomes an inverse current. All electrical currents may be said to be magneto-electric, for they emanate from magnets, (what is a voltaic battery but a magnet?) and they proceed in a direct course until they are reversed. This reversal is a repulsive phenomenon. To prove this fact, we must have recourse to the experiments of Mr. Faraday and of Signor Matteucci, on the course of the normal currents in the gymnotus, and in the torpedo. These were established to be direct currents from the brain to the caudal extremity. We know that when an animal sleeps, there is a normal current, proceeding from the brain to the caudal extremities. But if that current suffers a reversal, the animal immediately wakens. Is it easy to prove this fact? Very easy in the sleep of Animal Magnetism, and equally so in a great number of cases of profound sleep. All sleep is magnetic. All currents of air upwards have a tendency to produce in sleep reversals of normal polarity. A sheet of pasteboard or a large fan, throwing upward currents will soon waken a person slumbering in a healthy or normal sleep. It will do the same in a deeper magnetic normal sleep. We have to consider why it will not succeed in waking a person in the sleep of apoplexy. We know that the sleep in this case, results from a pressure on the brain. In health, the tendencies to sleep and wakefulness are regulated by normal amounts of force. In diseased sleep, the force of pressure may be supposed to be analogous to one of excessive attraction, not easily overcome by a repulsive force.* We must consider the opposite phenomena of

* The minds of many who converse and of some who write on Mesmerism are more confused on the meaning of the terms they are apt to use. We shall try to refer to some of the nosological definitions laid down by our predecessors. People are often making use of the words, Mesmeric Coma. Mesmeric or magnetic sleep is a healthy condition. *Coma*, a Greek word, has nearly the same signification as Cataphora, a term used by Galen and Hippocrates to signify an extreme condition of diseased sleep, (see Swediaur's "Nov. Nosol. Method Systema," Vol. I., page 308) from vitiated state of brain. Examples of this state are afforded by the Coma of drunkenness, by the sopor of opium, by the incomplete drowsy sleep accruing from those insufficient doses of ether or of chloroform, which fall short of their full intended effect of inducing complete anœsthesia, a state short of dead drunkenness; of carus, which is a more profound morbid sleep, short again of complete apoplexy. No doubt coma, sopor, carus, and lethargy, degrees of cataphora or morbid sleep, caused by loaded veins of the brain by serous effusion, or other diseased pressure on that organ, are stages in the symptoms of apoplexy. The reason of the present definition is that we may distinguish between healthy sleep and the degrees of apoplectic phenomena. Both are caused by the approximation of the cerebral molecules the one by a healthy attractive force, the other by morbid or by mechanical or peripheral pressure.

All states of anœsthesia, or insensibility, analogous to that morbid state induced by modern surgeons when they operate on a subject that has been removed from the region of the living into a condition very similar to that of a dead person when produced by ether, chloroform, or other medicinal agents filling the cerebral veins, are examples of the absence of feeling or sensation, sometimes without unconsciousness, in which the individual can experience no sense of pleasure or of pain. These result from approximation of the molecules of brain and nerves from forces of impulsion or pressure. But when the tonic or wholesome absence of consciousness of deep magnetic sleep is present, it results from central attractive forces. All narcotics act like crystals, having two opposite poles. One pole is attractive, acting centrally, but the other is repulsive, acting by pressure of loaded veins on the periphery of the brain, an organ inclosed in an unyielding vault.

sleep and wakefulness more minutely, in order to bring out clearly the fact that attraction belongs to the direct, and repulsion to the inverse current.

All sleep, whether morbid or healthy, is slumber, with more or less loss of consciousness, when dreams are not present, dependant upon an approximation of the molecules of living cerebral matter. This is easily made manifest from the events attendant upon the application of pressure to the living brain in any form, by the morbid fullness of veins within the vault of the bony skull; by the apoplectic effusion of blood, or of serum; by the palm of the hand applied over the surface of the brain, with sufficient force, after the calvaria have been carefully removed, in the skull of a living rabbit or cat; an experiment made demonstrative of sleep or wakefulness, alternately, by alternately pressing the brain and removing the hand with due dexterity.

In the present stage of our argument, it is of trifling importance, whether an approximative agency depends upon a force applied externally to contract the particles of a mass of matter into a more confined space, or whether the approximative force operates from causes finding their home in the interior of the mass. For the purposes of our illustration, the apparent result is the same. If approximation of the particles of the brain be effected in the living animal, whatever be the cause, impellent or attractive, there must necessarily be exerted an approximative force. That force may be an attraction from the centre, or an impulsion towards the centre. In the one case, it is an approximative agent by attraction; in the other, an approximative agent by impulsion. Whether the force arrive from a peripheral source, or from one central, the same *apparent* phenomena result.* The molecules are forced to become closer to each other. Now, the degree or amount of the approximative force determines the depth of the sleep; be it light slumber, or

* We reflect little that all impulsion results from a centrifugal or repulsive power.

deep anœsthesia. We now arrive at another stage of our argument; that sleep is healthy or morbid, according to its induction by central or by peripheral force, or in other words, by attraction from a central source in the brain, or by a pressure on its periphery. Healthy or normal sleep is the result of an attraction, towards each other, of the molecules of brain and nerve matter; while morbid or abnormal sleep is due to peripheral pressure, or impulsion. A fracture of the skull, causing pressure on the brain, effusion of serum, or of blood, or even the pressure from a loaded state of the veins, from whatever cause, may be examples of the presence of peripheral force. Normal sleep is characterised by the easy semi-flection of the limbs, and a deep free respiration, with the calm attitude and countenance of repose. Even rendered very intense and deep, by the attraction of a powerful internal magnetic force, it is not accompanied by snoring. Abnormal sleep, on the other hand, may vary in its characters according to many accidents, from the uneasy restless sleep attendant on weak digestion, undigested food in the colon, or on mental annoyance, to the deep stertorous coma of dead drunkenness, or of opium or of apoplexy.

I remember Abernethy to have stated in his lectures, that Ruysch had observed, in the case of a man, from whose skull he had removed a piece of bone, and had thus exposed the brain to view that when the man slept, his brain occupied less space in the bony vault of the cranium than when he was awake. Of course, an approximative force was present. When, in the year 1813, I was House Surgeon at Saint Bartholomew's Hospital, I had an opportunity of observing that during sleep, the brain of a boy from whose extensively fractured skull I had occasion to remove a large piece of bone, occupied considerably less space than when he was awake, and as I watched this boy with close attention and deep interest, for it fell to my lot to slice off portions of protruding brain, the case eventually recovering, I can fully corroborate the facts detailed in the following case which I copy* from page 48, of Dr. Macnish's work on the Philosophy of Sleep.

"Dr. Pierquin relates the following case. It fell under his notice in one of the hospitals of Montpellier, in the year 1831. The subject of it was a female at the age of twenty-six, who had lost a large portion of her scalp, skull bone, and dura mater in a neglected attack of lues venerea; a corresponding portion of her brain was consequently bare and subject to inspection. When she was in a dreamless sleep, her brain was motionless, and lay within the cranium. When her sleep was imperfect, and she was agitated by dreams, her brain moved and protruded without her cranium, forming cerebral hernia. In vivid dreams, reported as such by herself, the protrusion was considerable, and when she was perfectly awake, especially if engaged in active thought or sprightly conversation, it was still greater. Nor did the protrusion occur in jerks, alternating with recessions, as if caused by the impulse of the arterial blood. It remained steady while conversation lasted."

These general facts which could be corroborated, if our space permitted, by ample illustrations of similar cases, will suffice to make clear our meaning. We have adduced the proposition that sleep results from attractive, and wakefulness from repulsive forces. If words have any meaning, the truth of our proposition is clearly established. We have no occasion to overlay our reasonings with additional cases. It would be easy from notes written in Dublin, in the year 1814, to detail facts developed from my own experiments on many living animals in the Anatomy House of Trinity College, all converging to the point that sleep resulted from approximative, and that wakefulness occurred from divellant forces operative among the molecules of brain and nerve matter.

It is clear that the phenomena of sleep, as well as

* "Philosophy of Sleep," 8th Edition, 1854.

those of vigilance, are closely dependant to those operating among the molecules of the brain matter. If attractive or approximative forces operate, sleep supervenes. If, on the contrary, repulsion be present, and the molecules be separated at certain distances, wakefulness results.

There is another point of view, in which the application of attractive forces to the brain may be regarded. In the practice of Mesmerism or Animal Magnetism, many persons have been observed to be so sensitive, that the pointed termination of a rock crystal, directed to the face of such individual, has produced an ungovernable tendency to approach that crystal, and in not a few a deep sleep has, by the attractive force, been induced. Long ago, the Baron von Reichenbach observed that in persons of a certain degree of sensitiveness, there existed a strong attraction between the hand and a magnet. He observed that when he placed sensitive persons in a completely darkened room, they invariably observed streams of light issuing from the poles of a magnet.

Petetin, who practised as a physician at Lyons at the commencement of the worst times of the French Revolution, has, in a work, entitled "Electricité Animale," recorded the appearance of luminous emanations; and among our mesmeric somnambules these facts have been very frequently noted. More, we have repeatedly verified the Baron's observation that, in a dark room, streams of light issue from various crystals, as well as from the points of the leaves of healthy plants.* The Baron having made experiments with numerous crystals, I found that in these resided a certain attractive power, not previously noticed. He had observed that the power was concentrated in two poles, which were diametrically opposite, and constituted the poles of a principal axis.

Thus had the Baron established a claim to a share of the discovery, since amply corroborated by the facts

* See the Baron von Reichenbach's "Researches," Baillière, 1851.

developed by Mr. Rutter, with the aid of his Magnetoscope—the discovery that a law prevails in nature, constituting all crystallic arrangements of matter to be magnetoid, or be obedient to forces which, though not identical, are analogous to the forces engaged in the phenomena of Magnetism. My friend the Baron ranges these forces under the generic term *od*, and he deals with logical precision as to the facts he has been at the pains to discover. I have put forth the hypothesis that Nature deals with all organised arrangements according to strict crystallic laws; in other words, the magnetoid laws of Mr. Rutter, or the odic forces of the Baron von Reichenbach. The Magnetoscope in the hands of the inventor, as well as in those of his disciple, Dr. Leger, has shewn that all crystals are either magnets or magnetoid modifications of matter arranged to be endowed with forces analogous to those observed in magnets.

In a long experience of a quarter of a century, with many sensitive cases, and with numerous somnambules and clairvoyant persons, it has been my lot to verify frequently the accuracy of the Baron's results, not only with magnets, but with numerous mineral crystals.

The magnet has not only attracted the hand, but has, in very many instances when it had strong attractive force, so marked an influence as to exert a powerful attraction in the human magnetic centres, inducing immediate sleep. Some persons have been so susceptible to this attraction, that upon my sudden introduction of a large horse-shoe magnet, having a sustaining power of forty pounds, into the room, they have immediately fallen into a profound sleep.

The pyramidal end of an ordinary sized rock crystal, or an aggregation of three or four large crystals of nitrate of potass, have been to some persons very attractive, and in these, in a condition of full vigilance the hand has approached them with an *organic instinct*, or in entomologic language, with an *instinctive force*, grasping it. In very sensitive persons, if the hand be allowed to remain for three or four minutes in contact with the pointed ends of such crystals, sleep takes place, always refreshing—although in some deeper than in others.

The facility with which certain persons obey the influence of the attractive pole of crystals is very remarkable. Not only can they be induced to sleep by looking at the pyramidal end of a rock cyrstal, or at the terminations of an aggregation of several large crystals of nitrate of potass, placed conveniently upon a board, so that the terminations point to the face, or to the root of the nose, at the distance of an inch, but even when placed at the same distance from the pit of the stomach. Indeed, some have been found to be so very susceptible as to fall into a sleep under the influence of the attractive pole of a fraction of crystalline Epsom salt, so small that I have held it by the aid of forceps, or sometimes fastened it to a penstick. Of course, these susceptible subjects waken upon the opposite pole of the same crystal being presented to the face. This experiment I have very frequently exhibited.

All our facts on the attractive influence of magnets and of the attractive pole of crystals show only that sleep is a phenomenon dependent on attraction. We proceed to shew that attraction belongs to the direct currents. In very sensitive cases, sleep can be induced with great ease by passing a gentle direct current from the nape of the neck to the feet, the electrodes being zinc at the neck, and copper or platinum at the feet. Sleep induced thus, being proof of the presence of an attractive force. But place a person in a wooden or gutta-percha bath, properly insulated by dry glass supports, let the head be towards the north, and the feet towards the south. The electrodes I have used in such experiments have been of platinum. Forty pairs of Smee's elements have, under such circumstances, been in full action; the direct current taking its course from north to south, produced the most decidedly agreeable effects. The bath being at the temperature of 96°, I have felt extremely refreshed, though very sleepy while I remained in the bath. On another occasion I tried the effect of the inverse current, by placing the soles of the feet in the same bath facing the north, or source of the force from the battery. The result, in less than two minutes, was an intense headache. I was very glad to get quickly back to the former position, receiving the agreeable influence of the direct current passing through the warm water. I have tried numerous experiments with individuals of various constitutions, and I have arrived at the conclusion that currents of electricity passed through water placed in large glass basons, proper platinum electrodes being adapted to them, form a very powerful means of administering medicinal agents to act quickly on the system. Thus, salivation from mercury may be induced in most persons in a few minutes, by immersing the hands in an extremely weak solution of a mercurial salt. My friend M. Pocy's experiments in obtaining mercurial precipitation on copper cathodes by means of electrical currents from persons who had taken mercury in too large quantities, were very curious and instructive.

We proceed to additional facts in order to establish the attractive property of the direct electrical current. Mary Ann Douglas having been, with the exception of Jane Murrell, the most impressionable sensitive I ever had as a patient; when I placed a couple of small discs of zinc, each the size of a farthing, upon the insteps of her feet or upon the toes, and a copper farthing or a half sovereign at the nape of her neck, a current passed from the zinc to the copper or to the gold. It produced no immediate apparent effect, but in a few minutes she suffered intense headache, for the current did not take the course from the brain to the feet, but from the feet towards the brain. The direction was in fact inverse, and consequently repulsive. But when the gold was placed on each foot, the zinc applied to the nape of the neck, the natural or direct current was increased in force by the metallic direct current, and the girl slept immediately. If the metals were left for three minutes, the girl slept with a rigid condition of the voluntary muscles. All sleep resulting from attractive forces, it is clear that the direct current carried the attractive agent, and that the inverse current bore the repulsive agent. With more or less marked results, according to differences in constitutional sensibility, the same experiment never failed to be attended with the same results in twenty-two somnambules. It is hardly necessary to recapitulate the facts attendant on the relations between sleep and wakefulness, and the forces of attraction and repulsion. If in Magnetism there exist the two opponent forces, and the one leaves a magnet as a direct current until it be arrested by a magnet, when it assumes a return function, and that return function is characterized by a repulsive agency or force.

The direct current then, is an attractive force, and the inverse current is a repulsive force. Who shall say, then, that the whole philosophy of Nature does not lie in a nut-shell, if the Universe is planned upon the very simple basis of two antagonistic forces that account for all the possible phenomena of Nature.

We must next inquire what relations these facts have to the phenomena of health and disease. It has been asserted that the forces we have been considering are productive not only of sleep and wakefulness, but of health and disease. We must proceed to make out that we have been accurate in our language.

In ordinary life, how often have we witnessed the vanity of a man, displaying the vigour of his muscles, by appealing to the firm, hard condition of his arm, desiring the contracted state of his biceps to bear testimony to his manly power. He little reflected that he was illustrating a philosophic truth! He little thought of the approximative forces then playing among the molecules of his biceps muscle, and less perhaps of the relations of those forces to the universal force of gravitation, and less still to that force which the force of God's will conferred upon him, of exerting the motive power of will to contract that muscle, in order to display tonic force. Tonic force results in such a case from a direct current, issuing from the living brain, in obedience to the will. But let us inquire, what is tone? The word is derived from the Greek verb, τεινω, to stretch, or τονοω, to make firm. Turton (Medical Glossary) states its meaning to be "the natural strength of a muscle, or fibre." Spasm, from σπαω, to draw, may be tonic, or clonic. Now, the word clonic is derived from κλονεω, to agitate. The tonic spasm implies continued, sustained, direct force; the clonic implies indirect or inverse, agitative force, tremulous, unsustaining weakness. Tone is sthenia; clone is asthenia. Tone results from an approximation of the organic molecules of nervous tissue, and also of muscular tissue; clone, from a divergence of these molecules. Of course, in the one case, approximative forces must be present, operating among the molecules; in the other, divellant forces tend to weaken the connection between the molecules. Attraction governs the tendency to approximate molecules, while a divergence of molecules must be produced by a force of repulsion. The removal of all pressure from the living brain, in morbid sleep, awakens. Thus, it is easy to perceive that the stage of light sleep, deep sleep, and healthy, magnetic, rigid spasm of the voluntary muscular system, being stages in the degrees of attraction exercised among the molecules of certain forms of living organic matter, the inference is natural that the opposite manifestations are due to the agency of repulsive forces operating in the same spheres of living matter.

Wakefulness or vigilance is the opposite phenomenon to sleep. While there exists a due balance between the antagonising forces, attraction and repulsion, the wakeful state is not incompatible with normal or healthy tone. But if the normal or healthy relations be subverted, and the clonic tendencies overwhelm the tonic, the train of events is curiously corroborative of the principles now endeavoured to be illustrated.

We are not sure that our next consideration, naturally flowing from the proposition now established, will be well received by the medical world. It is this: whereas sleep refreshes and invigorates, and wakefulness fatigues and weakens, it is our purpose to assert that all invigoration is dependant on the same causes which produce healthy sleep; and that weakness or debility is the result of the same forces which produce wakefulness.

The one phenomenon, that of strength, has been translated by the word sthenia; the other, that of debility, by the word asthenia. The privative enables us to form the same idea as in electricity is expressed by the word negative. It will now be understood why I have thought proper from nosological considerations to introduce the two words, tone and clone, having the advantage of not being new, and at the same time in accordance with ordinary pathological views.

These words may appear fanciful to those who have not gone deeply into the nosology of diseases. I consider them important to elucidate the views I have adopted of the origin of all diseased conditions of the animal fabric.

Let us trace the origin of disease. We must first have a clear view of what we mean by the word Norma. A normal school is a fashionable expression for a model establishment, which may serve as a guide for future schools. When in my little book on Dentition, which appeared in 1834, the word normal occurred, many men, some of note, chose to find great fault with me for the use of such a term. I am not aware that it was at all inappropriate, but many like to be critical without being able to give the shadow of a reason for finding fault. It was a word in frequent use in works on anatomical science, and is now common enough, though it was in 1834 considered pedantic. The Norma of the Greeks was the tool which our carpenters call their *square.* It serves as a rule or standard. We wish to have some standard by which we may adjust our ideas as to what may constitute normal health, or perfect health.

This is a point to start from. Who that has thought deeply on disease could make up his mind on a definition of what constituted the idea, if he had not previously formed a basis for his superstructure. If perfect health can be defined, it does not follow that any aberration from its Norma shall be necessarily a disease. It may be only a disorder. We proceed to ask how we should define perfect health? The simplest idea we have of it, is that of a sound mind in a healthy body. But that again obliges us to proceed to the inquiry of what is a sound mind? We are at a loss to say whether our readers will allow us the privilege of discussing this question. We know what we mean by a healthy body. It is simply a living body in which each function is performed with normal regularity. We could add, in which the individual is quite unaware of his own existence, except from the consciousness of his own identity. But this would involve the destruction of our first definition; for many an insane man may be much diseased, and be quite unconscious that he is not sound in mind and body. What then is our standard? We must insist on the functions being normally carried on. Every aberration from health must be attended by some disturbance in the normal train of the functions. Each function is dependent on a due course of the electrical or odic currents of the body. When these run their normal courses, we have normal health; but if any reversal of the normal polarity of a current occurs, we must necessarily have a disturbance of health. Let us ask why we selected the word clone to designate the aberration from a normal current? We know that the first symptom of a state which introduces a catarrh, or a fever, or an inflammation is either a rigor or a shivering. In fact, an agitation. All convulsions are clonic or agitated spasms. We may be quite clear then that all disease must begin with a reversal of normal electric or odic polarity.

We must inquire next what are the circumstances which warrant us in the conclusion that these currents we have alluded to have their existence in the living body? The stupid school of imperfect, because not practical anatomical physiologists, become quite irascible with me when I endeavour to draw their attention to the incontrovertible facts which they despise. The Baron von Reichenbach has some observations in the preface to his "Researches" on, to say the least, the ill manners of Du Bois-Reymond. Birds of a feather are said to flock together. I shall make no other remark on the conceited translator into English of the aimless work of the dull German, than that he is to be pitied for imitating his friend.

It ought not to be doubted that Mr. Rutter has established the fact that man is a magnet, that the magnet Man is constituted of a number of minor magnets, which form him altogether into a large compound magnetic machine, with currents of magneto-electricity or odic forces issuing from his brain, which proceed according to the normal law established by Faraday and Matteucci. Stupid and foolish physicians may deny this fact. But it is a fact, deny it who may. We do not wait for the conviction of stupidities. Our reasoning powers are at least equal to theirs, and we have not the disadvantage of having repudiated Animal Magnetism and Phrenology. We need not only simply reassert the facts which Mr. Rutter has published, and has proved so many hundred times. We may do more, we may reason upon them as fully established.

I may allow that among those alluded to, some imagine currents of electricity cannot easily be either proved or inferred to exist in the living body. Was such a thing never heard of as sparks of electricity from the sudden pulling off of silk stockings. Place a gold leaf electroscope upon a table, then having prepared a sufficiently stiff iron wire by placing it in a handle of glass, so as to insulate the metal, which may be rubbed a few times on the hair of the head; if now the metal be slowly brought over the electroscope, the gold leaves will tell of the presence of electricity by separating from each other. If the presence of electricity be proved, may we suppose it to be quite static in the human body and flowing in currents elsewhere? What do the facts of Mr. Rutter's delicate instrument, or Du Bois-Reymond's exhibition with his delicate galvanometer tend to prove? Is it that the human will can set in motion something like a current of force that has no existence? Mr. Rutter, some years ago, on one occasion asked me to send him down to Brighton some cases of somnambulism, in order that he might, unbiassed by other influences, examine for himself the phenomena of Animal Magnetism. His request was complied with, and two young women were sent who resided for a while in lodgings in his neighbourhood. One was a young woman whose case has been noticed by me in the "Zoist," Mary Ann Douglas. She exhibited remarkably the phenomenon of the deep rigid sleep. When she was in this state, if I applied unmagnetised iron or zinc to her feet, and a copper or a silver coin to the back of her neck, it was marvellous to see how she went through various degrees of relaxation, until at last she became wide awake. So susceptible was she of the agency of attractive or of repulsive forces, that a rock crystal produced marvellous effects upon her. When pointed towards her through the wood of a large dining-room door, in the direction of her face, or in that of the pit of her stomach, while she stood outside of the door, and the operator, or the manipulator of the crystal inside, the door being held ajar, so as to enable the person holding the crystal to be completely out of her sight, and yet that her person should be observed by those who could see the crystal and watch its influence upon her; not all Mr. Faraday's wilfulness would be able to controvert the fact, that some unseen powerful influence attracted her towards the door. The pointed end being held towards her, the attractive influence made her sleep; and then if the opposite or butt end of the crystal were turned towards her, the force of repulsion operated, she started, and shortly became awake, retreating to a distance from the door, towards which she could instantly be attracted by turning the pointed end towards her. So very susceptible was she of the influence of crystals, that I have repeatedly, having previously fixed a small crystal of Epsom salt (sulphate of magnesia) to a penstick, offered the pointed termination of the little crystal to her face, and she slept immediately. Then I have wakened her very soon, by reversing the little crystal, by turning the penstick, and thus offering the butt end or repulsive force to her face. I have exhibited this very remarkable fact hundreds of times, in this and in several other very sensitive cases. If a tube of wood, or of paper, or of Wedgwood's biscuit porcelain ware, or of glass, thirty inches in length were presented, at a short distance from the forehead, and if I blew through that tube the sensitive would fall instantly into a profound sleep. If then I reversed the direction of the tube, turning the opposite end towards the forehead, and blew, an inverse force was present, rendered inverse in virtue of a change of polarity from the alteration of the molecular arrangement of the matter on the surface of the interior of the tube, she would in a very short time become awake.

Mr. Rutter made some very remarkable experiments on this sensitive young woman. One, made on repeated occasions, was not a little interesting, and established a fact on the existence of currents. The mesmeric passes with the hands in a direct course from the forehead downwards towards the lower extremities, easily induced sleep; while this sleep was a light slumber the Magnetoscope, though not so decided in the indications of force as in the condition of wakefulness, testified the presence of magnetic poles, but when additional attractive passes were made and the sensitive was reduced to that state of rigid sleep, which nosologists would call opisthotonos, no poles could be detected, though there remained evidence that the body in the rigid state, was a conductor of electricity, for when it was placed in a circuit, the current passed freely, establishing the curious fact that although electricity could be conducted, it had no longer the poles which could easily be distinguished in the full state of wakefulness.

We have to examine the next question, which we shall find relating to the forces directing the manifold functions of the various organs of the body. Dr. Prout was clearly of opinion that the operations of electricity were absolutely necessary to explain many of the phenomena he had to deal with, in what he designated as the primary and secondary processes of assimilation. These processes were attended by remarkable changes in the food, first in the stomach, and subsequently in the minute vessels known as lacteals. We are not very sure whether Dr. Prout in acknowledging the *power* possessed by the internal surfaces of these vessels did not admit *that* for which we have so strenuously contended. We mean the essence of Animal Magnetism, *the existence of a force endowed with the power of communicating vitality to the matured contents of these vessels.* He did not understand the subject as we have studied it; but he admitted the obvious inferences from the facts before him. We deplore the conclusion at which we are bound to arrive in the examination of the motives which actuate human conduct. There was a stolid pertinacity in my friend on the subject of Animal Magnetism, unworthy of a man who had approximated so near to the truth. It is to be much deplored that such a mind could not soar far above the wretched faculties characterizing the herd. He would, physiologist as he was, soon have traced out the analogies which it has been my lot to work out, and he would have abandoned his limited view of crystallography. If he had done this he would have done more. He would have accepted my views on the gradation and classification of forces.

Vitality has been the stumbling block of the physiologists. They are not aware that it is a magnetic phenomenon. They are not prepared by a previous knowledge of Animal Magnetism to believe in the fact that vitality is merely one of the forces in the infinite gradations of those forces which emanate from the great trunk force of gravitation. No reasoning would produce a conviction. What then is to be done? We must proceed to show how far facts will warrant us in proceeding to leave on record the trains of logic by which we have convinced ourselves; and these may, in time, alter the views of those who are to come after us.

Vitality is a subject of the deepest interest. Sir Benjamin Brodie, when he decapitated the animals, forming the subject of his experiments on artificial respiration, had not the sagacity or the grasp of mind to perceive that he was so near the point, on which the truth of the law of life could be defined. He could not discern the cause of the ebbing of life being dependant on a reversal of the normal current of electricity. It is to be said for him, that he could not foresee Mr. Faraday's discovery. But he could, had he been a man gifted with a scintilla of genius, when he heard of the phenomena of Animal Magnetism, have applied Mr. Faraday's fact to his dying rabbits. It is a remarkable fact that cunning most frequently destroys the higher intellectual faculties. What can be more preposterous than the whole of Sir Benjamin Brodie's conduct in the matter of Animal Magnetism? He has systematically, not only repudiated the whole subject, but he has taken pains to place himself on every possible occasion in a false position. If only a single act of indiscretion could be alleged against him, we would gladly overlook it, but he has mixed up with his repudiation of facts, so much unreasoning absurdity, that relying upon the authority of his position he has launched his whole character into the stream. We are not prepared to drown his reputation, because we believe that it is irretrievably lost. We shall, however, proceed to show, that no position, however exalted, can protect a man who abjures the cause of a holy truth. That Animal Magnetism is such a cause, the late President of the Royal Society might have been assured was an undeniable fact.

Vitality, and the numerous phenomena accompanying it, can be explained only by the facts of Animal Magnetism. We must recur to considerations on Tone and Clone, for they illustrate the subject of health and disease. But life, so little understood by Physiologists, is a force which requires renewing at every inspiration of air into our lungs. It is a force, not only derived from the great trunk force of gravitation, but is actually remitted to each living being, by the actinising rays of the sun's light. If the attractive force, productive of Tone, did not reach us in this way, we should soon cease to live. Nor is it immaterial whether we receive at one time a larger dose of the attractive force than is usual, and at another some share of the repulsive forces which are returning as portions of the inverse current to the sun. Sleep and wakefulness, tone and clone, health and disease are all so intimately connected with vitality, or the force to which we give the name of life, that we must endeavour to show not only that we are not confused on the important question of forces but that our views are clear and catenated. We know that if we be engaged in unravelling the intricacies of sleep, that sleep is not wakefulness. What are the causes of the difference between the two states? let us trace the phenomena. We know that in sleep, approximation of the particles of brain is a necessary condition. Then, in wakefulness, abundant facts concur to convince us that divellant forces are present. We need not mince the matter. Sleep results from attraction; wakefulness from repulsion. But we must make the world aware that we are no such shallow reasoners as to be obliged to say that we were aware of the repellent forces being sometimes productive of consequences which made more shallow reasoners detect our fallacies. We know very well that when we speak of sleep resulting from peripheral forces, we do not mean that these forces can possibly give the refreshing result of attractive forces. We wish it be understood that the consequences of a divellant or a repulsive force, even when performing the part of impulsion to produce approximation of molecules, can never be productive of the same consequences as those attendant on an attractive force. Never does a repulsive agency produce the same results as an attractive agency. These forces are necessarily opponent, not only in their functions, but in all their consequences. We are not bound to account for this fact, but let us on any occasion trace the results of either force, and as far as human health is concerned, we shall find that all attractive forces are beneficial, and repulsive forces are destructive. I have divided these into two. I call the one healthful, and the other lethal. We know well that the sleep induced by peripheral forces operating within the vault of the skull is not a healthy sleep. The sleep of apoplexy, the insensible sleep attending upon pressure of the brain from fracture of the skull, are examples of sleep from peripheral forces, or forces from the periphery of the brain, so is that from overloaded veins in the head from too much food or too much stimulating drink, and no one will contend that this kind of sleep is attended by healthful results. In these cases the human being, subjected to the influence of the approximative forces productive of sleep, has the symptom of snore, or stertor, which never accompanies healthy sleep. An abstemious man in health never snores. In the sleep induced by magnetic passes, unless the patient be out of health, no snore is present. Even when carried to the extent of great rigidity of the muscles and complete insensibility to external impressions, no snore characterizes the sleep. The assertion, then, is that all attractive forces induced by mesmeric or healthful magnetic agency, operating as they do to produce normal direct currents, are always attended by increase of tone, and consequently of health. But are there no circumstances in which the attractive force can be carried too far? Certainly not by human agency, which is in fact a spiritual force, the force of man's will, the extent of which is limited. A poison may produce death by hypertone. Cook, the victim of William Palmer, died from the hypertone, or opisthotonos, the extreme rigidity of attraction known as locked jaw, caused by strychnine.

Men do not reason from exceptions, we introduce them only to illustrate our position more clearly; that sleep and tone are the normal results of attractive forces; that central attractive forces do not share in the properties of peripheral forces, as these latter, though approximative of molecular matter, do share with repulsive agencies. The course of the currents, each characterised by its own force, must determine sleep or wakefulness, tone or clone, health or disease. When the whole voluntary muscular system is influenced by the normal downward magneto-electric current of Mesmerism, a delightful state of refreshed sensation follows. It is, in fact, a healthy general tone of the system. If rigidity supervenes, it refreshes amazingly, as sound sleep gives new energy and strength to the weary traveller, or to the hard-tasked labourer, so this magnetic-muscular rigidity conquers the weariness of debility, and restores health to even the most diseased subject. It was under repetitions of this deep rigid tonic spasm that Dr. Elliotson was enabled to perform the magnificent cure of cancer, which ought to hand his name down to posterity, if his noble courage, and his advocacy of the cause of Mesmerism had not established for him higher claims.

If enough has been advanced to show that the states of light sleep, deep sleep, and unconscious sleep, with rigid muscular system, are stages in the degrees of attraction, exercised among the particles of certain forms of living organic matter, the inference is natural that the opposite manifestations are due to the agency of repulsive forces, operating in the same spheres of living matter.

While there exists a due balance of the antagonising forces, attraction and repulsion, vigilance or wakefulness is not incompatible with normal or healthy tone. But if the normal relations be subverted, and the clonic tendencies overwhelm the tonic, the train of events is curiously corroborative of the principles now endeavoured to be illustrated.

With most persons, strong tea or coffee, at night, disposes to wakefulness. With some, of a more sensitive habit, these beverages produce not only restlessness, but jactitation, and the true clonic condition occasionally supervenes, of agitation and tremors. An example of the same clonic state is witnessed in a sensitive person, who, for the first time, smokes tobacco. In the advanced stages of typhus fever, when the asthenic state of the patient is reaching its climax, the clonic tremor and agitation are remarkably exemplified. The brain appears to refuse the influence of attractive forces, because the medical attendant is ignorant of the best means of restoring health, by acting upon the central attractive power. He gives opium in some form, perhaps, not being aware that the consequent peripheral pressure from loaded veins, loaded with unhealthy and unwholesome blood, may, and does frequently, aggravate the morbid condition he is anxious to conquer. Under these circumstances, however much the head is oppressed with heaviness, the refreshing sleep does not arrive. In some cases, night after night, and day after day, clone, or agitation, characterises the sensations and mind of the patient. The agencies of repulsion continue. Great debility of the muscular system, gradual deprivation of voluntary power, jactitation, with an anxious, wild look of the eyes, and an increasing agitation are observed. All the organic instincts of the body appear to have taken alarm. There is an undefined expression of depressed feeling. Then the muttering delirium comes on, and death results from effusion of serum, if all the polarities which have been reversed are not forced to resume their normal and direct courses, if the deep tonic sleep, by the aid of God's blessing of Mesmerism, cannot be brought to save the patient.

The pathology of clone, that agitative condition of muscle, which is the opposite to that of tone, is not exhausted in the study of fever. Clonic convulsions affect muscles with agitated, intermittent movements, not with that kind of spasm which is continuous, or sustained, as in tetanus. The convulsions of infants during dentition, the same kind of epileptic fits occurring in persons who suffer in consequence of dentition at more advanced years, or those who have had fits from worms in the alimentary canal, or from undigested food in the stomach, or colon, and the convulsions to which some adults are liable, either from hysteria or epilepsy, are examples of reversed polarities in the muscles.

Women, who die of puerperal convulsions, in some cases, exhibit, in a few hours, all the symptoms of severe epilepsy, followed by tonic opisthotonos, the extreme rigid spasms succeeding to the clonic convulsions. In such cases, death takes place from hypertone, and if the autopsy be practised in a short time afterwards, the internal surface of the uterus may be found to have a layer of coagulated lymph, while the brain and spinal cord are in a hardened, compact state, evincing the agency, during life, of the extreme attractive forces which have constituted that stage of this malady a real tetanus. In such cases of death by tetanic spasm, the patient has not necessarily been killed by poison, although the symptoms have not been remotely dissimilar to those furnished by a case of death by strychnine, and the whole illness may not have been of more than three hours' duration. I witnessed a case, to which I was suddenly called, of a lady in whom the combined violence and suddenness of both reversals of polarity, 1st., from the normal or direct current to the inverse, producing clone; 2nd., from the inverse to the direct, producing hypertone, having determined the severity of the symptoms. I have attended a case, in which puerperal convulsions, at the end of the fifth month of pregnancy, have commenced with severe epilepsy most suddenly, blindness and deafness succeeded immediately. Consciousness was complete for two hours after the access of most severe clonic spasms; then, after a short interval of perfect calm, occurred a sudden reversal of polarity, to produce a disease of hypertone, or tetanus with opisthotonos, concluding with a reversal to the clonic convulsion, and sudden death by apoplexy, accounted for by irregular reversals about the heart. In this case, the changes occupied about three hours.

The death by hypertone may be attended by an empty state of the cavities of the heart, but the blood is not completely fluid, as in the case of hypertone when an animal has been killed by lightning.

Death by lightning occurs in two ways. If the electric force strikes the metallic furniture of the harness about the horse's head, it kills the animal at once by a powerful direct current, inducing hypertone, if the removal of the high force of vitality, an exalted force in the ranks of magnetic forces appertaining to forces directing the functions of high grades of organisms, have not at once killed the animal. But the electric force may have entered the nervous matter of the brain, by an inverse current from the spinal cord, and the death would be from an inverse current, and, consequently, from hyperclone. The shock, in either case, removes the high order of magnetic force we know as vitality, or the vital force. If a shock, powerful enough to form a strong battery, be administered to a steel bar magnet, the magnetism is discharged, and is dissipated to join the all-pervading force of nature, the force of gravitation, permeating our magnet, earth, and streaming its magnetic aura around it.

But, suppose when the electric force struck the horse, a gentleman, seated on the box of the carriage, was holding up an umbrella to shelter him from the rain, and the electric force, attracted by the point of the umbrella, passed in a direct, but much milder stream, through his body, rendering him, for a time, rigid and insensible, from which paralysed state his recovery, though slow, was in the course of time complete. In this case, which actually occurred to a near relation of mine, his watch had stopped, and it was found that all the steel in the works had become magnetic.

A lady, travelling in Switzerland, was killed by lightning, which had first struck the earth, and thus reaching her as an inverse current, was the more sure to destroy life, as a clonic or lethal force. The difference was remarkable in these cases. My cousin survived, the current being a direct one, received and weakened by a pointed conductor. In the lady's case, there was an inverse current from the earth; rigidity was present in the case where the current was direct. In the other, the inverse current precluded the possibility of such an event. In death from lightning, whether the current be direct or inverse, the vital force being at once removed, the blood globules die, and the blood is not found to be coagulated. In fact, the removal of the vital force has been the result of the polarities of the system being so completely destroyed, or extinguished, as to demagnetise or devitalise the living spheroidal magnets we know as blood globules. Hence, the death of that in which was the life. The death by lightning is a most appalling idea, for the vital force has in an instant fled. The science of the positivists cannot tell us where it has gone. The students of Animal Magnetism can, if they have pursued their investigations boldly into spiritualism.

During a voyage to Bombay, in the early months of 1823, I witnessed a case of locked jaw, strikingly illustrative of the reversal of vital polarity. A young lady, of unusually vigorous mind, who was on board the ship, was accosted one day, in the southern tropical latitudes, by the captain, in anger, on the quarter-deck. He used unwarrantably some very brutal expressions, which so shocked her, that her agitation was apparent in her excessive tremors. She hastened down the companion gangway towards her cabin below. At the last step, her foot slipped. She was seized at once with a sardonic grin, and the next instant with opisthotonos. I caught her in her fall. She was rigid, and in unconscious sleep, her back so much curved that her head tended towards her heels; the urgency of her spasm was appalling. She was placed upon her bed. The pulse, at first full and hard, was becoming small and feeble. The face was swollen and flushed. The veins were opened in both arms, and the patient, having lost a prodigious quantity of blood, was relieved of her alarming symptoms. Although much weakened, she had no hysteria, but she had the dangerous tremblings, from the mental shock, which were characteristic of reversal of polarity, preceding another reversal to hypertone, and must have shortly died, if the bleeding had not relieved the congested veins, and the coincident collapsed arteries.

These considerations lead us to reflect upon the consent which must prevail in the animal economy between the nervous and vascular systems. While tonic and clonic states of the nervous and muscular organs are manifestations of two such opposite causes, is there nothing remarkable in the tendencies of the vascular system during the operations of these opposite forces? This is a question involving the study of facts, that belong to a wide range of diseases;—involving too the solution of problems on the influence of the nerves on the blood and secretions, for which we are not sufficiently advanced to offer replies and explanations. But facts enough exist to make certain general propositions quite clear. We have proof of a tendency to a normal balance of the circulation in veins and arteries, in the tonic state of the system; while in the clonic, there is a decided unbalancing of the circulation, by the veins becoming larger, more full and distended, and the arteries smaller, and consequently containing less blood. As a rule, then, congestion waits upon clone, while the normal state of the vascular system attends upon tone.

The tendency of the loosening of vigour is to reverse the normal nervous currents of the body, and each reversal of a normal current, be it ever so partial, must necessarily be productive of some aberration from normal health. Under healthy rule, every organ of the body has its own polar conditions, and consequently its proper currents; (Leger; The Magnetoscope, p. 47, et sequentes). Any reversal in these polarities must induce first, an accumulation of blood in the veins, and subsequently, those changes to which the idiosyncrasies of the individual may lead. Disease, then, in any organ, must arise from a more or less persistent state of reversed normal polarity, or reversed direct currents of the parts. The relations of the reversal may be bounded by changes produced in the organism of parts. Now, organic magnets may, by reversal, lead temporarily, even to increased tone and activity, changed again by the laws regulating morbid currents to a contrary result—to a loss of energy. Dr. Leger was wont to exhibit illustrations of these facts, by shewing, that under certain morbid conditions, increase of power was evinced in some organs, by the effect on his magnetoscope, a wider scope of indication exhibiting the fact; and on the other hand, the pendulum could produce evidence of a loss of energy, by its narrower extent of oscillation or rotation.

Whatever may be the train by which we arrive at the facts we have considered, it will not be denied that the direction of the electric or odic currents is important, that if one course, the direct, be coincident with the phenomena attendant upon tone, and the inverse courses be coincident with clonic signs, that, if sleep coincide with a compact state of brain, and wakefulness with one in which repulsive agencies are present; that if approximative agencies be impulsive or peripheral, the brain, or rather the vital force sleeps unhealthily, but if the approximative forces arrive from an interior or central attraction, the sleep is healthy or normal; that if good reasons are offered, tending to the proof that sleep and vigilance are closely dependant on forces operating on the molecules of the living brain;—that if attractive forces existing in certain substances, as magnets, or crystals, have an agency in causing the central nervous attractions to operate;—that if these in operating, produce a deep healthy sleep, followed by a rigid or tonic spasm, which has been found to be endowed with a marvellous health restoring power; that if the contrary phenomena, coincident with repulsive agencies on molecules, be such as to produce clonic and debilitating forces, ushering in low fevers and other trains of asthenic disease, or even if these repulsive agencies, arresting suddenly the normal currents with extreme violence, cause lethal convulsions, we have then sufficient motives to examine questions relating not only to theories of sleep, but to the practical application of the facts, at which we have arrived, to the prevention and cure of disease.

When a healthy man is considerably exhausted by fatigue, we are aware that the best means of restoring his lost tone is to allow him to indulge in a sound refreshing sleep. Common sleep, in one's bed at night, is the more refreshing in proportion to the degree of tone which the system has acquired. If this tone could be multiplied, one should have proportional increase of vigour. A healthy sailor, when he turns into his berth, often packs his sleep of four hours, with so much tone, that he obtains the refreshment of a mesmeric sleep, which, in fact, is only common sleep, multiplied, if the expression may be allowed, by a factor of a larger power than itself. The sailor, having had strong exercise, feels the fatigue inducing a reversal of polarity from an inverse to a direct current. Breathing at sea a pure air, charged with health inspiring ozone, his clonic forces yield to the tonic, and he sleeps with an intensity in a ratio to his previous fatigue.

Is it necessary to ask the question, whether it would be desirable to know the means by which this multiplied sleep could be practised? Is it desirable that artificial means should be adopted to secure the blessings of tone and health, when the multifold accidents of life have deprived us of them? Desirable or not, to those who oppose themselves to progress, events will march, the flood-tide of knowledge will advance, and the obstacles set up by small minds will be demolished by the forces called into operation from the energy and earnestness with which the cause of truth proceeds.

Allowing for the very numerous differences of constitution in human beings—for their physical idiosyncrasies, as well as their mental, and regarding each difference as a modifying force—although it may be to some hypercritical and sceptical minds most difficult to reconcile certain apparently stubborn facts, it is yet in our power to classify them, and to deduce laws. The relations of attractive and repulsive forces to the phenomena of the living body are very striking. We have but to watch and to trace some of these, to arrive at most interesting conclusions. If one pole of a crystal be found to encourage sleep, and if the opposite pole have the effect of producing wakefulness;—if Mr. Rutter's Magnetoscope have established the fact that the two ends of a crystal have opposite poles, and that, in fact, every crystal is a magnet; and that every portion of a living vegetable, and of a living animal be a magnet; conclusions fully corroborated by the patient philosophical researches of the Baron von Reichenbach—there can be conceived hardly any termination to the magnetic relations of crystallic and organic bodies, and in prosecuting our inquiries, in order to harmonise the influence of various forces with the vital dynamics of our systems, we should endeavour to arrive at some principle, which may be unerring in practice for establishing a due amount of tone in the frame. The influence, as far as we yet know, of the attractive pole of crystals, would appear to be tonic. Tonic forces operate to produce sleep. Magneto-electric currents, from the brain downwards, favour sleep and tone.*

The means and appliances to obtain normal sleep, and its consequent tone, then, must consist in those arts, which have a tendency to effect our object. It is well known that sleep may be induced in some persons in a very short time, by passes slowly made with the slightly curved open hands of another person, at the distance of from half an inch to an inch, from the crown of the head, downwards, along the face and chest to the pit of the stomach. This may be done either with each hand alternately; with both hands employed together at the same time, or with one hand singly.† So very susceptible have some subjects been found, that I have repeatedly seen persons who have thus been put to sleep by one pass. There is, perhaps, no end to the varieties in the degrees of susceptibility in different individuals to the influence of these magnetic manipulations, known as passes. For the purposes of illustration, the very susceptible cases are undoubtedly the most striking. With those not so susceptible, the patience of the mesmeriser may be sorely tried. But where to the sufferer the importance of the cure is vital, no effort should be spared to accomplish the end in view. Though I am certainly an advocate for the deep mesmeric sleep in the treatment of all very serious and grave maladies, such as typhus and typhoid fevers, nay, in all fevers, especially in acute rheumatism, in scarlatina, and in small pox, and do not believe that cancer can be cured without the deep sleep accompanied by rigidity. (I quote

* See note at page 264 of 4th volume of the Zoist.

† See Townshend, Facts in Mesmerism, Deleuze, Teste, &c., &c.

my own words from a paper on the cure of a large polypus of the uterus) : "The result of many cases makes it sufficiently obvious that sleep is not necessary in order to effect a cure by mesmeric agency." "As far as we hitherto know, a great deal would appear to depend upon the relation between the peculiar force of the mesmeriser, and the peculiar aptitude of the patient to receive the impression of that force." In hundreds of cases of rheumatic pains in various parts of the body, Mr. Capern, now of Weston-super-Mare, has removed the suffering in ten minutes, and I have known him, on many occasions, to chase away pain of a severe character by a single stroke of his hand. "In like manner, I have myself repeatedly removed pain in various parts of the body from persons of a susceptible nervous fibre. Major Buckley has frequently, to my knowledge, done the same thing without the induction of sleep. It is a strong fact, bearing upon the question, that at a time when I was a sceptic as to the power of his influence in this way, in the year 1845, he removed an intense head-ache for me ; and when I told him I was certain it would recur in ten minutes, for that it was dependent on a condition of stomach, with which I was familiar, and which would be removed by a dose of calomel, he assured me that, if I would allow him to make a few passes along the abdomen, I should not require the medicine. Having no faith, I ridiculed his proposal, but from a feeling of civility towards a most good-natured man, I submitted, and in a few minutes the passes had the same ample effect as if I had taken a moderate dose of castor oil. The experiments of sub-mesmerism, to which the ridiculous compound word electro-biology has been applied, are illustrative of the same general proposition, that sleep is not a necessary condition for the cure of disease by mesmeric agency. The wonderful efficacy of local mesmerism in the cure of severe acute inflammation without sleep is an additional proof tending to the same point. In whatever mode, and under whatever circumstances, an ample and powerful development of

Baron von Reichenbach's odic force positive is made to affect a patient labouring under pain or disease, whether sleep accompanies its presence or not, the cure is commenced, and if the forces can be kept in action for a sufficient time, the cure is effected. The more I reflect on the subject of sleep, the more I become assured of the practical advantage of the theory, which I have proposed in the fourth volume of the Zoist. All sleep results from an increased compactness of brain and nervous matter. Sleep results in different individuals from different quantities or intensities of the attractive forces. In one, the amount of power necessary to produce the sleep may be very different from that in another. But the same force attractive which brings on sleep is the odic force of the Baron von Reichenbach,—the sleep producing power of the pointed end of the rock crystal, of sulphate of alumina, of sulphate of soda, of sulphate of magnesia, of nitrate of potass, of the north pole of the magnet, of the downward mesmeric passes, of the fingers pointed a sufficient length of time to the eyes of the patient, or of the gaze of a mesmeriser, whose mind is occupied with benignant feelings, or of the will of a mesmeriser directed by the same class of feelings. The mesmeric force attractive may operate on the nervous system of individuals in a more wide and clumsy manner by causing the eyes to look upwards strainingly, squinting at an object held or fixed at a few inches from the forehead. This coarse and unpleasant mode of hypnotising is, like the silly and unsatisfactory explanation, which has been given of its *modus operandi,* characteristic of the low tastes of the badly organised brain, from which many mean and unenviable feelings have emanated. The mode of inducing enlarged pupils of the eyes and the sub-mesmeric condition—the first stage towards the approach of sleep—by the zinc and copper discussed by the American electro-biologisers, belongs to the same class of agencies. All the phenomena of electro-biology can be produced on any individual, who is susceptible of being brought to the requisite impressionable condition by the practice of any of those modes of mesmerising which have been known to us for a long time. Secrets are out of the question."

But though sleep is not an essential condition for the cure of a long list of diseases, there are some maladies which we might in vain hope to cure, without the induction of very sound and deep sleep. Epilepsy, not dependent on dentition, is one of these. I had a case of this malady under my hands, with my fore and middle fingers pointing to his eyes daily, for one hour, and sometimes two hours, for upwards of two years before I could succeed in dilating his pupils sufficiently to make him somnolent, and thus making him drowsy and sleepy, a condition attended by extraordinary benefit to him; for, whereas in the beginning, his severe fits were in duration five hours and a half to six hours, he was at the time, his weak-minded mother removed him from London, never afflicted with insensibility for more than three minutes.

In our attempts to produce sleep, we observe a great variety of phenomena. These vary in the order of their appearance, according to difference of constitution. There are persons who, remaining awake, may, upon being mesmerised, exhibit a complete want of power in the trunk and limbs, coupled, perhaps, with insensibility in these parts. These are, for the most part, the cases with phenomena analogous to catalepsy, for, if the limbs be moved by the mesmeriser, and placed in any extraordinary position, they remain as they are posed, until they be again altered, and made to assume another pose. In many such cases, however, if a few downward passes be made along the limb, so placed in a condition of catalepsy, it becomes rigid, and then it can no longer be moved about; it is stiff, and no longer amenable, unless it be blown upon, or otherwise relaxed, by demesmerising passes. The difference between catalepsy and rigidity is very striking. In catalepsy, the muscles are not rigid. The tone is not so far advanced. I had a case of a young woman, a few years ago, who was so susceptible of mesmeric passes, that I could induce her to sleep by making a single pass. Six or eight passes produced rigidity, and twenty to thirty curved her body, and made her a subject of opisthotonos; a condition I could always produce by affixing a couple of sovereigns to the soles of her feet, and then placing a disc of zinc upon the skin of the nape of her neck. A powerful direct current passed from her neck to her feet, and a rigid sleep instantly seized her. By applying zinc to the feet, and gold to the nape of her neck, we had the inverse current, and it invariably wakened her and released the spasm.

The train of events, in the case of a person susceptible of being influenced by the ordinary mesmeric passes, is, first somnolence, shortly afterwards, a gentle sleep, becoming deeper and deeper, as the passes proceed, becoming more and more placid, according to the temperament of the patient, and often modified, or much influenced by the benevolent or attractive character of the mesmeriser. Some sensitive patients in this stage become sleepwalkers, or somnambules. Some become clairvoyant, and furnish deeply interesting subjects for the study of the philosophy of mind; for now the varieties of phenomena, more or less associated with the magnetic properties of the preponderating phrenological organs of the brain develope themselves very curiously. In some, the trains of symptoms indicating errors in bodily health, are marvellously exposed to our knowledge. By acting upon the obvious suggestions afforded by these indications, many patients have been strikingly benefited.

But where no mental phenomena are manifested, and where the changes, upon repeated or long-continued mesmeric passes, pursue their normal course uninterruptedly, the stages are few. Deep sleep succeeds somnambulism, and a tetanic rigidity succeeds the deep sleep. This is the best course for the health-restoring agency of Mesmerism to pursue. Perhaps none of the diseases, that have not wholly disorganised or destroyed important structures is able to resist the influence of repetitions of the powerfully toning force. It appears to re-establish the original order of Nature; and obliges parts which have not been too much altered in structure, to resume their healthy forms and relations. It normalises abnormalised organs. It replaces disease by health. This language is very deliberately, very earnestly used, from a thorough conviction of its truth, and is not in the least too strong, considering the facts on which I rely.

CHAPTER VIII.

A REITERATION ON SLEEP.

Voluminous as are the works which we consult, when we propose to give a condensed view of the facts relating to any subject, we are at a loss to find the points at which we desire to arrive. We are told that we must wait until more facts arrive in order to elucidate those points. But all our reading amounts to an acknowledgment that our predecessors have only a very limited range of facts to deal with. This is well illustrated by a reference to all the works on physiology. We desire to know more than they have yet recorded on the facts of sleep. Not one of the physiological labourers of the age knows anything more of sleep than the nurse who watches the babe in its cradle, or the attendant in the wards of the hospital, who witnesses nightly the most interesting phenomena of healthy and diseased sleep. No physiologist has yet given any account of sleep and wakefulness, philosophically considered. This arises from a neglect of the study of Animal Magnetism. Surely, if no other plea could be offered for the study of this magnificent range of facts, this one should be a sufficient reason for laying aside brutal prejudices, and for studying the phenomena which are the surest guides to the cure of intractable disease. If we survey the scope of these facts, we are in danger of losing the thread of our discourse. We purposely avoid wandering from the subject of sleep. This is our present topic. We have already seen that the phenomena of sleep, and of wakefulness are dependent on two opposite manifestations of magnetic force. We have now to trace the relations of health and of disease to the same cause.

Many will affirm that no possible relation can be traced between these phenomena, and that all our distinctions, based upon magnetic agencies, must be fanciful. We do not pretend to argue with sciolists. It is so easy to affirm or to deny a proposition, that unless we know that a man, who objects to our experience, has had the opportunity of testing the facts fairly, we must decline to place him in any position as a judge. It is simply absurd to expect a man to know what he wishes to decide upon, when he has not been at the pains of studying the subject. If such a man as the late Sir Benjamin Brodie were to offer himself as an umpire, we should decline to accept him, because he more than once said, that if he were to "witness the facts himself, he would not believe them." Fancy a man refusing to believe in Sir William or Sir John Herschell's facts, because he would not believe them, if the telescope made them as clear as the sun.

That the facts relating to sleep and to wakefulness have a close relation to the phenomena of health and of disease, we shall proceed to demonstrate. We have already made use of the following illustration. When a rope is twisted to a certain degree, and stretched between two points, the particles of hemp approximate to each other, and the rope is said by the sailors to be taut. Now, let us compare a muscle to this rope. In a relaxed state, it is flabby, but when the will influences its contraction, it is said to be tonic. Taut and tonic convey analogous ideas. Suppose, now, we have a contrary state. The muscle, or rope, is relaxed. Necessarily there is not the same tension, the particles are wider asunder. If we now examine the condition of the particles, we shall find reason to conclude that they are not only divergent from each other, but unable to return to their approximated state, without the application of the force which originally produced their compact state. This force is an attractive agency. It is a magnetic force. Will any one venture to deny this proposition? We really do not know what our sciolist might venture to deny. Our next question is, will the muscle obey the stimulus of the will of another person? This is a point that may admit of a question by those who have never witnessed the facts relating to the force of the human will. Hundreds of persons have witnessed all the phenomena of the human will operating upon sensitive male and female subjects. Many a time have I made persons travel considerable distances, when desired to do so by the force of my silent will. Many a time have I, in public carriages, obliged persons I never saw before in my life, to place their hands in mine, or to take hold of my shoulder, or my elbow, or my knee; of course, not being aware of the cause which influenced them, they have either looked uncommonly awkward, or have sometimes fallen asleep, under the powerful attractive agency of the force. My friend, Mr. Henry Stafford Thompson, of Fairfield, has frequently adopted the same experiments, and with similar results. I do not know how to condense all the facts on this subject, into a small compass, and shall therefore refer my readers for them to the notes I have published in my edition of the Baron von Reichenbach's Researches, and to certain papers on the human will, which will be found in the several volumes of the Zoist.

The facts being established, all cavil is absurd. The inference from them will lead us to the conclusion that sleep and wakefulness may be produced by two opposite magnetic forces; that these forces may be set in operation by the human will, which thus becomes a motive power; and that if this motive power be exerted to produce sleep, it is an attractive force, and if to produce wakefulness, it must be a repulsive agency. The human will can and does exert these opposite forces. If certain phrenological organs are called into operation, during the direction of the will, it becomes an attractive force; but if certain other organs come into play, the influence is repulsive. There is a law regulating the attractive and repulsive organs of the human brain. This law is as invariable as any other magnetic law, and is the cause of our being agreeable or repugnant to our neighbours, in our intercourse in society. Many are the occasions in which every person must have experienced, without being able to assign any reason for the fact, the abhorrence they have entertained for particular individuals; and, on the other hand, the readiness with which they have been glad to accost an interesting stranger. Superficial thinkers believe themselves warranted to refuse attention to this fact. They are to be allowed to flatter themselves they are far wiser than their neighbours, and respected accordingly.

All magnetic phenomena relating to sleep and wakefulness have not been embraced in these considerations. Some are of a nature so curious, that we may well supply our readers with a discussion on them. When a man sleeps in his hammock at sea, he is most frequently amazingly refreshed by his deep slumber of four hours. The sailor, on board-ship has, upon the whole, a very agreeable life. He eats and drinks heartily; he is decidedly fatigued, but when he turns in, he sleeps profoundly. Generally, this sleep is a magnetic or mesmeric phenomenon, for he enjoys a more sound sleep than he would on shore. The reason is obvious. We know that at sea the air is impregnated with a large quantity of ozone, and this agent has a tendency to deepen sleep. Some of my mesmeric sensitives, upon a single sniff of ozone, have fallen into a profound rigid sleep for several hours. This sleep is always refreshing, and in some cases, so invigorating, as to lead to the complete restoration of sound health in persons who had long been ill. There is more in this sleep than we are apt to imagine. If we endeavour to probe the causes of our invigoration, we shall find that as the muscular system approaches the condition of rigid spasm, the particles become more and more approximated. We have already seen that tone means magnetic approximation of the particles of nervous fabric. Muscular structure follows the same law. We might go as far as to say that all the animal tissues are, to a more or less remarkable degree, influenced by the same law. Health then is the result of tone.

We must now inquire why we have deferred the consideration of disease, until we could establish in our minds some clear idea of the cause of health. We perceive that sleep and health result from currents having the same direction. They are derived from the existence in the system of attractive forces. These are produced by direct currents of Magneto-Electricity. How, may we ask, are the phenomena of disease produced? This is not so easy to make clear. Notwithstanding all our efforts to persuade men to look into the subject of Animal Magnetism, we have failed to convince any considerable number of persons that disease is dependent on magnetic laws. Disease leans upon clone. Clone, we have proved, is dependent upon reversals of normal or direct currents. But we may be told that all cases of reversed currents do not necessarily produce disease. That is very true. But it is nevertheless true, that all health and all disease must be dependent on the degrees of their respective phenomena. Dr. John Clarke, the celebrated predecessor of his brother, the late Sir Charles Clarke, was a man of shrewd apprehension. He used a good expression, when he said that the most ephemeral ailment of the body was a temporary derangement of organisation. If in his time, these present ideas could have been put forth, he would at once have adopted them, for like his brother, he was a thinker. A temporary disturbance of the structure of a part necessarily results from abnormal polarity. Disease then, varying in degree, may range from slight disturbance to very serious disorganisation. What can induce any man to deny this proposition. It is a clear statement of a fact, and the changes accompanying different degrees of clone must be considered stages in the different forms of disease.

The word Clone is adopted in reference, principally, to asthenic states of system. It is, however, applicable to all the degrees of illness; for all catarrhs, all the forms of rheumatism, all inflammations are invariably preceded by reversals of polarity, and the severity of the attack seems to be always in some ratio to the intensity of the magnetic change. We are thus led on to consider the immediate connection between reversals of magneto-electric currents, and the circulation of the blood. This is a subject as difficult to explain as that of its intimate connection with disease; for, though the reversal of polarity may be complete, and severe rigors result, time is required for other changes. The whole body is a complicated system of living magnets. Life implies a vast variety of instinctive forces. Each tissue is distinguished from other tissues by being regulated by its own force; a force generated from the great trunk force emanating from the sun. This is not so easy to understand, for how do separate forces reach the separate tissues? The actinism of light is a force derived from the great trunk force. It finds its way into the body by various channels. It is respired by the lungs, associated with oxygen. It finds its way through the eyes;—nay, through the whole body, by the agency of dioptrics and catoptrics. It is not a question how each instinctive force is generated. We must be satisfied with the fact, that no form of matter can act in generating force. The *modus operandi* of life in the refinements of exquisitely delicate forces is at present too subtle a subject to occupy us. We must be content with the fact that they are forces regulating inert matter.

The influence of light as an agent in producing health is of great importance. We shall be at no loss to understand the operation of sun-light in restoring colour to anœmiated subjects. Those who have been pent up in close dark rooms are placed in circumstances very unfavourable to the development of health. They are subject to a condition known as etiolization. We are obliged to confess, that we are accustomed to regard these wan denizens of poverty with extreme pity. Even a restoration to the light of heaven often fails to restore them, and they sink victims to consumption, or some other form of scrofulous disease. Light, then, regarded as a sanitary agent, is of vast importance in the economy of nature. We do not know how we should do without it. It is the invigorating principle in the air we breathe. It is the magnetizing agent which pervades the whole body. From it is derived the principle of life; the vital force which radiates through every part of the system. It is the nidus of the soul of man, and if this be doubted, we must proceed to lay before our readers the reasons for the assertion. These reasons are but common-place truths; they relate to the highest flights of philosophy; they belong to a science which is, of all others, the most holy. We shall now proceed to lay the facts before our readers.

Summary must be our progress in detail. We have to shew that from magnets and from crystals, as the elementary forms of all organised matter, we must trace the progressive steps of higher development. We are not bound to go back to the year 1819, when these ideas first occupied my mind, but we may be assured that no train of investigation, once entered upon, is so captivating as that of tracing the gradations of nature's various forms. All the highest philosophers have given way to the tendency to classify and to generalize. The most comprehensive minds are those gifted with the memory for minute details, while they are enabled to look beyond these to the powers which are exerted by a superior being in arranging, and in reducing to the law of series, the largest as well as the most minute objects. We are now upon the great law which operates in fashioning atoms to arrange themselves, or rather to be arranged according, first, to palpably geometric forms, and then to more apparently fantastic shapes, which, however, still retain a shape subject to geometric rule. No crystal however minute, is without form. All is law, all is order. As the severely geometric form is modified into one which, it would be wrong to say, was aberrant, for what may appear to us at first as aberration, is found always, on deeper investigation, to be strictly obedient to rule or norma, so, we may observe in all cases, that what we suppose to be law is often only aberration from norma. We are gifted with neither prescience, nor omniscience. Our faculties are too limited to enable us to understand how a straight line may be undulated, without being aberrant from the norma of straight lines. It may be only a temporary manifestation of the presence of force which produces undulation. This force, having expended itself, is no longer operative in modifying the form of what was essentially the same object, whether in a strictly straight, or in an undulated form.

We have now settled, in our minds, the idea of aggregations of atoms according to laws regulating the various forms of matter. We have all of us, or we ought to have, the knowledge of the great fact, that the sun is a magnet. If we do not know this fact, we are unprepared to understand Sir Isaac Newton's great proposition that the sun is the centre of the solar system. His view of gravitation was derived from facts dependent on the sun's attractive force. We are none of us obliged to feel that we are quite sure of what Sir Isaac Newton thought, when he said that gravitation was the force which regulated the movements of the planets; but we must give him credit for having reflected on all the facts in science known at the period of the world when he made his great astronomical discoveries. He must have reflected deeply on the phenomena of the magnet. No man, so profoundly versed in physics, could have been ignorant of the necessary connection of the meridional and equatorial forces with the sun's central power of attraction. What, then, is the inference? That he was quite ignorant of the laws which regulate attraction and repulsion? On the contrary, he scanned all their properties at a glance, when he boldly announced that gravitation was a compound force, and that it consisted of two forces having opponent tendencies. He could never have, otherwise, used the terms centripetal and centrifugal. We are not prepared to assert that he had anticipated the modern discoveries of Oersted. But there is no doubt that he fully appreciated the influence of his own two opponent forces, or he would not have thought of their *universal* application. He was right in his generalizing conclusion. He felt that he had made a discovery universally applicable. When he found that his two forces accounted for the whirling of spheres in their concentric orbits, he had only to apply his law to the minutest spheres, and he could equally deduce the laws necessarily resulting from the existence of his opponent forces. Was it necessary for him, then, to try how far his new theories could be adapted to the confusions which were to enter into the heads of his non-mathematical, or even of his mathematical successors. Too many of these, ignorant of the laws regulating phrenological science, have set themselves up, as judges, on affairs of which they are content to know very little. We so often find men pretending to understand subjects they have never studied, that we need hardly be surprised at exhibitions of ignorance. We are careful of exposing our own ignorance, not venturing on subjects beyond our depth; but we know enough to be able to detect a sciolist when we meet him on a subject we have carefully studied. Many such exist in the fields of science. Prone as most men are to surround themselves by interested advocates, few are able to detect the fallacies of ignorance. One of these consists in concluding that we are good judges of our own minds. We feel a pleasure in accepting praise, let it come from any quarter. We forget that it may not be disinterested. Many are the errors into which this love of approbation has led men of even superior minds. The organ indicating the existence of this propensity is larger in some men than in others. To be the idol of a coterie, is a folly that few can resist.

We are not sure that we should be able to resist it ourselves. Fortunately, our lot has not been cast among coteries. We have instinctively avoided them, and hence perhaps the mutual repulsion between us, and those who delight to be the idols of the sciolists. There are few who can resist the temptations of flattery; fewer who can manage to steer clear of parasites. We are not advocates of misanthropy, but it would be more agreeable to become the slave of some despotic prince, who could appreciate manly vigour of intellect, than to be surrounded by lovers of the ingenious and the chimerical. All phrenologists can appreciate this preference. The one class is prone to flattery and imaginative adulation. The other delights in truth.

Our present occupation is in a search for truth. We propose to limit ourselves to this healthy pursuit. What is truth? A question often asked, never satisfactorily answered. It is, nevertheless, very easy of reply. Truth is illumination from Heaven. We cannot use a stronger expression than God's truth. We may be sure that our definition is a very just, and a very cogent illustration of that which puzzled Pontius Pilate.

We proceed to shew that if our definition be founded on fact, we shall find our future course more easy.

We all know, or we ought all to know, that magnets and crystals emit light. Those who do not know this simple truth, should lose no time in reading my edition of a translation of the Baron von Reichenbach's work on the Dynamics of Magnetism, &c., in their relations to vital force. It is a most interesting book, and may well occupy the attention of persons desirous of becoming versed in the facts that now are exciting intense interest in Europe, as well as in America, and which, in spite of all the obstacles thrown in their path by the sciolists of England, will force their way into the channels of public opinion. We are sure of our position. All the world, holding together, cannot conquer the truth. The illumination from heaven is like the rays of the sun, and great must be the power of the monster emerging from the depths of darkness, who shall be able to extinguish that light.

Sir David Brewster is little aware of the error he has committed. When men of science allow their temper to blind them, they are then sorry specimens of our intellectual race. My good friend had an excellent opportunity of distinguishing himself. I had hoped when my friend Dr. Leger wished to dedicate his little but most important brochure of the Magnetoscope to me, that Sir David would have appreciated the compliment I caused to be transferred to him. He has, however, shewed no sign of recognising the great facts developed in the book to which his name is prefixed. I am sorry for him, for the effect which will be produced on his reputation may not be considered very enviable. Free thought is very rare among men of science. It is not to be understood that free thought implies a leaning to infidelity. In that sense, it is unhappily too common. We should then denominate it cramped thought, for it is the result of narrow assumptions and limited views of natural facts. Free thought leads to high aspiration, and high aspiration belongs to large minds. It is not now a question whether such men as Sir David Brewster can arrest the progress of knowledge. He has done his best to throw discredit on most important truths. This conduct will injure him, but cannot have any influence on the truths. It is a pity men cannot perceive this when they allow their love of ruling the thoughts of the world to overcome their love of truth.

We are now to consider how far the world will be willing to receive our account of the forces regulating the arrangements of magnets, crystals, and the higher grades of organic apparatus, necessarily obedient to the laws of magnetism. We are quite aware of the difficulties we labour under, even when we hint at a preference for a classification of facts which forces the introduction of natural orders for any accepted system of artificial arrangement. Those who were students of natural history, in the days of my college life, may remember the opposition made to Jussieu's views, where they clashed with the favourite system of Linnæus. Time is required to mature a healthy growth of opinion; and there is no doubt that, in time, the views I advocate will fully supersede the present artificial conclusions of the men of science. It is not requisite that men should yield their stubborn convictions in a day, but when they have well digested the new facts introduced by numbers of investigators, they will find that they must expand their systems of knowledge. We are about to enter upon a new era in science. Those who are not ready to believe in the facts which have been witnessed by numerous persons of average intellect, will find themselves compelled to account for phenomena they do not comprehend. Time passes with great rapidity, and events succeed each other in such manifold ways, that we have little idea of the importance with which we are surrounded when we survey the absurdities, as we may consider them, which are daily pressed upon our notice.

We are upon the eve of new prodigies; we know not what extraordinary scenes may succeed to those we have already witnessed. Many are the sceptics of to-day, who will be the believers of to-morrow. Hence we may conclude, that amidst the ever-changing aspects of science, our present idols may yield to a race of still more daring sciolists. Such considerations are worth reflection; and few men will say that we are not bound to make amends for the vituperation we have lavished on the real lovers of truth.

Time hastens on, and we feel that we should be wanting to our own dignity if we permitted ourselves to waver in the course we have for years pursued. Mesmerism was assailed for the imposture and improbability of its pretensions. Nothing daunted by the assaults of its enemies, we continued to study its phenomena, and we found ourselves amply rewarded in the continued development of new and marvellous facts that were afforded us in the course of our researches. No man can question Nature without receiving such a reward. But many mistrust their own powers of observation. Some imagine themselves incompetent to believe the evidence of their own senses; while others are downright scoffers. Some think it creditable to laugh at all philosophy which does not emanate from the brains of their own patrons. Others expunge from their memory the vestiges of all that can militate against the preconceived resolution to deny the most palpable truths. We are not now advocating a system of *gobe-moucherie*. We are not believers in impossibilities. It is sufficient to indicate the course of our opponents, which has been characterised by a folly, and by a stupid want of faith in the evidence of their own senses, discreditable even to the lower agricultural boors. There are no terms too contemptuous for the degradation to which many of the opponents of Mesmerism have subjected themselves.

Prone as we are to wander through mazes of facts, without giving ourselves the trouble to arrange them in any kind of order, we cannot regard the new truths of Animal Magnetism as unworthy of classification. Whether we consider them as a part of the arrangements which have been made since the beginning of the solar system, or of the world we inhabit, we shall find ourselves equally at a loss to become convinced of the position, to which they ought to be assigned in the order of our knowledge. There can be no doubt that truths so obvious require, at all times, our best attention. We are obliged to give them that attention, if we wish to be considered more than commonplace thinkers. No man likes to be thought of as a fool, and yet many act as if they really were no better. No man likes to be considered as a poltroon, and yet his want of courage to tell the truth he is fully aware of, is the most egregious cowardice. The dilemma to which many reduce themselves is so awkward, that they are obliged to yield to absurdities, rather than acknowledge a respect for truths they cannot gainsay. We do not fight against weak minds. We do not wonder at the stupidity of idiots. We are the champions of truth against sciolists and men of genius, who pretend to have a knowledge of which they deem themselves proud. Rather than succumb to the truth, the men we select as our adversaries would throw all their vaunted science into the sea. We must grapple with the science of England, simpletonically so called, which tells us that it will never yield one iota to the absurdities we advocate.

There is not much to be said, when we are voted to be so plentifully absurd as not to be worth a moment's consideration. Yet we defy the science of England to gainsay our facts, or even to find a flaw in the fabric we are now erecting to the glory of God. All that we see around us warrants us in the efforts we are making to connect real science with the existence of a deity, and we are bold to say that no series of facts which the science of England can adduce, are so cogent, and so convincing on the evidences of natural and of revealed religion as those we have undertaken to advocate.

However multifarious are the sects of the Christian world, however diversified may be the tenets which characterise these sects, the fact is still the same. There is but one true Christian religion. Its essentials are to be found in one book. That book is our code of morals. No man, however elevated his rank, need be ashamed to own that his conduct is governed by that code. Whoever he may be that scoffs at that code, cannot fail to be weak in intellect. He may not be wanting in acquirements, but he must be wanting in common sense. Who shall advocate the good taste, the good feeling, the honest sentiment of an educated person of the nineteenth century, who shall scoff at the plain morality we may find in the New Testament? If this be so, where is the person who shall gainsay the facts on which that morality is based? There is no doubt of the truth of the basis of the morality of the New Testament. We are not now contending for idle words, upon which learned sciolists may split hairs. We are engaged in advocating the truth, and we say it is based on the foundation of all morality? What is the foundation of all morality? It is nothing more than thé holy work of the same hands that fabricated the glorious universe we see around us. We may say, in other words, it is the science of God.

CHAPTER IX.

CHENEVIX, ELLIOTSON AND WILSON.

If we desire to explain the causes of the phenomena of either mineral, or vegetable, or Animal Magnetism, we must inevitably recur to the antagonisms of the universal force of gravitation, or to forces centripetal and centrifugal; for, allowing ourselves the use of figures of speech, all Magnetism may be said to resolve itself, under divine law, into attraction and repulsion.

We can hardly reflect upon any phenomena in nature which do not relate to these antagonistic powers. Their ramifications are distributed everywhere. Nature is pervaded by them, and we can conceive of no change that is not dependent upon them. They prompt and regulate all the chemical, physiological, and mechanical impulses, incessantly productive of the perpetual changes proceeding around us in all directions. Nay, we may proceed further, and trace their operations into the phenomena of mind, for psychological forces, capable of being measured and estimated by the Magnetoscope, must evidently obey the laws of Magnetism.

Can we imagine a pursuit more captivating than an inquiry into the phenomena of a science that takes so extensive a range, and which involves such exquisite considerations? Dr. Elliotson says of it, in a deeply interesting paper, in the first number of the Zoist, "Mesmerism has always been true. Dimly known for thousands of years, in barbarous and semi-barbarous countries, known as to some of its high results to many of the great nations of antiquity, though the knowledge was confined to the chosen few, it is only beginning to be seen in its various aspects and ramifications, and to assume the character of a science, a science of the deepest interest and importance, inasmuch as the phenomena of life transcend those of all inanimate matter, and the faculties of the brain—the mind—are the highest objects in the universe that man can study; and inasmuch as its power over the faculties of the body at large, and especially over the whole brain and nervous system is immense, and therefore capable of application to prevent and to remove suffering, and to cure disease, far beyond the means hitherto possessed by the art of medicine."

This paragraph, with permission to abate the sense in which the mind is made subservient to brain is characteristic of the large thoughts of the author, whose flexibility of mind has not failed to land him on the domains of spiritual philosophy, with strong convictions of the holy nature of the high mission in which his life has been for so many years engaged.

Dr. Elliotson, when he wrote the paper of which I have quoted the introduction, was little aware that his contribution to the history of Mesmerism, and its expansion into spiritualism, was so important in establishing an epoch in the progress of the truths he has so nobly advocated, and of the science he has so assiduously and so successfully cultivated.

If we reflect on the earlier history of Animal Magnetism in England, we shall find, after the ineffectual efforts of Mesmer to interest the College of Physicians, efforts in which he met with no success, that the facts remained with Dr. Dumainaduc, and a few adventurers, who made no great progress here. A remarkable hubbub succeeded upon the effort to introduce a new method of removing pain by a pair of metallic tractors, which a Mr. Perkins brought, I believe, from America. They consisted of pieces of small, tapering, wedge-shaped metal, gold, or compounds of gold and copper, which were used by passing the points gently down over the seat of the pain. It was said to be a speculation in charlatanerie. I will not be bold enough to deny the fact, for I have not sufficiently investigated the matter; but this I know, that Dr. Haygarth, of Bath, in his alleged triumphant exposure, proved only that wooden tractors answered, with some patients, as well as gold, *ergo*, gold could do nothing! A sequitur worthy of the medical mind! I do not pretend to advocate the cause of a quack. I only contend for the possibility of two pieces of pointed gold, gently passed down the skin over a painful part, might possibly remove the pain, for they might possibly encourage the normal direction of currents of force, the polarities of which, in order to constitute pain, had been completely reversed. Thus, eager for the notoriety of being considered sharp and acute, did this Bath logician flounder.

Dr. Elliotson has pointedly insisted upon the truth of Mesmerism, and has thus established not only the superiority of his own mind over the minds of the small fry of those would-be-considered philosophers, who oppose the progress of a kind of knowledge fraught with holy blessings to mankind. It is well that an epoch has been fixed for the introduction of Animal Magnetism into this England of ours. Honour to all who have, at any time, brought in portions of real truth into this land, and whoever has brought in any facts relating to Animal Magnetism, has conferred an important benefit on our country.

"Animal Magnetism is true," wrote Mr. Richard Chevenix, in his essays, communicated in 1828, to the London Medical and Physical Journal. And he proceeds, "In the whole domain of human acquirements, no art or science rests upon experiments more numerous, more positive, or more easily ascertained." All this either is, or is not the truth. If it be the truth, then in what a predicament are those who, without due examination—it need not be said, but it may be affirmed, without any examination at all—have set their faces against the truth.

The Magnetoscope gave to Dr. Leger the means of arriving, with mathematical precision, at the conclusion that the mental capacities of human beings were as various as their physiognomies, or as the configurations of their persons. Reviewing the workings of mind in the various characters that have acquired for themselves, by their stolid opposition to the truth of Animal Magnetism, an unenviable notoriety, we should arrive at the conclusion that our deeply interesting subject had been projected into the world at a period of its history, when scientific minds were devoid of scope, incapable of expanded views, wanting in depth of thought, and deficient in the faculties that can be excited to high aspiration. It might well be said of the minds of those most distinguished for their opposition to Animal Magnetism, that they were with, perhaps, one or two exceptions, belonging to a class that never rose above mediocrity. Dr. Elliotson says, in the essay already quoted, that "Accident determines the knowledge, the opinions, and the pursuits, both of individuals and of most nations. Where any cerebral organ, or group of organs, preponderates over the others, it will impel the individual, with little reference to the influence of surrounding circumstances, or even in spite of them; but the generality of men are without such preponderance, and take the kind of knowledge which is offered them, imbibe the opinions of those around them, and fall into the habits and pursuits of their countrymen." This is so correct an estimate of the motive impulses regulating the opinions and knowledge of the herd of men, that we can wonder little at the retarded progress of any great truth, when the minds which lead opinions at any given time, rise not, for the most part, above the common-place standard, and are content, like the monkeys congregated in a zoological menagerie, tc expend their activity in exercises of cupidity and envy.

Sheridan, in the *Critic*, tells us that envy is the most powerful feeling of the heart, and our experience would lead us to add that it exerts a most energetic influence in retarding the progress of truth.

Dr. Elliotson must often have experienced the force of this fact, since the period when he tells us that "the establishment of Mesmerism in England sprung in 1828 from the visit to London of an Irish gentleman, named Chenevix, who had fixed his residence for several years in Paris. His visit was short, but during his stay here, he prevailed upon several persons to see him make trials of mesmerism; and afterwards prevailed upon the editors of the *London Medical and Physical Journal* to publish the report of all that took place."

It is right to say that Mr. Chenevix was for some years a sturdy, and an unflinching opponent of Animal Magnetism. Nay, he had been in the habit of ridiculing mesmerism. "At length," I again quote the words of Dr. Elliotson, "after nineteen years he condescended to witness mesmerism in the person of a young lady in Paris. 'I went to laugh,' says he, 'and I came away convinced.'

"To suspect anything like a trick in the parties concerned, was impossible. They were of the highest respectability and distinction, and some of them I had known for many years. The magnetiser was, indeed, in the frivolous French metropolis, called a charlatan, which made me suppose he was not so; and the event proved I was right. He was, indeed, poor; he exercised his art for money; he gave public lectures at three francs a ticket. Many young physicians had as fair a claim to the title as he had. But from the hour above alluded to till the period of his death, I remained acquainted with the Abbé Faria, and never knew a man to whom the epithet impostor was less applicable.

"No sooner had the Abbé Faria begun to operate, than the countenance of the lady changed, and in two seconds she was fast asleep, having manifested symptoms which could not be counterfeited. The sitting lasted about two hours, and produced results, which, though I remained a sceptic upon some of the most wonderful phenomena, entirely convinced me of the existence of a mesmeric influence, and of an extraordinary agency, which one person can, by his will, exercise upon another. The Abbé Faria offered every means to dispel my remaining doubts, and gave me all necessary instructions to obtain total conviction from experiments of my own. I most zealously attended his labours, public and private, and derived complete satisfaction upon every point relating to mesmerism, even upon those which appeared supernatural. Many of the experiments I repeated, not only upon persons whom I met at his house, but upon others totally unacquainted with him, or with his studies, and was ultimately compelled to adopt the absolute and unqualified conclusion announced above,—*Mesmerism is true.*" And, Dr. Elliotson asks, "what was his conduct? Did he content himself with saying, 'It's very wonderful, certainly.' 'I don't understand it,' and then think no more about it, turn to something by which he was making money, or which was pretty or popular. Did he content himself with saying, I should like to know the use of it? as though the sight of any of nature's wonders were not in itself of use—a high, an invigorating, a noble, intellectual, and truly delicious and improving occupation—far beyond the occupation, however necessary, of procuring food, and raiment, and property, the administration of which to the more humble wants of our nature constitute the only view possessed by many of utility—as though every fact of nature were not a part of the mighty, the universal whole—as though a knowledge of any fact in nature could fail directly or indirectly, sooner or later, to give power to good. Did he content himself with saying 'It cannot be doubted but it is a dangerous power, and may be turned to mischievous purposes!' as though whatever is a power is not a power to evil, as well as to good, and greater power to evil in the very same degree in which it is a greater to good; as though heat, without which we die, without high degrees of which, the arts administering most to our advantage, and comfort, and propriety of all kinds cannot be practised, is not converted to the most destructive purposes by the ill-disposed, and a source also of incalculable, unintentional, accidental mischief;—as though the steel instruments always on our tables, and in our hands, and in all parts of our dwellings might not in one moment be made implements of injury and death; as though half our medicines in hourly use may not be made instruments of death. Did he content himself with saying, 'Well, I am sure the medical profession will see its truth and importance, and I shall leave it to them!' as though any revolution in science or institutions had ever come from those in whose hands the subject was placed, and had not always been forced upon them, and forced upon them with toil and anguish, and persecution to those who effected it. He had read that Christ, for his efforts to substitute, in the room of long public prayers and sanctimonious scrupulousness about indifferent things, and the pomp and ceremony, and mechanism of worship, and nicety of doctrines, and priestly assumptions—to substitute humility, disinterestedness, and universal benevolence, saying, 'It was said, of old time; but I say unto you,' was vilified by all the Jewish establishment, and the so-called respectable Jews, and then nailed on high in the open air to two crossed pieces of wood. He knew that the discoveries of Newton were long excluded from the university of which he was a member, and were introduced, through stratagem only, by Dr. Samuel Clarke explaining them in notes, without any appearance of argument or controversy to the book of Descartes, used as a text book by all the tutors, though, like the notable fellows, now forgotten and in the dust, unseen and unremembered; and that the exploded and unfounded system of Descartes kept its ground more than thirty years after the publication of Newton's discoveries. He knew that the medical profession laughed Harvey to scorn, and made the public think him too great a fool and visionary to be fit to physic them, when he taught the circulation of the blood, and was so perversely obstinate that not one doctor who had arrived at years of discretion, and had attained his fortieth year, when the discovery was announced, ever relented, or admitted the truth of the circulation, or allowed himself to the day of his death to be the wiser for Harvey's discovery, but lived on perversely and piggishly boasting of error; though probably many noisy professors of unbelief did, like a host of noisy living doctors who declare their unbelief at this moment in mesmerism, and like gentlemen of a dark hue in the basement story, 'believe and tremble.' Mr. Chenevix knew that beneficial changes in bodies of men, united as an institution or corporation, or even in a profession or occupation, and standing up to their vested rights of ignorance, unimprovability, and undisturbed motionlessness, required always heavy blows and pressure from without, before these bodies bestir themselves and think of moving onwards. Did he think a truth less important, because neglected and despised? No. Mr. Chenevix, surprised at the pusillanimity of the French Academy, which could not deny, and yet had not the manliness to avow the facts which one half of its members declared they had witnessed, resolved, with all due humility, yet without shrinking from the task, to devote some time to the collection of facts, and to offer the result to a much more enlightened public than that to which the art is compelled to appeal in France.

"Mr. Chenevix had no opportunity of renewing his trials of mesmerism till May, 1828, when happening to be on a visit in Ireland, he inquired for some patient among the peasantry, no matter what the disorder." He was not long in obtaining the object of his wishes; our statement is abridged. The case was a woman of 33 years of age, who had suffered under severe epilepsy for six years, and had, under this affliction, fallen into the fire, and dreadfully burned her leg. Besides a tendency to paralysis of the left leg and thigh, the poor woman had spasmodic contraction of her hands and feet, accompanied with racking pain, which sometimes lasted twelve hours or more. Epilepsy is often attended by absence of mind and loss of memory, and these symptoms were present here. She seldom slept two hours at a time, had constant thirst and bad appetite. She was in the eighth month of her sixth pregnancy. Her first attack came on after her first confinement. Mr. Chenevix mesmerised her for forty-five minutes, and produced only slight drowsiness. The second day, after being mesmerised forty-five minutes, she was drowsy till the next day. On the third day, she was mesmerised forty-five minutes, without sleep. She was not mesmerised on the fourth day; on the fifth day she fell asleep, after being mesmerised nine minutes; on the seventh day she slept in three minutes; her health improved marvellously. On the sixteenth day, she was mesmerised without being aware of the presence of her mesmeriser, for she was operated on at the distance of fifteen feet, a door being interposed. She slept in fourteen minutes. On the seventeenth day mesmerism was omitted, and she had a severe attack of spasm in her left leg and thigh, for six hours, followed by coldness and numbness; on the eighteenth day, she slept in half a minute, kept asleep and left limb mesmerised. Pain was gone, and strength and heat were restored. Under a pursuance of the treatment she had no return of the symptoms of disease. She continued to improve in health rapidly. Her neighbours, ignorant of her having been submitted to mesmeric treatment, were struck with her improved appearance. On the 28th June she was delivered. Mr. Chenevix called on her on the 6th July, and found her up and well, except rheumatic pains in her left shoulder, which, by mesmerism, he removed in ten minutes. On the 17th July she went to thank Mr. Chenevix for her recovery; and neither then nor afterwards was he able to affect her again. "From the very first day she was mesmerised," writes Mr. Chenevix, "the symptoms were alleviated, and decreased regularly as the treatment advanced. In less than a week, thirst, sleeplessness, shivering, and pains to which she had been subject for six years, ceased; the paralytic tendency diminished, and the spasmodic contractions were entirely removed after the twelfth day of mesmerising. Although none of the extraordinary symptoms of lucidity occurred, although this patient awoke the instant she was spoken to, her cure is interesting, as being completed so rapidly. Twenty-one sittings sufficed. Even at the period when she used to be most afflicted, the touch of a finger, so slight as to be almost imperceptible to myself, roused her from her state of mesmerism, and with a sensation which she described as like the prick of a pin.

"Between May 23, 1828, and January 20th, 1829, Mr. Chenevix mesmerised upon 164 persons, of whom ninety-eight manifested undeniable effects, some in one minute, some not till the operation had been repeated several times. There was hardly one instance where disease existed, that relief was not procured, and many of the patients offered phenomena as extraordinary as any recounted in Germany or France." While prosecuting his experiments, he had the good fortune to meet with many benevolent and zealous persons—not of the faculty—who made trial of the art with entire success, having hardly ever failed to procure relief for their fellow creatures, at the same time that they produced phenomena which highly surprised and gratified them. He counted fifty who had become both converts and practitioners. However, three physicians of public establishments in the neighbourhood of the place where his experiments were made, attended to him. Drs. McKay, Peacock, and Cotter, the first of whom kindly lent his assistance upon all occasions, and testified the truth of the wonders he saw to everybody. We have not, however, heard of these gentlemen having prosecuted the object, or stood up for it, even when it was experiencing the furious hostility and roguery of some Irish journalists, and medical men. At any rate, had they profited, and felt it their duty to see that their profession profited by what Mr. Chenevix taught them, Mesmerism would now have been perfectly established in Ireland. Dr. Cotter yielded to evidence more slowly than the other two gentlemen, and though he saw two epileptic patients sent to sleep in about half a minute, a fit of one instantly arrested, one struck motionless by Mr. Chenevix's will only, while walking across the room, and set at liberty by the same power as instantaneously, a suspicion of connivance still lurked in his mind, and he was therefore requested to take five patients of his own to Mr. Chenevix for experiment, never before seen or heard of by Mr. Chenevix. He accordingly took a female, whom he had been treating four years for indigestion, costiveness, and headache. Her usual aperient was ten grains of calomel, and thirty of jalap. She had no idea of what was to be done to her, and was suffering at the time under violent headache; after being mesmerised for three minutes, her headache was better, and in five minutes was gone. In eight minutes she was in the soundest mesmeric sleep Mr. Chenevix ever witnessed, and continued sleeping for thirty-five minutes, when he awoke her. During her sleep, Mr. Chenevix was informed by Dr. Cotter, in Latin, that her bowels were at that moment particularly bound. He, therefore, directed his attention to procuring an evacuation, passing his hands before the abdomen, without, however, touching it, or approaching nearer to it than three or four inches. In less than an hour after she had left the house she had three evacuations, and, for some days, her head was considerably relieved. The treatment was, unfortunately, not continued, or we have no doubt, from abundant experience in such cases, that both her head and bowels would have been completely restored to health. Dr. Cotter now practised himself in a similar case, of a young lady, aged fourteen. The pain here was in the left side, instead of the head, and, for a length of time, Dr. Cotter had given her medicine without any permanent benefit, so that he was desirous she should relinquish so injurious a habit. She had no idea of what he was trying, when he mesmerised her, but in four minutes she was completely asleep. He mesmerised her only three times, yet she had no return of the pain, and no longer required aperients, for which, previously, there was a continual necessity. It is a common effect of Mesmerism to render the bowels regular, although aperients have before been habitually required."

This very interesting paper might be wholly transferred to my pages, so full is it of important facts. Mr. Chenevix goes on to cure a case said to be one of consumption. He cured seven cases of worms by Mesmerism, some of them complicated with other serious disorders. Dr. Whymper, Surgeon-major, and Mr. Smith, Surgeon to the Coldstream Regiment of Guards, permitted Mr. Chenevix to make trials on the men. The first was affected in twenty minutes. The second, Richard Ireland, went to sleep in six minutes; after he had slept five and twenty minutes, transverse passes did not awaken him, but he awoke as soon as he was called by his name. Mr. Chenevix, to show that the power of mesmerising was not confined to himself, taught a serjeant the simple art, and then gave the two surgeons illustrations of the effects of his will, in producing hot or cold sensations on at least eighty cases.

It is painful to find such lengthened series of facts as Dr. Elliotson states in this most interesting paper offered to men like Mr. North, in whose Medical and Physical Journal, Mr. Chenevix condescended to publish his valuable papers. Dr. Elliotson remarks that this man never followed up the subject, and that he pretends —and it must be pretence—not to believe in Mesmerism. It matters very little what men of Mr. North's calibre of mind believe, or pretend to disbelieve. The great matter is, the amount of success which the astounding facts that Mr. Chenevix demonstrated before the persons he names, had in rivetting the truths of Mesmerism on the attention of men of a higher class of mind than that which belonged to poor Mr. North. It seems that the facts demonstrated by Mr. Chenevix were brought to the notice of Mr. Brodie, afterwards Sir Benjamin Brodie, of Mr. Faraday and Dr. Hargood, of the Royal Institution, Dr., now Sir Henry Holland, and the two late Drs. Babington, father and son.

To those who are at all familiar with the study of the mental capacities of their neighbours, and have had extensive opportunities of guaging the powers of mind of such of their neighbours as have attained to the notoriety which the accidents of life have obtained for them, it will be a matter of no wonder that the persons whose names appear above as individuals appealed to by Mr. Chenevix, were unable to appreciate the importance of the facts offered to their notice. They were, one and all, not only men of average, commonplace minds, but some of them really of very poor minds. I may be told that some of them have achieved reputations. I am quite aware of the force attaching to character, and of the sneers which hero-worshippers have ever ready for those anxious to estimate men according to their real deserts. Every person who could receive the facts presented by Mr. Chenevix, without being impelled by a desire to inquire into them, must have a miserably poor intellect. Let the history of the life of each of these men be well scanned, and his claims to immortal fame will vanish to very nebulous dimensions.

Names are no guarantees for virtue or talent. People may assume that they are so, and often unwisely become irascible, when they are made to reflect on the fact. If they suffer from having consulted a blunderer, with a certain reputation, they console themselves with the idea that their own wisdom was infallible, in trusting to a fool, who had the care of certain numbers of other fools.

How does the world estimate the talents of medical men? We know that they are mostly led by fashion. We know that most men seldom inquire into more than the source whence they receive the recommendation. So many become the victims of clumsy stupidity, that they are bound to acknowledge the probability that nature was as much in fault as the practitioner. When, in early life, the celebrated physician, Dr. Gooch, assured me that the majority of my competitors would be found grossly ignorant of the principles which ought to guide them in their operations, I could hardly believe the assertion. I found, however, abundant cases of suffering produced by malpractice, which the undoubted skill of such men as Thomas Copeland was quite unable to relieve.

It is not necessary to advert to specific instances. The subject is alluded to only for the purpose of guiding our readers to reflect on the absurdities to which the world is apt to give its assent, when proposed by men who have acquired a name for some specialty. Sir Benjamin Brodie was one of the names celebrated for a specialty. If his knowledge could have been tested, it would have been found to consist of those half-fledged conclusions which go by the designation of dogmas. In consultation, the Baronet was always dogmatical, always authoritative, never argumentative. He would spurn a French consultation, in which each man should be bound to give, in the presence of the relatives, his reasons for the opinions he entertained, and for the practice he recommended. Sir Benjamin was never ready to submit to superior minds. He was never the equal in intellect to Thomas Copeland, whose reminiscences of the blunders of his contemporary, would have formed an interesting, though a melancholy chapter, in the history of London surgical practice. Dr. Elliotson has called Sir Benjamin, when deprecating, in the Zoist, the poor creature's conclusions on Animal Magnetism, a carpenter-surgeon. Dr. Elliotson meant to say that he could make use of his handicraft tools, but that he had not the capacity of mind to allow him to judge on such a subject. He was right, and many will hereafter back the acumen displayed in the remarks on that occasion. The Baronet became President of the Royal Society, and is a proof of the low ebb of the science of England. How many would be glad to know the grounds upon which such a sciolist could rest his claims to the distinction which placed him at the head of the scientific body of our country. Some are apt to believe that a course of life devoted to hospital intrigue, has been equally successful among the toadies of science. How this may be, is not our question. We would denounce the man as unfit for the chair he occupied, on the specific grounds that being the first to suggest the experiment of keeping up artificial respiration, with a view to determine the cause of animal heat, he blundered in the obvious philosophical conclusion, at which he would have arrived, had he prepared himself to be a true physiologist, by the study of Animal Magnetism.

We are loth to be considered personal in our remarks, and nothing should induce us to drag a man, and one who is now passed away from our sphere, before the public as a culprit in the cause of science, were it not that a holy sense of duty obliges us to be no respecter of persons. Sir Benjamin Brodie not only abandoned the true path, along which a physiologist should travel, but he, on more than one occasion, sought opportunities of publicly declaring his absolute ignorance of the facts of Animal Magnetism—a science of which no man, who pretends to acquirement, should be so grossly ignorant as he was. The President of the Royal Society should not despise facts, but he was proud to shut his eyes to truths which are familiarly known to many a mechanic in the manufacturing districts of Britain, to many a poor seamstress, and to a numerous body of labourers from the country, who have been taught to produce physiological phenomena in Animal Magnetism. The fact that an infirmary had been for years in existence, where the President of the Royal Society could witness physiological truths of vast importance, made not the slightest difference in his desire to diminish his amount of ignorance. Ignorant he was determined to remain, and he declared that he would not believe the facts of Animal Magnetism, if he were to witness them with his own eyes. No man can blame me for spitting and roasting this fat fowl after such a declaration. It will remain a glaring discredit to the Royal Society to have had such a President.

Few are able to understand why a man of such slender capacity should volunteer to write a duodecimo volume on psychological inquiries. It is a volume replete with very superficial facts. It could have been written by a vain, smart collegian of eighteen or twenty, and is quite unworthy of a man who had lived to the years of one whose opportunities should have led him to conclusions very different from those that occupied the minds of the prigs of Edinburgh nearly fifty years ago. Phrenology is now an accepted branch of knowledge, and it is absurd for such a smatterer in psychology as Sir Benjamin Brodie was, to tilt a lance, in knight-errantry, against the truths developed by the genius of Gall, and corroborated by the labours of Leger. The Royal Society may have been a refuge for the poor Baronet, but it will hardly cover his shame with the whole ægis of its reputation. Shame that though they were jockeyed by his career of cunning, such a man was suffered to vault into the chair of that august body.

Gall's merits are not in question; his genius is too well acknowledged to be blighted by the paltry pen of the Baronet. Leger, however, is a name not so well known. It has been my endeavour in the fifth chapter of this work to give my readers some idea of his important discoveries. Let us ask what has the Baronet done? His first essay was a series of physiological experiments on living animals, to endeavour to determine the source of animal heat. He did determine that this heat was not dependent upon the process of respiration. This fact was important, but there he stopped. Has he ever made another discovery? Has he ever added a new scientific fact to our fund of knowledge? Not one! We cannot allow his experiments on poisons, and ill-digested mass of surgical observations, so full as they are of bad logic; nor his trifling and really trumpery psychological inquiries, in which he condemns the magnificent discoveries of Gall, perhaps the greatest genius who has enlightened our century, to be at all worthy of consideration. Then let us review his one discovery; he found that when he deprived a living animal of its brain, he deprived it of its source of heat. It would have been natural for so very deep a philosopher to have thought a good deal on this fact. I once drily suggested to him that he was the discoverer of the cause of mesmeric phenomena; my friend looked at me with the sardonic grin which such a profound philosopher has always a right to assume. He felt I meant something, but his deep wisdom could not tell what. I could not persuade him to be generous. He had then the opportunity of gaining golden opinions. He might have said, "I have wronged you all who have so patiently studied the facts of Animal Magnetism. I see now my error, I made a great fool of myself at the Royal Medical and Chirurgical Society. I am unable to make any other reparation than that implied in a recognition of your claims, and the fact that you have flashed across my brain, that Animal Heat and Animal Magnetism are one and the same thing." Would Sir Benjamin Brodie have been degraded by such an acknowledgment? He is infinitely more degraded by not having made it. But poor man! How could one with so petty, so insignificant a mind have dreamed that I proposed so simple and easy a mode of getting himself extricated from his dilemma. The probability was, that his mind did not comprehend the obvious corollary to his discovery. He was an active little man, but no thinker, and would always require prompting to make out any new fact. He never had the remotest pretension to scope. The probabilities are that Matthew Baillie was the suggester of his experiments.

It is weary work to give estimates of the truth in regard to such small minds. We feel the task to be a most disagreeable one, and were it not for the necessity imposed upon us by a sense of justice to the holy cause of Animal Magnetism, of which this poor creature has been so determined, so inveterate, and to borrow a good analogy from the lower animals, so really pig-headed an opponent, it would not have been worth our while to expose the shallow pretensions of so veritable a sciolist.

Time and space hardly permit us to go on with criticism. We pity the poor creatures we feel obliged to lash. They are the favourites and the demigods of their own cliques,—men who little reflect upon the infinitely small space they occupy in the great universe; men who think themselves somebodies, when really in less than forty years they will be nobodies. Such is the vanity of science, hardly more durable and almost as frivolous as the vanity of the crinoline. Both are hollow, both are emblems of falsehood, of hypocrisy, and of insatiable love of self-glorification.

When Mr. Chenevix condescended to offer so interesting and so grand a subject as Animal Magnetism to the notice of such men as those whose names appear above, he could not have been aware that not one of them was gifted with scope; and without the endowment of scope, no mind is aught but ordinary in its capacity. Mr. Faraday's repudiation of Animal Magnetism sufficiently measures the area of his calibre. Sir Henry Holland's capacity may be estimated, when Dr. Leger's Magnetoscope shall again be able to declare the mathematical value of the mental forces emanating from the diminutive *petit maître* and fiddle-faddle mind. Then, Dr. Hargood and the Babingtons; they wanted the faculty to perceive that they were not destined to set the Thames on fire. To be wanting in the power to estimate the phenomena of Animal Magnetism, when they have been demonstrated, stamps the Lilliput scope of any man's mental capacity.

And these are the men to whose wondrous judgment Richard Chenevix, having vast truths of transcendent importance to publish in England, was recommended to submit them.

And these are the men whose immortality is to be associated with that of Richard Chenevix, whose introduction of Animal Magnetism into England forms so important an epoch in the history of the science. Chenevix was fortunate in his reception by two real lovers of truth, —John Elliotson and John Wilson. The former, an object of mortal envy to a host of pismires. Nevertheless, a man of no common intellect, whose appreciation of the phenomena of Animal Magnetism cannot fail to immortalize his well-earned fame. He has worked hard to store up a huge mass of most important facts. The Zoist and the Mesmeric Infirmary are the records of a fame which is undying; but as I have before indicated, his chiefest glory is in the sincere and fervent worship he has always devoted to the "truth," which is for "ever the type of the great Supreme."

My friend, the late Dr. John Wilson, gave an *accueil* to Richard Chenevix, which must have gratified him, from its honesty and heartiness. Wilson was Physician to the Middlesex Hospital, but the world never estimates such a man. He loved knowledge for the delight it yielded him, as he loved pictures, because his pleasures were of a high order. He was a man of taste and of pure heart. He worked at various subjects. Mathematical science was a delight to him. He shewed intellectual acumen in his studies in Animal Magnetism, by some experiments on its effects on the lower animals, and he published a very curious as well as interesting pamphlet, which he entitled, "Trials on Animal Magnetism, on the Brute Creation." He magnetised fish, as well as savage brutes. I was with him on one occasion at the Surrey Zoological Gardens, while honest Mr. Cross was proprietor of the menagerie. The great male elephant was put into a deep sleep by the strenuous and energetic passes of my friend and colleague, for two hours and twenty minutes. The keeper told me that the doctor off with his coat, and wrought like a Trojan, and got the old animal into a sound sleep, and no mistake. Mr. Cross had a very savage irascible hyena. Dr. Wilson mesmerised him, and it was amusing to see the delight of the savage animal at the doctor's approach. It was the attractive power of benevolence communicated to the animal by the force of Dr. Wilson's will.

It is to Dr. Elliotson and to Dr. Wilson, then, that we are indebted for the vast physiological importance which accrues to the science of England for their encouragement of Mr. Chenevix, and for their subsequent exertions in introducing a knowledge of Animal Magnetism and of its phenomena to this country.

I may be permitted now to enter a protest against the language which Mr. Chenevix has used in writing of the French metropolis, and of his opinion on the frivolity of the national character. It does not become a man of science to sneer at a nation which has really done so much in every branch of science. France has not only contributed her full share to the world's civilisation in the advancement of science, but in the encouragement of literature, she has excelled in deep thought, in every branch of philosophy, in her published works of pure mathematics, in her works in poetry, in her exquisite taste for style in language, in her taste in the appreciation of all that is grand and high in art. She has produced men of marvellous genius, and all this betokens the contrary of frivolity. It evinces large intellect and deep sensibility, those characters of soul without which her grand revolution could not have thrown off so much vice, and have given to the future so many sure pledges of a glorious progress in all that is pure and excellent in the human character. It is the fashion, and particularly among the Irish Celts, to depreciate the Celtic race. The Irish are themselves as frivolous as any nation may be. They little dream how centuries of ignorance have diminished their heads. Even the delightful Catholic teachings of Thomas a Kempis have not been able to remove from them their savage disregard for the sacred respect which every advanced Christian thinker should entertain for human life. Frivolity is the child of ignorance that has never learned to concentrate ideas. It cannot be said of the French that they do not concentrate their ideas. They are not only deep thinkers, but they are men of heart, and if their character is now undergoing a phase of stormy moral or immoral rioting, time will land them on a shore where they must find a sublime restoration to the noble principles spread over the works of so many of their magnificent authors.

Mr. Chenevix had, perhaps, not studied the works of those Frenchmen who have left such rich treasures of facts on Animal Magnetism. To enumerate these would fill a volume. The names of De Puységur, Tardy de Montravel, Petetin, Foissac, Jussieu, Deleuze, Mialle, Charpignon, Frapart, Cahagnet, Bertrand, Gauthier, Ricard, La Fontaine, and a long list of writers, who have distinguished themselves, not only by powers of keen and correct observation, but by exalted sentiments of cordial benevolénce. Our subject is an exalting one, and perhaps no writer has treated it more eloquently than M. Alphonse Teste, whose work on Animal Magnetism is a most interesting set of lectures, which should be recommended to every one desirous of examining this deeply interesting subject. In the natural anxiety to hasten the condensation of my voluminous notes into the press, I feel a strong desire to transfer passages wholesale from Teste's book into my pages. I agree with him in all that he has written on the claims of the great Van Helmont to be considered as the modern father of scientific Animal Magnetism. I agree with him in numerous sagacious opinions he has published, but I particularly agree with him in those noble sentiments which give such a charm to his book, and which communicate to his pen "une style si fluente et si chaleureuse." It is an old English saying that an honest man is the noblest work of God. It is hardly necessary to show the high appreciation in which honesty of purpose is held by Teste when he respects the English Elliotson for his thorough disinterestedness in preferring the cause of truth to a large practice, and at the same time remarks that some French magnetisers, less eminent for their scientific position, but not less courageous, have braved public opinion, and have abandoned their profession, to become the missionaries of a proscribed truth! "I shall cite only one man," says Teste, "the most intrepid of all, Marcillet, whose passes have been more precious than all our writings, a hundred times over, and who has more effectually served Magnetism without comprehending it, than did, for thirty years, the elegant pen of Deleuze, and the numberless publications of that great, good, and ardent philanthropist. Marcillet is a man of heart, convinced even to the point of fanaticism, and who consequently fears nothing, doubts nothing; his somnambules participate in his calm and confident attitude. The resolution of their magnetiser screens and shelters them against extraneous influences, for the most part, and generally against malevolent influences. There is no general rule, however, without its exceptions. When Alexis Didier, and Marcillet visited London in the year 1849, I was requested to persuade them to give a *séance* at the house of one of the mulish physicians, who had ranged himself against the holy truth of Animal Magnetism. Phrenologically, this poor creature of a physician had a miserable head. I at once said that no phenomena would be obtained. I was confident that Alexis, with such an *entourage* as he would have at Old Burlington Street, would have all his clairvoyance smothered and burked. I was weak enough to be overruled, and accompanied the worthy Alexis Didier and the brave Marcillet into the anti-mesmeric vortex. Of course, the result might have been anticipated. The party assembled had scarcely a good head among them. Marcillet was not hustled, but the tone and manner of the assembled sceptics was quite sufficient to depress even Marcillet. I felt thoroughly disgusted, not at the failure, but at the causes which had ensured it. But these poor creatures could not crush the clairvoyance of Alexis everywhere.

Those who had the good fortune to witness the magnificent phenomena that accompanied his somnambulism will not soon forget them. In the third volume of the Zoist for 1845 and subsequent years, will be found very interesting details of facts relating to these. But there are many important facts relating to the clairvoyance of his brother which are not published. My excellent friend Adolphe Didier is distinguished for this wonderful and extraordinary faculty, and I have witnessed his clear lucidity in the diagnosis of disease in many instances. He has on several occasions, as respectable officers of the Life Guards can testify, exhibited most marvellous clairvoyance and prevision; it is to him that I am indebted for his act of kindness in furnishing me with a translation from La Harpe of the very interesting account of the prevision of Cazotte, that happened in the year 1789, four or five years before the events described.

It is very gratifying to find M. Teste so honestly upholding high principle, and a thorough devotion to the cause of truth as he does in his advocacy of the virtues and of the precious services rendered to our cause by M. Marcillet. Well may M. Teste rejoice over the results of the mission of M. Marcillet, when he can assure us that numbers of persons, not only in Great Britain, but in France and Spain have been convinced of the truths of Animal Magnetism by the zeal and energy of this man, and his extraordinary somnambule, Alexis Didier. It appears, too, that in spite of the venomous envy of a certain set of small-minded practitioners in France, the cordial acceptance of the truth of our subject has been endorsed by the higher class of modern physicians and surgeons of Paris. Fouquier, Rostan, J. Cloquet, Adelon, Orfila, Trousseau, H. Royer Collard, professors of the Faculty; Bousquet, Ferrus, Pariset, Londe, Husson, Mare, Reveillé-Pariset, members of the Academy of Medicine. In all probability, many more names of high note would be found to acknowledge their belief in the truth of the phenomena which have been investigated by so many highly distinguished French writers. The works of character and value written by our neighbours are numerous, and are creditable to the cultivators of medical science. The best books in our language are no doubt very good. Two are by reverend gentlemen. One a deeply philosophical work by the Rev. Chauncy Hare Townshend, entitled "Facts in Mesmerism." The other was originally a modest little book to which its author, the Rev. George Sandby gave the appropriate title of "Mesmerism, the gift of God." This little book written in 1843, was soon out of print, and the author in a subsequent edition, recast his work and published it in a much enlarged form under the title of "Mesmerism and its Opponents," an admirable and comprehensive work. Mr. J. C. Colquhoun, a very learned advocate of the Scottish bar, published in 1833 an interesting and very highly important work on Animal Magnetism, under the title of *Isis Revelata*, which passed through several editions. Mr. Colquhoun has since written another work evincing considerable power of thought, and great reading. It is a pity that a man who can write so well should be so rabid on facts, which it is quite impossible to overturn; Animal Magnetism is true. Mr. Colquhoun does not hesitate to admit this, but he must needs deny facts I have witnessed at least a thousand times. Is there no way of looking at a truth, except through a wilderness of learning? There must be some awful confusion of ideas in any man's mind, who thinks that the high class of force involved in animal magnetism is to subvert all belief in the existence of a deity. In reality, Dr. Leger's discoveries in Phrenology, which may owe much of their importance to the undoubted truths, easily demonstrable to the power of the human will, which is not only an immaterial mental agent, but is even a mechanical motive force, capable as the New Testament instructs us, of moving mountains. I believe Christ told us a *truth*, and I believe, moreover, that no two truths in the universe are inconsistent with each other. If Dr. Leger's truths are incontrovertible, all reasoning against them will never prove their tendency to materialism, and if they be revelations of truth through a poverty-stricken worthy Roman Catholic, all the bigotry of all the deacons of Scotland united to that of all the bishops of England will not be able to shake the truth that has been established through the revelation vouchsafed to a truly pious, religious man in the discovery of the Magnetescope. All due respect may be accorded to Mr. Colquhoun for his learned work "Isis Revelata," and his history of "Magic, Witchcraft, and Animal Magnetism," but if he believes that bigotted dogma is to demolish God's holy truth, he is awfully mistaken.

It was not the intention here to notice all the English writers on Animal Magnetism. Honour to those who have supported the truth, for they have fulfilled a pious and religious duty. A number of small but good works in English on our subject claim our praise. "Vital Magnetism," by the Rev. W. Pyne, is a very good and useful book. "A Compendium," by Mr. Lang, of Glasgow, is an excellent work. All persons interested in Animal Magnetism should read the works of as noble, energetic, and laborious an investigator as ever lived to adorn the medical profession—the late Dr. Esdaile, of the Bengal Medical Service. His "Mesmerism in India," and his "Natural and Mesmeric Clairvoyance," published in England, will well repay attentive study. The reports of his marvellous and astounding surgical operations for the successful and painless removal of stupendous tumors from patients during the health-bestowing mesmeric sleep, are to be seen in the Zoist. They were published in Calcutta, but it is to be feared are not to be had here.

The contemplation of such a high character as that of the late Dr. Esdaile, so absolutely moderate in all his wants and desires, so unselfish, so wanting in the self-seeking that has characterised those small minds who have set themselves up in opposition to the truth, induces one to point to his memory with the deepest veneration. He was remarkable for the natural ease, and frank simplicity of his character; his sincerity and directness of purpose, his sturdy advocacy of Animal Magnetism, and his virtuous self-abnegation completed a character of rare honesty and faithful integrity.

CHAPTER X.

VAN HELMONT.

When our admiration for noble and rare virtues, and for depth of character is suddenly excited by unexpected appeals to our judgment, we are seldom able to scan very minutely the origin of the errors in which we have lived for years. I may confess, that no biography I had met with before the eloquent lectures of M. Teste, which is entitled the "Magnétisme Animale expliqué," I was under the hazy errors of my teachers at several universities. I remember the vulgar ridicule cast by an able professor of chemistry in Scotland upon the now venerated name of Van Helmont. The amiable and clear-headed Deleuze is the man to whom we are indebted for leading our thoughts into correct channels on the character of this great man. I may refer my readers to the remarkable lectures by Teste, above alluded to, for a condensation of the biographical essay by Deleuze, while I take the liberty of incorporating in the present chapter most of M. Teste's ideas.

Thus, I shall be able to do some justice to his estimate of the debt we owe to the real father of modern Animal Magnetism.

Without pretending to enter into a polemic on the respective claims of Van Helmont and Mesmer, we cannot do wrong in tracing some of the facts relating to the former, and in holding up his character to the admiration of my readers.

Whatever may be the merits or the demerits of individuals, we have to shew that our subject does not suffer wrong from them. Yet, when the spirit of truth is in the ascendant, we cannot be wrong in evincing our admiration of those who have fervently worshipped that spirit of truth. We have established, as far as these British Isles are concerned, the epoch at which Animal Magnetism was introduced here. The sun of that morning gilds with peculiar lustre the names of Chenevix and Elliotson, and the late Dr. Wilson; as to those who were stolid and could not perceive the nature of their position, it may be as well to quote a paragraph from the work of Dr. Elliotson which I accidentally met with in a periodical under the title of the Mesmerist. "The opponents of Mesmerism merely act the parts of puppets, not knowing why they so act, and blindly obeying the general laws, by which a supply of opponents to every truth and improvement is always provided. The statistics of opposition to good things would shew that their courses always obey fixed laws, and they are to be pitied for being destined to the parts which they so eagerly perform."

These chapters are not intended to trace out any history of Animal Magnetism. If they induce people to think and to reflect on the stimulus which an abstract love of the truth has in different epochs of the world's history, given to men of generous and simple minds to devote deep attention to a kind of knowledge at present either disregarded or tabooed by sciolists—they fulfil their purpose.

Of all those who have hitherto studied Animal Magnetism, the greatest man was undoubtedly the celebrated Van Helmont. In every respect, he had the greatest claim to our admiration. The more deeply we reflect on the principles constituting unusually extensive intellect, large moral virtue, and ardent desires for the amelioration of our race, the more thoroughly we become convinced that the individual whose mental capacities enable him to indulge in the aspirations of such intellectual, moral, and philosophical character must have large scope in order to embrace and grasp the objects on which to engage the thoughts of a large mind. The human race has seldom produced a man gifted with a wider scope of intellect. If we regard him in any point of view—physician, physiologist, philosopher, thinker or writer, this was, says M. Teste, a wondrous character. In his faculty for observation, and for his power of induction, he was unsurpassed.

In the hands of Van Helmont, the knowledge of medicine, regarded as science or art, was completely revolutionised. The study of the laws of vitality superseded the routine of the Galenists, and the disciples of the Arabian physicians.

Jean Baptiste van Helmont, Lord of Merode, Royenboch, Oorschot, and Pollines, was born at Brussels, in 1577, of a noble family, of high consideration. His father died in 1580, leaving him under the charge of the boy's mother, and of an uncle, who both struck by his precocious intellect, felt it a duty to cultivate it with care.

A singular and a curious child, anxious to be acquainted with every thing. No doubt, Gall could have told us that this boy had an unusually large, healthy head, endowed with numerous and powerful cerebral organs. With a highly nervous temperament, of course, the child apprehended eagerly and rapidly all it was taught. During the tender years of his childhood, the boy remained with an intelligent mother; but when his guardians considered it expedient that he should commence an academical course of study, he was sent to the University of Louvain, where he remained until he had completed the full curriculum of studies which entitled him to take the degree of Master of Arts. But the young man declined the usual academic honours, simply, as he said, because he regarded these titles as the playthings of vanity.

At that time, the courses of philosophy at Louvain were carried on by Jesuit instructors, and it was one of these, the Father Martin del Rio, who wished to instruct our young sophister in the mysteries of magic, which he taught. The regular professors of the university appear occasionally to have been much annoyed by the proceedings of the teachers of philosophy, who, in these days, at our university and hospital colleges, would be considered as outsiders. Like some of our more modern intruders, they tended to keep up a stimulus, which was favourable to the ventilation and improvement of original thought

From all we learn from the works of Albertus Magnus, and from the curious materials gathered together by Cornelius Agrippa ab Nettersheym, of Vallemont, and of several other authors, who have written on occult physics, the leading phenomena of Animal Magnetism, especially those of the powers of the human will, and of the influence of the hazel wood on the human being, probably, too, of somnambulism and clairvoyance, must have constituted much of the subject-matter of the mysterious studies connected with the knowledge of magic. But too much confusion must have existed in the facts of the science to captivate such a mind as that of Van Helmont, for he gave himself up to study the works of Thomas a Kempis and of Tauberus, and it is said of Van Helmont that these studies led his ardent imagination into the obscurities of mysticism, and that, in order to obtain the favour of participating in the influence of the divine grace, he abandoned, from deep humility, all his wealth to his sister, and renounced, without regret, all the privileges of position to which his birth had entitled him.

Here were facts that were quite sufficient to prove this man to have been a most extraordinary character. If we believe his own statements—and why should we not believe them ?—he soon gathered the harvest of his complete self-abnegation, for he enjoyed the contemplation of Theophanies, and like Socrates, felt the happiness of mental intercourse with a guardian angel, or spirit, which, on all important occasions, communicated with him.

To those who have been similarly blessed with the great Van Helmont, there is nothing wonderful, though grandly admirable, in the course he pursued. His humility prompted him to believe that the abandonment of his fortune was but trifling, when he contemplated the example of Christ, and, resolving to imitate him in works of beneficence and charity, he determined to study medicine. He commenced by poring over the works of Hippocrates, of Galen, and of their immediate successors. When we know that he had a guardian angel to induce his thoughts to proceed in correct channels of reasoning, we need be at no loss to account for his want of enthusiasm, and for the ardent desire he felt to reform the errors which he detected.

Anecdotes are not wanting to show his reasons for the abandonment of a conjectural science, when he went to travel in France and in Italy.

It is said that, ten years later, he heard a voice, that of his familiar spirit, urging him to overthrow the system of the Humorists. Accident had thrown him in the way of an empiric, who had fascinated him with the idea of experimental chemistry. Dreaming on the discovery of a universal remedy for the cure of all diseases, he pursued the science of chemistry. About this time, it would appear he married a rich heiress of Brabant, and by her had several children, and, among others, a son, François Mercure, who became very celebrated. Van Helmont resided in the neighbourhood of Vilvorde, and devoted the remainder of his life to chemical operations, and to philosophical inquiries on the physical and intellectual organisation of the human race. For thirty years, his laboratory afforded him constant and assiduous occupation, notwithstanding his very conscientious attention to several thousand patients whom he annually treated most successfully. All this time, devoted to the good of his fellow-beings, he most cheerfully dedicated gratuitously from a high sense of duty. He never received any compensation for the time he devoted to acts of kindness to his fellow-beings. The Emperors Rudolph II, Matthias, and Ferdinand II endeavoured, without success, to attract him, by handsome offers of riches, dignities, and high position, to remove to Vienna. Van Helmont was proof against all the blandishments placed within his reach. He would not quit his retreat, where he completed most of his works. Here he died of bronchial flux, in the year 1644, at the age of sixty-seven.

No biographer, except the excellent and accomplished Deleuze, has done anything like justice to this extraordinary man. I may translate his words: "Van Helmont was a man of genius; he forms an epoch in the history of physiology and of medicine. He it was who first drew attention to the epigastric forces; he recognised the powerful action of the stomach upon the other organs; he perceived that the diaphragm was a powerful centre in the economy of the living body. In reflecting upon the reciprocal influence of living beings, he perceived, in all bodies, a principle of movement in their nature—a force by which they mutually act and react, and to this principle he assigned the name of *Blas*.

The writings of Van Helmont are replete with profound views on the uncertainty of medical science; on the means of improving it; on the insufficiency of the logic of the schools; on the true foundations of our knowledge; and of the necessity for an alliance between the metaphysical and moral sciences with those of the physical and natural sciences. These ideas were sufficient to evince the wide scope of intellect which characterised the genius of this extraordinary man. He seems to have had almost a prevision that mental science would one day be cleared of all the rubbish of metaphysical disquisition, and become a branch of real physical or natural science. The researches and discoveries of Van Helmont in chemistry were of vast importance. The first lights on aeriform bodies were struck by him, and the appellation he bestowed on these substances, that of *Gas*, holds good to the present day.

The absurd outcry of the class of common-place minds (and this class I have, in my notes to the Researches of the Baron von Reichenbach, designated as stupidities,) against Van Helmont's imagination, is of a piece with the frivolous objections raised against Animal Magnetism.

"That which gives so peculiar a charm to the writings of Van Helmont," says Deleuze, "is the elevation of soul, the complete absence of all desire for reputation, and of all earthly interest." Van Helmont was no self-seeking envious hypocrite, cunning as a fox, perpetually intriguing to rule his neighbours. His large soul was imbued with an ardent love of truth, and a strong desire of being good, and of doing good. Charity, a wide tolerance, was with him a religious sentiment, governing all his other sentiments. If his style be in some passages, obscure, these never fail to interest the reader, and to originate new thoughts.

This great man has been accused of credulity and of superstition. Are we to attach these words to every pious belief a man may entertain on the Holy Scriptures, which does not accord with our own convictions? Then they apply to Van Helmont, for he regarded the Scriptures as the depository of all truths. But if these words be regarded in the injurious sense often attached to them, their tendency to intolerance and fanaticism, in weakening the empire of the reasoning faculty, then never had they a greater enemy.

Superstition, in truth, would appear to be the property of the thoroughly ignorant mind, and as its etymology would imply, should stand above the reasoning faculty, that power which is the most valuable gift of God to his creature, man; which enables us to distinguish those ideas that are injurious to the happiness and well-being of man, and those that tend to destroy the tranquillity of individuals, and the harmony of society. Superstition would lead us to attribute unholy powers of acting, and of producing mischievous and even mechanical evil effects to the devil. But the belief that angels, mediators between God and man, watch over us, inspiring us with ideas of peace; inciting us to virtue, and consoling us, when we are unhappy; that those who have preceded us in this life can read our thoughts, can rejoice at our good resolutions, and can communicate holy thoughts to us; that there exists a correspondence between heaven and earth,—between time and eternity,—that God favours those who fervently pray to him, by encouraging their hearts, and enlightening their spirits, in giving them knowledge which they never could have acquired by study. These, and other so much ridiculed opinions—how do they injure mankind? What disorder do they introduce into society? How can they possibly lower our intelligence? seeing that all men of great genius, from Plato downwards, have indulged in such thoughts.

Van Helmont yielded to them, and if he explained phenomena, in his time, otherwise inexplicable, by referring to the will of God, but which since his time have been accounted for on the principles of natural philosophy, the real difficulties have not been overcome, they have simply been postponed.

It is so common a fault with the world, or herd of Stupidities to ridicule and condemn that which they have never taken the trouble to comprehend, or even to study, that it has been customary in some lecture rooms for professors to play off their small wit in detraction of the greatness of Van Helmont, on account of his hypothesis of an Archæus, or intelligent agent in the living human being. Few can realize the idea that men of genius can conceive larger thoughts than can possibly enter into the narrow capacities of actual Stupidities; few can believe that a man of genius can think thoughts that far transcend his own capacities of intellect, and hence the self-complacency and envy of paltry minds.

It is not only curious, but very surprising, that, in society, we are obliged frequently to succumb to the Stupidities by whom we are surrounded; and nothing can account for this curious result, but the existence of forces in nature dependant on the phrenological organs of those who, in the accidents of life, are thrown up, into the more influential positions of society. It is to be hoped that these accidents are regulated by infinite wisdom. With this hope we must bow in deep humility to our fate. Why should the wondrous anticipations of the knowledge of our present age, by the illustrious Swedenborg have been allowed to pass unheeded. What man of science ever refers to them? Why should all the anticipations of the knowledge of the present age by the marvellous Van Helmont pass by unheeded? The truth is, that the accumulated stupidity of successive years drowns the valuable facts developed by men of great genius, who have been very insufficiently and imperfectly understood.

Both Swedenborg and Van Helmont, seers into Heaven, and prophets by *constitution*, anticipated in their works some very important discoveries of modern scientific labourers.

Van Helmont absolutely anticipated some of the ideas of the greatest men of the age in which we are now actually living. The ideas of Bichat, and of a man quite equal to him, though tempered with greater caution, with ideas quite as comprehensive, with capacities to estimate facts quite as minute, and a power to classify and to generalise, more extensive, the underestimated Professor Macartney, of Dublin, promulgated ideas and held opinions very analogous to those of the great Van Helmont. Dr. Prout, in his Bridgewater treatise, not dreaming that Van Helmont had put forth ideas of intelligent agents in the animal economy, shocked some of his admirers, while he failed to induce them to consider him a man of genius far beyond the comprehension of his successors.

Modern knowledge is shocked at the idea of an Archæus, but does not hesitate to land us in the expression of instinctive forces. In the sixteenth and seventeenth centuries, the pretty expression of instinctive forces did not exist; it is quite evident that our ancestors could not entertain the same ideas in which we now indulge, but we may marvel that genius had penetrated nearly 250 years ago into the lights of our own time. They had not telescopes with penetrating power, but they had clairvoyance, or something analogous to it.

We may well wonder at the mind of such a man as Van Helmont, who could anticipate the ideas of 250 years. We do not sufficiently reflect on the powers of a man who could anticipate the gradations of thought implied in the lapse of time during such a period. There is a difference, that should be noted, between the Archæus of Van Helmont and the instinctive forces of the present age, which evinces the possession, by this extraordinary character, of what some now call mediumistic force. He was enabled to assert that his Archæus consisted, not simply of instinctive forces, but of an union of vital force with a spiritual kernel. In fact, this great man was a spiritualist of a high order. He had revelations analogous to those vouchsafed to Emanuel Swedenborg. The faculty of imagination is, probably, a condition necessary to the prophetic, or mediumistic temperament. As in the case of the Swedish seer, it may have had a tendency to run wild, and, considering the Roman Catholic bias in the mind of the Belgian philosopher, it may account for some of his fallacies. (It is clearly to be understood that in the word bias, no offensive meaning is implied to any religious sect; it is simply used in the strictly logical sense in which it is defined by Locke in his Conduct of the Understanding.)

In endeavouring to defend Van Helmont for the idea he promulgated, in the use of the word Archæus, which he intended for, as indeed it was, a deeply philosophical thought, the object is to point out the analogy to the thoughts of the philosophers of our own times. Bichat, Macartney, and others. One cannot help, in tracing the ideas of this great mind, being struck with the spiritual connection which continually obtains between vital and magnetic laws. Magnetism, according to this philosopher, acts everywhere. "It is a paradox for those, only, who deride all things, and who attribute to the power of Satan what they cannot explain. "*Magnetismus, quia passim viget, præter nomen, nil novi continet; nec paradoxus nisi iis qui cuncta derident, et in Sataniæ dominium ablegant quæcumque non intelligunt.*"

"What then," asks Van Helmont, "is Magnetism?" "An occult influence exercised on each other at distances, either by attraction or repulsion. *Sic vocitamus eam occultam coaptationem qua absens in absens per influxum agit, sive trahendo vel impellendo fiat.*'

"The mode of this occult influence obtains from Van Helmont the name of *Magnale magnum*. It is not a material substance, capable of being condensed, measured, weighed. It is a purely ætherial, vital spirit, or force, penetrating all bodies, and agitating the mass of the universe. It is the moderator of the world, because it establishes a correspondence between all its parts, and between all the forces, with which they are gifted. The light of the sun, the influence of the stars, the shocks communicated by the torpedo, &c., are spiritual agencies, or forces, that is, unlike material emanations, but like imperceptible light, traversing by radiation from one suitable object to another."

These paragraphs, from one of Van Helmont's works, cannot but remind us of some portions of Mesmer's works. His universal fluid is undoubtedly here foreshowed, if not anticipated. In fact, all that men may fashion out on the subject of forces must finally resolve itself into the grand idea which Sir Isaac Newton wrought out into the grand trunk force of universal gravitation.

The whole of Van Helmont's system of anthropology rested on the basis of Spiritualism. He distinguished man into exterior and interior, assigning certain faculties to each. Exterior man is animal, using reason and will, in virtue of blood, the interior man is not animal, but is the true image of God.

"*Dico hominem externum esse animal ratione et voluntate sanguinis utens; internum vero non animal sed imaginem Dei veram.*" It is certain, from many of his remarks and propositions, that Van Helmont was familiar with the phenomena which have occupied the attention and studies of De Puysegur, of Tardy de Montravel, of Deleuze, and of other students of the higher developments of Animal Magnetism. He knew well the phenomena attendant upon the exercise of the silent, concentrated will of man, a force which is the essential agent in Animal Magnetism.

"When the imagination is strongly excited, the soul engenders a real or an essential idea, which is not a sure quality, but an intermediate substance between the body and this spirit. When this idea has thus clothed a material substance, and then an entity or specific existence, the intelligence recognises it, the will attaches itself to it, and directs it, the memory recals it.

"When this ideal entity spreads itself externally as vital spirit, it requires but a slight excitation to travel to a distance, and to execute the behest forced upon it by the will.

"Bodies are but the half of the world, spirits are also everywhere existent. Thus, spirits are the ministers of Magnetism. Not the spirits of heaven and hell, but the spirits which are formed by man, and which are in him as the fire in the flint. The will of man divests itself of a portion of its vital spirit, which, uniting itself to the ideal entity, acquires an intermediary existence between what is material and that which is not, and spreads itself like light.

"The will sends forth and directs this substance, which, once launched, like light, and not being material, is not arrested, either by distance or by time. This substance is not a demon, nor is it produced by the devil, it is an action of the spirit which belongs to our nature.

The material world is ruled by the immaterial, and other bodies are in subjection to man."

Little are men of science aware that in the sixteenth and seventeenth centuries lived a man who, had he been on earth at this day, would not have talked nonsense on table-turning, but would at once have seen, in the phenomena connected with it, a meaning fraught with blessings to mankind. Van Helmont had a mind large enough to know that man's happiness depends on the belief in those phenomena that are so well described in the Old and New Testaments, and that those phenomena have been investigated, of late years, by humble, modest, and deep-thinking, and closely-reasoning logical philosophers—of such men as the venerable Baron von Reichenbach, and the cautious and scientific Mr. Rutter, whose galvanoscope experiments on the will-force—are of a scientific value, which his modesty will not allow him to estimate.

Van Helmont held that the word magic had, in scripture, two senses, the one good, and the other evil. That which resided in man was decidedly good. All magic virtue has need of excitation. In the brute body, this excitation takes place by an interior heat, and awakens the movement of corporeal spirits—in Magnetism, it takes place by an interior contact. The magic virtue of animals is developed by intellectual conception: that of exterior man by a strong imagination, a continuous, deep, and concentrated attention; that of the interior man by the Holy Spirit.

Van Helmont continues in this strain: "I have, until now, deferred the unveiling of a great mystery, it is that there is in man an energy such, that by his own will, and by his imagination, he can act out of himself, and impress a virtue, exercising a durable influence on a far distant object. This explains what we have said of the ideal entity which proceeds to execute the orders of the will of the Magnetism of all things, produced by the imagination of man, or by the spirit of other things, and by the magic superiority of man over all other bodies.

"*Ingens mysterium propalare hactenus distuli: ostendere videlicet in homine sitam esse energiam qua solo nutu et phantasia sua queat agere in distans, et imprimere virtutem, aliquam influentiam, deinceps per se perseverantem et agentem in objectum longissime absens.*"

"This power we possess of acting out of ourselves by the sole force of our will is undoubtedly incomprehensible, but do we better understand how our will acts upon our own organs, how it moves our arm? The union of the soul and of the body, the action of the one upon the other are phenomena of which the cause is impenetrable.

"Man is the image of God, not by his external form, but by his soul, by the faculties with which he is gifted. But God who has no corporeal organs, acts by his will alone. It is by his will alone that he impresses motion on all his creatures; it follows that man can likewise act by his will.

"The human soul, continues Van Helmont, being the most perfect image of its creator, in it resides, in the highest degree, higher than in all other animals, the power of the will; the soul transmits it to the vital spirit, which harmonises with it, and which externally reproduces its faculties; but this power of acting outwardly does not belong to it exclusively—it shows itself, although much more feebly, in all the beings endowed with life and sensation. These have a portion of will, more or less active, more or less powerful according as they approximate more or less to man, who dominates all. And this ought to be, because God is the principle of life, and His spirit pervades all nature."

Van Helmont asserts that we can attach to a body the virtue with which we are ourselves gifted; that we can communicate to it certain properties, endow it in fact with magic or magnetic qualities, and avail ourselves of it, as a vehicle, for conveying salutary effects to our neighbours. This is quite true, and has in more modern times been amply verified by the Marquis de Puysegur, by Deleuze, and by a vast number of others who have been practical magnetisers.

(My friend Mr. Henry S. Thompson, Mr. Capern and I myself having been gifted with will-power in a high degree, can corroborate all the assertions of Van Helmont, above quoted, and almost copied from the eloquent work of M. A. Teste, entitled "Le Magnétisme Animal expliqué.)

This great man, Van Helmont, arrives on the subject of the Will at this conclusion. Since man possesses the force of acting by his will upon a distant object, it is clear that this energy has been a gift from God, and that it is natural to him."

"*Probato nunc eo quod homo habeat vim per nutum agendi, satis confirmatum est homini istam energiam a Deo datam et naturaliter ipsi competere.*"

CHAPTER XI.

CASES TREATED BY ANIMAL MAGNETISM.

We have related the steps of our own researches up to the point of conviction of the existence of a series of forces in nature, all and each determining a law of formation, established upon geometrical forms, and attended with a mathematical certainty, the proof of the existence of a perfect scheme, under the title given to it by La Place, of the "*Grande Formula.*" It is true, quite true, that there does exist such a scheme, based upon an intelligence which directs the creative force of universal nature. If Animal Magnetism had done nothing more than to rescue from the oblivion to which the philosophers of modern times seem eager to hasten it, the ingenious hypothesis of a grand mathematical formula, this would have been an ample benefit conferred on science. But it has done, and is doing much more. We are bound to leave out of the question all collateral issues. We have to worm our way through facts not familiar to the world. We have to shew that the facts which have been observed by myself, are of an importance second to none in nature. What are these facts? The obvious result of making the mesmeric passes is to put persons to sleep. We have long known this. But though various treatises have appeared on the phenomena of sleep, no strict theory of sleep has ever appeared in the works of the physiologists. Why is this? The facts I propose to lay before my readers have all been known for a length of time. No man addicted to deduce principles from such facts has taken the trouble to draw from them the only inference they are capable of yielding.

We know that there are phenomena attendant upon sleep, which render it a deeply interesting subject. Shakespeare has told us that it resembles death. We say that when it is very deep, it is death. We do not mean *that* death which is attended by absolute decomposition of the particles of the organic body; but that kind of death which means a separation of the soul from the body. This is a startling proposition, and, of course, is not satisfactory without proof. We offer this proof. When a man sleeps the sleep of death, he is quite unconscious of all that passes around him. You bawl in his ear, you endeavour by loud noises to waken him, you may burn him with hot iron, you may nip him, and try various expedients of a painful nature to rouse him to consciousness. Nothing succeeds. The man is dead, and you cannot, with certainty, predicate that he will return to life. We are not satisfied with your assertions of a probable issue. We say, you cannot to a certainty predicate the issue. The man may recover. He may eventually die in that state. You tell us you know he is alive because he breathes. This is no argument. It only shews that he must be more vivid than a corpse; but what we contend for is, that a man without consciousness, who for several hours may be submitted to the tests we have enumerated, is not a man possessing the principle which constitutes an animated being. What is that principle? It is a force. Here again we recur to the properties of the great trunk force, for it is a force having attractive and repulsive powers; a force which we designate the soul. On what authority do you call it a force? We have no need of authorities. We base our reasoning on facts. It is a fact that no man recovers from the sleep of death, unless we can bring his soul back to his body. We shall presently give very striking examples of this proposition.

We are not contending for a trifle. We propose to go on with our train of argument. Are our statements true or false? If not false, who can gainsay us? Can those who fight against us produce as well-ascertained facts as those we are now publishing?

The cause of Animal Magnetism requires a champion who shall be quite fearless.

Will the idea of restoring a soul to a dying body, exasperate the false reasoners? We have more close reasoning for them.

The soul of man is a force, notwithstanding the prejudices of our opponents. It stands forth as the highest force in animated nature, which, irrespective of the matter forming each *nidus* of force, is simply a vast series of forces, each having its determinate place in the "grand formula." Man is but the head representative of the forces governing animated nature. The forces of his body, like those of all other animals, are in series. They follow laws, analogous to magnetic laws, and are defined in all their creative functions, in the production and regulation of those nucleated organic structures which constitute the living molecules of man's frame.

Man is a congeries of forces all under the regulation of magneto-electric laws. He is, in fact, a magnet, having poles in each section of his organization. This alone is sufficient to demonstrate the course of his currents. The brain is the prime magnet, and the vital currents travelling thence to the caudal organs, supply each portion of the frame with its necessary powers. We learn this from Mr. Rutter's Magnetoscope. We learn it from the Baron von Reichenbach's facts. We learn it from the abundant truths acquired from Mesmerism, and, moreover, from the analogies offered by Mr. Faraday's conclusions, and those of the ingenious Matteucci. Time and space are wanting to go into demonstrative details. It is not only a fact as we state it, but it is a fact derived from the blunder committed by Sir Benjamin Brodie, in not recurring to the experiments he performed, when he established that the source of animal heat was the brain, the prime animal magnet which he removed, when, notwithstanding his artificial respiration, the subjects of his experiments became cool, and he lost the opportunity of the just inference from what he might have obtained from them. He might have done worse than to obtain the assistance of some superior mind to suggest how he could retrace his steps, and gracefully read a paper on his amended views on the source of animal heat to that Society which had placed him in the false position of its chairman.

We are anxious to proceed. The thread of our observations leads us to the next question. If the magneto-electric currents be in the direction we assert them to be, then what is the inference, when a number of passes made by our magnet-shedding influence places a patient in the condition of sleep. Necessarily we give that patient tone, health must depend upon tone, and, consequently, the state we know as congestion yields to the attractive force we introduce into the system, and health and strength replace extreme debility, and that condition which we have called clone.

When one looks through the ranks of that which should be a liberal profession, we cannot but ask, why are they all banded together against that which should teach them to cure diseases, which, by all the resources and appliances of their art, in its present state, are quite incurable. "Do you mean," said the late Dr. Chambers to me, one day, "to send us all back to school?" More than this worthy man have put the same question to me. There are men in the profession who do not put such questions, for when the subject on which they are profoundly ignorant is alluded to, they think it more dignified to tell a falsehood. Such, for instance, as this: "Madam, your husband has an incurable disease of the liver, if you insist on his being mesmerised, he will inevitably become insane." No return to school would suffice to polish the heart or the moral apprehension of such a physician as this.

The man has not in him the stuff of which a gentleman is formed. Let him be content with his honours as a fellow of the College of Physicians.

We could mention the name of this sordid being, but we forbear from contempt for such a character, and our course now leads us to illustrate the propositions we have adduced in the preceding pages, by a few striking cases.

An endeavour has been made to shew that the facts of Animal Magnetism may be so arranged as to afford principles guiding to the development of very important knowledge. Let individuals of heart and conscience reflect upon the following case:

Mr. Rixworthy was the keeper of a billiard-room in Piccadilly, situated between the Burlington Arcade and Bond Street. My friend, Mr. Joseph Hands, had been called to see him whilst he was labouring under symptoms of advanced typhus fever, and very soon perceived that the medicinal remedies commonly employed in such cases, could do, and had done very little or no good, and as the man became rapidly worse, he stated his views to the friends of the patient. Mr. Hands was urged to call in further advice, and I was by him requested to visit Mr. Rixworthy. Mr. Hands had had great experience in the practice of Mesmerism, but neither he or I could indulge in any very great expectations of recovering a case in which a complete want of consciousness had come on. The patient had a small thread-like pulse, so frequent and weak as to be hardly perceptible to the fingers, and indeed it was hardly possible to count it. Lying on the bed in that insensible state, with effusion on his brain, for no amount of light could, in the slightest degree make his very dilated pupils contract. He might be said to be dead, although he still breathed. If his arms or his legs were lifted up, they fell from the hand like lifeless matter, quite powerless. The mouth was encrusted with a black dry hard matter, and the tongue was dry, black, and hard. Of course the poor man's venous system was universally congested; was it worth trying to bring this patient to life by practising Mesmerism? Life seemed to be ebbing fast. What was there that could possibly be a source of hope? his age, for he was only a little past thirty, and it might be added, a certain confidence in the energy of those who would undertake to mesmerise him. I sent for the best Mesmeriser I then knew, Mrs. Skelly, who came in less than two hours, during which interval I continued making slow passes from the crown of the head down the body and legs. When Mrs. Skelly arrived, that energetic woman continued her health-restoring passes for six hours more, and full consciousness returned; at my ensuing visit, I found Mr. Rixworthy sitting up with his clothes on, conversing. It was very unwise, but on my remonstrating with him, he promised to return to his bed. It appeared, however, that he felt too well to remain long there, and upon their insisting on my directions being obeyed, he threw off his coat, neck-stock, and placed himself on the bed, with his trousers on. Falling asleep, the shank of a button pressed upon a small spot on the skin of his back, near the spine; the next morning, I found him awake, but awfully altered in countenance, the pressure of the metal upon the skin of the back, on one side of the spinal column had produced mortification, for the congested capillary veins had not had time to recover their tone before they were exposed to the injury, and the patient was now beyond the reach of our aid.

All the medical experience of the Fellows of the Royal Colleges and of the Royal Society may be defied to produce such a successful case of the treatment of typhus fever, up to its point of recovery by the energy of Mrs. Skelly; but I propose to adduce another case, not very dissimilar, although not coming strictly into the category of typhus fever, it was, however, so thoroughly typhoid that I do not choose to submit it to the bed of Procrustes. In the month of June, 1846, a gentleman and lady who were well known to me, called on me one morning, while I was breakfasting, and requested me, without a moment's loss of time, to accompany them to visit their son, a boy, at that time at school a few miles out of London; the distance was traversed by railroad, and, on our arrival, we waited a short time for two medical men who had been in attendance upon the case. We proceeded to the bed-room, and found the little patient muttering in low delirium, he was lying in bed, on his back, and knew nothing of what was passing around him. With a dry, black tongue, and lips as dry and parched, he was in what I have called a complete state of clone, the lips muttering, the fingers busy in picking the bed-clothes. The mother, who had never before witnessed such a case, seated herself disconsolate by the side of the bed. It is difficult to convince many people, who have had some experience among persons who have died of typhus fever, that in such an instance as this before us, with all the worst symptoms of typhus, when it is running down to its fatal termination, that it was not a true case of genuine typhus, but, to all intents, this was so analogous to that formidable disease, that, in all essential particulars, the patient might be said to be in his last stage, and certainly not the most distant chance of recovery existed by the ordinary means of practice usually adopted by medical men. It has already been said that the boy was insensible, and his father and mother were deeply affected; he had been well only four or five days before, and the change to the typhoid delirium, with picking of the bed-clothes and subsultus, was a very sudden affair.

There are in the medical world certain rules for the treatment of such cases, and the physician and surgeon had left none of the resources of their art untried; they were men of good talents and of sound information, in their profession their practice was extensive, and it was clear that they had done their best. It was necessary to communicate to the parents that there was nothing further to be done. I ventured to suggest Mesmerism;

the two doctors stared at me with a smile of derision, the father was half inclined to be angry, and was at first not contented to adopt the suggestion, but he eventually yielded upon my firm assertion that there remained nothing else to be done.

I mesmerised the boy with slow, downward passes for four hours, at the end of that time, he slept sufficiently to enable me to remove him from the bed on to a mattress, and he was soon placed, sound asleep, on his back, into one of Willoughby's carriages, his mother and I entered the carriage, and I continued to mesmerise him for two hours and twenty minutes; at the end of which time he was in bed, at his father's house in Hyde Park Gardens; a mesmeriser continued to operate; he slept beautifully all night, and we were assured of his recovery. After that he slept nightly very soundly, although a very large abscess was found in the right hip-joint; the original cause of this abscess, which was in time succeeded by many others, owing their existence most probably to a kick from a school-fellow in a game of foot-ball. The condition of constitution productive of these abscesses need not be too closely alluded to, except for the purpose of illustrating the pathology of venous inflammation, which arose from a state of venous and lymphatic congestion; all these cases of congestion, whether in veins or absorbents, analogise very closely with certain states of gout and rheumatism, and consequently also of typhus and typhoid fevers, for each and all of these diseases owe their existence to a full condition of veins, to an absence of tone, to a prevailing state of clone, and necessarily to inverse currents of vital force, or negative currents of Animal Magnetism. A succession of abscesses took place, which placed him in great danger, and were attended by solutions of bony matter, and separate parcels of exfoliated bone, which were sometimes a cause of great pain and inconvenience, as they occasionally passed from the bladder.

I was called to him on one occasion, in the dark hours of night, or rather at two o'clock in the morning, when I found that large quantities of blood and of purulent matter had passed from the bowels. Spiculæ of bone from the hip of the right side came away, and carious bone in front of the right tibia had declared itself, in this condition the peritonæl inflammation was very urgent, the boy could not bear the weight of the bed-clothes, and the whole of the abdomen was so extremely tender that the point of the finger touching its surface in any part was agony to him. Of course, his pulse was small, hard, and frequent, the face was almost hippocratic.

Most men would have considered such a case invincible; I had had equally well-marked cases of puerperal fever at the Queen Charlotte's Hospital, exhibiting this painful condition of peritonæum, which I had cured by steady and indomitable persistance of downward passes, combined with the force of will; and I determined to adopt this course in the case of this boy; I worked with determined energy and will for two hours, and had the satisfaction to find my efforts crowned with success. The cries and moans became less marked, and my little patient fell into a sweet sleep; he slept for four or five hours, and awoke free from pain. I had great difficulty, after this attack, in regulating his spirits, he was occasionally quite boisterous in his good-humour, and could hardly be kept quiet enough to produce the cure that we so ardently wished to effect. I never despaired in this extraordinary case. That deep, refreshing sleep assured me of a recovery.

A large abscess was formed under the fascia of the thigh, a restless desire to get the poor little fellow more rapidly well, induced the parents to call in additional advice. My old friend and colleague, the celebrated Thomas Copeland, urged very sensibly that as the patient had made a good progress towards a cure under magnetic treatment it should steadily be persevered in, so great an authority whose mind was untramelled by the slightest love of intrigue ought to have reassured the boy's friends, but they must needs call in Sir Benjamin Brodie, he was consulted first at his own residence in the absence of the patient. I never was more struck with the active restlessness of this man's small mind, his anxiety seemed to be to show the father and myself that he was quite familiar with cases which were characterized by rapid formations of abscess in the hip joint, accompanied with a typhoid fever, he repeated that these cases were well known to him, and quickly running to his book shelves, brought down a work in which he had treated of them. He showed me two cases, I remarked that these both ended in death; he would not be instructed even when he found that not only the boy had not died, but was doing remarkably well; in a fidgety manner he agreed to call and see the patient; and when he had seen him there was not a word about the astonishing fact that the recovery was due to Animal Magnetism, he seemed to be impatient to shew that the holy science against which he had resolved so foolishly to commit himself, should not have any kind of encouragement from him, all he had to suggest was that some mechanical contrivance should be adopted to straighten the limb, now nearly anchylosed, by the patient's inability from his exuberant high spirits to attend to rules, and the anticipations I had ventured to state, added to my serious warnings, were realized. I believe about seventeen openings, resulting from abscesses, took place, he was treated *secundum artem*, and I might add *ignorantiæ*, and though the influence of magnetism on his system continued to exert such a healing efficacy as to enable the climate of the south of France to gain the credit of his cure, he still after the lapse of several years remained a valetudinarian victim of the ignorance of the effects which Animal Magnetism could produce in such a case upon the human system. The men who had subsequently the charge of this case were quite ignorant of the means of using this wonderful agent, which has under God's blessing given him the amount of health he at present enjoys. Measures were pursued in this case which did no earthly good, numerous abscesses burst, the power of mesmerism had controlled in the worst periods of his malady, incurable by any other agent.

Having witnessed a recovery from the jaws of death, when two practical men had evinced their inability to save him, and having seen that the mesmerism of six hours and twenty minutes had placed him on the high road to recovery, still the parents could not satisfy themselves of the emptiness of the sounds arising from the tinkling cymbals of false fame, and the empty bubble of reputation too often acquired by intrigue; but were led away by fashion, thank God, however, that the patient was in the hands of wise spirits who conducted him to a cure, and who thus insured that he has become a useful member of society. What, then, has this case added to the experience possessed by Sir Benjamin Brodie's clique? Has it led their leader or any one of themselves to strike out a new principal of practice? Did it lead their leader to reflect? No! he adhered to what he had so often said on the question of Animal Magnetism, he could not, and would not believe his own eyes, nor perhaps the evidence of any of his senses. But here is another case which might have reminded this rash and unreflecting surgeon of the number of times he has lost patients from opening feculent abscesses. We were in consultation at Kentish Town on the case of a very interesting and beautiful lady, who was in great danger from an abscess of this nature; he proposed as the abscess was beginning to point, to open it with a large lancet, I protested against the treatment and Brodie immediately lost his temper, and sharply reminded me that he was the surgeon to advise, I told him that after no small experience in these cases, corroborated by the judgment of our friend Copeland, and by the wise instruction of my old master Professor Macartney who had written perhaps the best book of modern times on some important points connected with inflammation, he turned upon me and said. "*It is a book full of lies.*" I remarked that there was not the slightest room for a lie to creep in, it was the most philosophical book I knew on the subject, and it was written by a man who is now gone to another world, but of whom I had often heard Sir Benjamin Brodie himself boast, as the dear and intimate friend from whose instructions he had derived more sound knowledge than he had ever acquired from any other man; the result was, he said "Well I insist on opening this abscess in my own way," and the lady died in a few days. Now for the case of a person who had feculent abscess treated upon very different principles; when the Commandeur and Chevalier Marques Lisboa was at the head of the Brazilian legation in London, I was called on to visit Mr. Pereira, then an attaché to his Excellency, this large-minded man freely placed his name and case at my disposal for publication, in any way in which it might serve the cause of truth, of course it was published in the Zoist, September, 1847.

I perceived that my patient was afflicted with bronchitis, which was accompanied by low typhoid fever; in a few weeks, he became much better, and, indeed, was so far recovered that he was about to leave London for Sussex, on a visit to his friend, Mr. Abel Smith, in the beginning of January, 1847. I was led to infer the existence of a feculent abscess, he had a bad complexion, and a serious return of typhoid fever, which prevented him from fulfilling his intention of leaving town; my representations alarmed his warm friends, Mr. and Mrs. Vanzella, to whose house he was removed, and they were anxious for further advice; my arguments did not satisfy my good friend, the physician, who came to assist me in the conduct of the case, and he could see only an ordinary case of typhus; on the 14th of January, just forty-two days after my first visit to this gentleman, it was found that he had been three nights without sleep, notwithstanding the administration of rather large doses of opium, and of other narcotics, administered after consultation held in due form; muttering delirium, and a small, thready pulse of 120; jactitation, twitching of the fingers and hands, feet and toes, became very troublesome, and his vacant stare and constant efforts to pick the bed-clothes, left no doubt in the minds of his kind host and hostess that a fatal termination was at hand; they had called in the ecclesiastics of the Roman Catholic Church who had administered to him in good time the consoling rites of their religion. I visited my patient at ten o'clock at night, and was asked if I had the slightest hope, for, with such delirium and the impossibility of procuring sleep, it was thought idle to expect a favourable change; the scepticism and astonishment of Mr. and Mrs. Vanzella were strikingly manifested when I said, "Will you object to my trying Mesmerism?" at once they thought me a fool, but opposed no obstacle; they retired into another room, and left me with the patient and his servant; at half-past ten I commenced making very slow passes, from the crown of the head to the pit of the stomach, continuing to work incessantly, without relaxing for a minute for three hours and a-half, accompanying the concentration of the mind in a prayer powerful and efficacious.

I succeeded in obtaining for my patient a very sound and calm sleep, which lasted, without interruption, until past seven o'clock in the morning. From that day his principal medicine was mesmerised water, occasionally, but seldom, assisted by a very gentle dose of castor-oil. A mesmeriser attended him daily for some weeks, during which time the fecal abscess declared itself externally; my old friend and colleague, Mr. Copeland, then attended him, and saved him with his usual skill from a serious operation, and I can testify to his philosophic delight at the progress of Mr. Pereira towards a rapid and complete recovery, under the influence of a daily tonic mesmeric sleep, which secured as well an excellent refreshing slumber every night.

Here was a case attended by a very different man from the late President of the Royal Society, here was no rash and mulish plunge of a lancet into a fecal abscess, but a cautious, skilful, and philosophical attendance on the process of nature, an inquiry into all that he could not at once comprehend in the phenomena of Animal Magnetism, and the calm, investigating spirit of a large mind.

My patient, Mr. Pereira, had the noble courage to desire that his name might be published in connection with his case, for the benefit of mankind; it is only by the aid of such enlarged minds that the narrow and stupid prejudices against Mesmerism must give way. I have just pride in learning that this distinguished gentleman has become a chevalier and a commandeur; he has married, and become the father of a family, he is in the favour of his sovereign, whom he has represented as a minister at Berlin, and at other courts of Europe.

Typhus is perhaps as striking a form of disease, to illustrate the fact of abnormal reversals of polarity, and its attendant, a loaded condition of the venous system, as any we could select, and, happily, there have been numerous cases of fevers of an aggravated typhoid type, in which the downward mesmeric passes have restored the magneto-electric currents from the brain to their normal direct courses, and have thus saved life.

Passing over the atrocious consequences of envy at the cures performed in cases of typhoid puerperal fevers, by mesmeric agency, and these have been numerous, we may notice the case of a young lady in Regent Street, whom I was called to visit in the beginning of April 1849. She had a stiff neck, with some fever; a couple of doses of gentle medicine removed her illness, and I ceased to attend her. On the 16th I was recalled by her father, and found her suffering from low fever, with a dark tongue, and feeble pulse of ninety-six, her countenance was indicative of the disease she laboured under. Experience having warranted me in a conviction of the value of calomel, sennæ, rhubarb and neutral salts in such cases, recourse was had to those mnch-decried medicinal, and really magnetic agents, in such doses, and at such intervals as the symptoms of the disease and the effects produced justified; I did not neglect the use of mesmeric passes, which are so valuable in the treatment of fevers, especially fevers of a typhoid type.

On the 18th of April Dr. Elliotson was requested to join me, and he suggested the application of a blister to the hypogastrium, to obviate a slight tenderness or pressure about that part. The pulse was rising and had reached one hundred and ten, delirium set in and I had the doctor's full concurrence in the reliance I placed in mesmerism as an efficient agent in arresting the progress of the symptoms which would most probably otherwise have led to a fatal termination. The influence of the power was remarkable. On the 19th of April we had daily up to this time conferred and deliberated on the treatment, and had adopted the course of remedies, which in addition to mesmerism, we deemed necessary; we had found, however, that the chief agent in calming the nervous symptoms, in allaying the fever, and in controlling the delirium, was mesmerism; its efficacy in sustaining the forces of the patient during the 16th, 17th and 18th of April, while she was losing frightfully profuse quantities of blood from the bowels was most remarkable; on the 19th Dr. Addison was asked to visit the young lady.

This person, so ignorant of the health-restoring soothing efficacy of mesmerism, prescribed in its stead an opiate which so aggravated the delirium of the patient that she passed a most wretched night; the mother distracted by her anxieties, was well nigh delirious herself; she called me up early in the morning, and imploring my pardon in the most affecting terms begged me to overlook the results of a maddened state of mind. Dr. Elliotson and I having retired on being superseded by Dr. Addison, again attended, and again adopting the practice of mesmerism, succeeded in marvellously calming, soothing, strengthening, and finally curing our patient.

The mesmeric passes gave her an appetite to enjoy her food, even during the height of the fever, as well as the power unusual in such circumstances of perfectly digesting it when swallowed. It might here be remarked that it is high time for such men as the late poor Dr. Addison to turn their attention to Animal Magnetism; it is to be feared that there are dozens of such persons now existing in our profession, they would do well to know that opium will not at all times sooth, and that there are many cases in which it will very much aggravate the delirium it is desired to cure. At page 464 of my edition of the translation of Baron von Reichenbach's researches, I have said that certain proofs should be offered only to minds that can admit of the truths of mesmerism. To others, as they are incapable of appreciating the subtleties of fine reasoning upon subjects of an extremely delicate nature, it is loss of time to waste facts.

I have not the slightest hesitation in declaring my conviction that the young lady, who, on the night of Monday the 19th of April, 1849, swallowed that ill-judged opiate, must have died, had not the practice of mesmerism been assiduously adopted. I have Dr. Elliotson's full concurrence in this opinion. I well know the relief in my own person afforded by Animal Magnetism. Had it not been for my friend Mr. Henry S. Thompson of Fairfield, I must have died of a complication of three serious deseases under which I laboured in the year 1847. I was suffering agonies from severe and painful bronchitis, and an attack of pneumonia, which Dr. Robert Ferguson and Dr. Elliotson both agreed yielded the distinctive crepitus of inflammation, besides this I laboured under hepatitis or inflamed liver; there could be no mistake upon the matter, for my medical attendants were persons who knew their business, they prescribed such remedies as are usually given in such cases, calomel, squills, tartarised antimony, ipecacuanha, and I had leeches applied, blood removed by cupping twice to the extent of thirty ounces each time, mustard poultices, and blisters.

The kindness and undoubted skill and attention of my medical friends did not in the least relieve my agonising pains. Propped up in bed, unable, for upwards of sixty hours, to lie down, I was visited by my magnetic friend, Mr. Thompson, at nearly ten o'clock at night. He made passes for four hours and a-half, fixing his unswerving gaze on my eye, and willing me to sleep. At the end of four hours and a-half, a part of which time I was watching my sensations as he impressively and successfully removed one pain after another. I was in a sound sleep, having drowsily and almost unconsciously removed a hillock of pillows from about me, and having assumed a natural horizontal position. The sleep continued more than five hours. All coughing and pain had ceased, and I awoke in the morning with a delightful feeling, refreshed, eager for nothing but nourishment. I ate a hearty breakfast, got up, and dressed. Mr. Thompson and his daughter were kind enough to visit me daily for an hour, until I was restored to health. I do not forget their kindness, but do them honour by devoting the life they have been the means of saving to the continued strenuous and conscientious advocacy of the cause that must, sooner or later, take its firm hold on public attention, as the great curative principle where medicine is now administered, and as a far more powerful force when it is accompanied by the essence of human kindness.

After the striking facts we have just noticed, it is hardly necessary to recur to the subject of typhus, for the influence of mesmerism in my own case of gouty bronchitis, complicated as it was by inflammation of the substance of the right lung, and of the liver, sufficiently evinces the value of the fact that in forcing the abnormal currents, productive of disease, by a reversal of polarity, to assume the normal courses required to restore health, we do not fail to accomplish our object by a truly scientific process. If it were necessary, however, it would be very easy to enlarge this chapter by details of cases cured by Mesmerism, in which lethal currents were making head against the commonly prescribed medicinal agents. These would embrace successful cases of cure in typhoid puerperal fevers, in typhoid scarlatina, and in typhoid small-pox. It is not our purpose, however, to prolong details in this department of medicine, although some of them might illustrate the causes of the envy, the ignorance, and the love of money arrayed against the progress of truth; the reiterations of the fact that dropsy, general and insisted, gout, rheumatism, diseased heart, tendency to fatuity, puerperal mania, and several other forms of insanity, have been completely cured, would fail to convince medical coteries that Animal Magnetism was not intended as a sly means of poaching upon their manors; they have now a grasp of practice, and that is their manor. No innovation can be admitted, especially one implying the necessity of going back to school, in order to learn those portions of natural philosophy of which they are disgracefully and grossly ignorant. After the facts which have been adduced, and the numerous cases made known by those who have had the industry to search hundreds of volumes published since the year 1783, on the phenomena of Animal Magnetism; can any other imputation rest on our opponents than that of an utter disregard to the cause of truth, suggested by the meanest motives actuating human beings.

It will be perceived, in the remarks on the causes of dropsy, which we have ventured to assimilate to those productive of rheumatic affections, of certain compact textures, that the reduction of electro-magnetic currents to their normal courses, would be our indication for the cure of these generally untractable affections. A striking case occurred to me. In the year 1843, when I was living in Wimpole Street, I was called to the wife of a master-carpenter, residing in Fitzroy Market; she was a stout woman, of short stature, and her *embonpoint* had become remarkable of late years; now, however, there could be no doubt of the presence of anasarca, she was pregnant with her fourth child; her age was about forty-two; her respiration was laboured, and her secretions were very unhealthy. I prescribed for her alteratives and expectorants. She had a slight cough, and her liver was slightly enlarged; the heart did not give sounds characteristic of valvular disease. I suggested to her the advisability of being mesmerised, as an experiment, for the purpose of toning the system. In the course of a very short time, four or five days, the mesmeric passes induced a very profound sleep; just at this time she began to suffer from pains in the region of the left kidney, and these were irregular in the times of their accession; soon the legs began to swell, and both limbs were ædematous; the same condition, in the course of a fortnight, characterised the upper extremities; the abdomen, the face and the eyelids were full of water, and both hands, to the tips of the fingers, were remarkable for the same condition. In this state, she was taken in labour, and was delivered of a fine, healthy-looking girl; the anasarcous state remained for nearly a week after her delivery, when suddenly the water very rapidly decreased in quantity. I had not taken the precaution, at the time of her delivery, to ascertain the girth of her body, but on the second day I measured round her abdomen with a tape, and found that she measured fifty-four inches in circumference. Mesmerism had been daily practised for an hour at a time each day; suddenly, on the tenth day after birth of her child, a large abscess, containing very offensive pus, was discharged from the bowels, which, upon examination, was found to contain the *débris* of a mulberry calculus, weighing nearly three ounces; this woman was soon able to walk to my house in Wimpole Street, where she was daily mesmerised for twenty minutes, the practice was continued for three months, and the measurement of her abdomen then gave us the circumference of thirty-two inches; she was in extraordinary good-health for some years, her son had learnt to mesmerise her, and even for the slightest cold this treatment was practised, to her great satisfaction and comfort.

It would appear, then, that the tonic agency of Mesmerism is calculated to alleviate and to cure diseases of many other descriptions than those we have hitherto noticed; general dropsy has repeatedly been wonderfully benefited by the adoption of this practice, but there was a case too tempting not to relate, of a large ovarian cyst, which occured in a washerwoman, who was employed at a celebrated school in the neighbourhood of Hoddesden, in Hertfordshire, superintended by the clever and charitable Mrs. Ellis, whose well-known works on education have established for her a great reputation. The woman had been tapped for an ovarian dropsy, and the fluid had again accumulated in the cyst to such an extent as to impede her breathing; the operation of tapping was urgently recommended by the surgeon who had attended her. A gentleman and lady who resided in the town being willing to undertake the task of magnetising her regularly, I was requested to state whether I could have any hope of benefit accruing in this case. I replied that I had known Mesmerism applied in such cases with success; she was regularly mesmerised, and at the lapse of some three or four months I heard that she had been entirely cured. My first case of ovarian disease was in Jane Love, whose right ovary was considerably enlarged, and the tumour vanished at the end of three months. My friend, Mr. Capern, whose reputation as a mesmeriser is well known, had a similar case of successful cure by Mesmerism of a large ovarian dropsy. We might relate numerous cases of a similar nature, but I prefer here to introduce a striking instance of the cure of a disease as intractable as ovarian dropsy. About the year 1853, a worthy old gentleman, who was very much attached to a charming daughter, had called in several eminent practitioners in the city to attend this young lady, under an attack of mesenteric disease. She had a good deal of fever, and the ordinary medical attendant, who was a zealous and assiduous friend, residing in Queen Street, Cheapside, finding that he could make no progress towards amendment in the treatment of her case, called in Dr. Jefferson and Dr. Pereira; after some consultation, they recommended her removal to Croydon; but her health did not improve. After she had been there some time, I was called upon to visit her, to give an opinion as to whether the practice of Mesmerism would give her any chance of recovery from her perilous position, for the father had been told that she had not more than two or three weeks to live. On examining the case as thoroughly and completely as I was able, in every particular, I could not agree in opinion with her former attendant. I mesmerised her, and finding she was susceptible of the influence of my passes, by the drowsiness she exhibited, I recommended a healthy young woman to act as a daily mesmeriser, to keep her as much as possible in a state of deep sleep. I need hardly add that, my advice having been followed, this very handsome young lady recovered completely in less than three months, and was married to a gentleman to whom she had been engaged, and is now a happy wife and mother.

This is only one of numerous cases I could cite of persons labouring under mesenteric disease, or, as it is often called, Marasmus, that have been cured by the holy magnetic agency of the higher forces of Animal Magnetism.

We might continue to indulge in relations of very remarkable cures, brought about by the healing efficacy of Animal Magnetism. If we be right in the general proposition that all disease is more or less dependent upon clonic conditions of system, however brought about, it is clear that the induction of continued normal currents of animal magneto-electricity, from the head downwards, continuously pursued by a healthy and benevolent person, for a sufficient consecutive period of time, would cure all diseases. We may refer to Dr. Elliotson's case of cancer (see Zoist, No. 23, vol. 6, page 213) as a striking example of the power of Mesmerism, in conquering, under circumstances of proper energy and perseverance, the most intractable of diseases. We do not hear of those who, for their own purposes, create a great bother about hospitals for incurables, and cancer establishments, having the common sense to lead the attention of mankind to that agent, which has hitherto been the only one known to effect this object. Subscriptions are got up, which find their way into the pockets of certain interested parties, certain secretaries, or others, but we hear of no cures of these cases; if people want to bestow their money where it would be usefully applied, they should go to No. 36, Weymouth Street, over the door of which house they would find painted up the words Mesmeric Infirmary, and an active, intelligent gentleman, whose honest face at once gives them the assurance that their charitable bounties are not thrown away, and the records of cases in the books of the Institution, form a guarantee that there is more than empty sound and profession for their money—in truth, the Infirmary is the home from which will spring the full practice of the most glorious science which can occupy the attention of mankind, for it leads not only to charity, but to the true worship of all that is beautiful and good, in the worship of our most holy and beneficent Father. Those connected with that Infirmary have often proposed that the doctors engaged in the treatment of insanity should practise Animal Magnetism for the cure of their afflicted patients, but have not yet been successful in introducing the treatment to the notice of men who consider themselves as psychological physicians, which seems to imply that they are far too learned, and too self-glorifying to bow to the opinions of those who have written in the Zoist. As Mr. Faraday might tell them, they had better "look sharp," lest the Zoistical writers may, before they are aware of it, convince the public that the great cure for insanity is Animal Magnetism.

A very poor, but very beautiful, girl was admitted some years ago at the Queen Charlotte's Lying-in Hospital; she had been seduced by a scoundrel of an attorney, who is said to be living at Hampstead, he had not treated her with the consideration and tenderness which her case required, and soon after her confinement, she was seized with delirious frenzy. I mesmerised her for four hours, and calmed and soothed her to a great extent; as soon as she could be removed, she was taken to her sisters, who lived in the neighbourhood of the Hospital, where I attended, putting her daily into a deep sleep, and, in a fortnight, had the happiness to find her perfectly recovered; she very soon married well, and is now in good health, a respectable mother of six children.

It has been in vain that I have urged medical men, practising in the specialty of insanity, to study Animal Magnetism, they are generally either wondrous wise, or too much addicted to their own vain conceits to believe that the great law of the forces regulating the universe can have aught to do with their specialty. No one need meddle with insanity but a man educated in the absurdities of Sir Benjamin Brodie's psychology, or in the metaphysics of the Scotch schools of philosophy. A gentleman had heard some statements that I had made respecting the discoveries of the late Dr. Leger by the aid of Mr. Rutter's Magnetoscope, and I was requested by him to see a cousin of his who was labouring under mania of a very distressing nature; she was a young lady of about twenty, whose father was a clergyman in the country, who had an excellent living, and who had three daughters, extremely good-looking, and very amiable, one of whom was the patient I was desired to visit.

The story related to me was that one of the three girls was on the point of being married, when the case of insanity suddenly became a source of great anxiety to her relatives. She was in the habit of attending her father's church, and the eyes occasionally wandering, met those of a young man who was said to be particularly handsome, but whose social position did not suit the views of her father; at the end of about two years, during which the young enamoured couple kept their affections a profound secret, the father discovered that his daughter, a highly nervous and sensitive girl, was in love; fearing the shock to her system of his disapprobation, he kindly and tenderly communicated with the father of the young man, a wealthy farmer in his parish, and arrangements were entered into between the two parents to provide for the young couple. Returning to his study, he summoned his daughter to communicate to her the arrangements which he thought would gratify her. She fell on his neck, and, in a few minutes, her joy being too much for her, became a wild maniac.

The medical men who had the charge of the case, were eminent in their calling, but after many vain efforts to do her good, had, with all their psychology and opium, been unable, at the end of ten months and a-half, to be of the slightest service to her; when she arrived in town, she was placed by her relations in a convenient cottage near the Alpha Road, St. John's Wood. On my first visit, I found her walking in the garden, with two attendants, and she received me with violent gestures, and spat in my face. I mention these particulars to show the degree of aberration of mind with which this interesting young lady was afflicted; it being unnecessary to relate all the details of her case, it is sufficient to remark that she was very insane. The wiseacre physicians, who had talked so much nonsense before a Committee of the House of Commons, on the subject of clairvoyance, could assuredly have been able to give no better account of the causes of certain phenomena in this remarkable case than I could myself. If they had had the humility of real philosophers, they might have been enabled to trace the thread in the tangled skeins of those mysteries which relate to the subject of clairvoyance; to show their supreme contempt for those facts which God wills to be existent in the world, is not the way to prove themselves to be either learned practitioners, or deep, wise-thinking physicians. This insane young lady was a very fine clairvoyante, and could read aloud passages that her father was transcribing from the sermons of Archbishop Tillotson, and other eloquent divines, at a distance of more than 170 miles. I mesmerised this lady for upwards of an hour each time, during eight days, at the end of the fifth time she was perfectly restored to her senses, and I continued mesmerising her three times more for the purpose of insuring the permanence of the cure. Mesmeric agency had in this case not only deprived her of her clairvoyance, but I have the firm conviction that it had been the means of dispelling the influence exercised upon her brain by evil spirits.

Mr. Fradelle, of the Mesmeric Infirmary, in Weymouth Street, has had some very striking cases illustrative of the principles which guided me in this cure; he has been very successful in expelling evil spirits from their possession of insane persons.

We have already said that we do not profess to be adepts at miracle. What are hallucinations? for these, in fact, constitute the essence of insanity; if we say that we are convinced we have a leg of mutton hanging to the nose, when none of our neighbours can see any such thing, we labour under an hallucination. How many times has Dr. Elliotson, in Conduit Street, and have I, at my own house, before numerous persons, exhibited instances of hallucinations of various kinds to which individuals were subject, when in what Dr. Elliotson very properly designated as the sub-mesmeric condition, that is in a state of neither sleep nor wakefulness, and yet in a hazy state of mind, between the two. We have made some patients believe that they were made of bird-seeds, and they have been wonderfully alarmed lest they should be swallowed up by birds; we have made them believe, at other times, that they were made of glass, and that they were so fragile, that they felt obliged to call out to us to beseech us not to break them to pieces; they have imagined that we presented them with brandy, soda-water, with sweet liqueurs, sweetmeats of various kinds, although the object presented to them at each time has been simply a glass of water, silently willed to be the object indicated.

Can we not persuade the world to look upon the subject in a common sense point of view, and induce them to train their physicians into grooves, which shall enable them to perceive that we have exhibited in these our victims of temporary hallucinations, specimens of insanity which they could, if they pleased, learn to cure with as much facility as we have cured them by the processes of Animal Magnetism, or Mesmerism. It is quite evident that hallucinations, in order to constitute permanent insanity, must be themselves permanent, but if that condition of sleep which constitutes somnambulism be induced by mesmeric passes, by the force of the human will, or by the other various modes of inducing this sleep, we necessarily remove the cause of the hallucination, and thus cure the insanity; this may require longer or shorter periods of time, but mesmerism may be said, when applied with sufficient perseverance and sufficient energy, to be a complete cure for all these cases. I could not here repeat, for it would obviously not be proper, all the cases that have occurred to me in private practice in this department, but I may assure my readers that Mr. Barth did not at all exaggerate the completeness of the cure he performed as stated in the 9th Volume of the Zoist for September, 1851.

It may be remarked that all these aberrations from the healthy standard of mind, occurring in India, have more or less some connexion with the causes producing Gout and Rheumatism; no set of causes are more prolific of disease in India than those dependent upon fullness of the venous system; and that fullness is to be traced to errors in the primary and secondary processes of assimilation; of course, then, the most rational mode of treatment, and certainly the most philosophical in such cases, would be deep sleep mesmerically induced. Dr. Esdaile informed me that he never suffered himself to be conquered by the obstinate non-receptivity of a patient. He would on some occasions employ as many as fifteen mesmerisers to work an hour each with passes, until he had established not only a condition of sleep, but anæsthesia; it is just this condition which is so desirable in curing obstinate disease dependent on congestion. There are many cases of severe congestive disease which require the absolute deep sleep, there are many other cases in which the high magnetic force of human magnetism applied to a patient labouring under severe Gout or Rheumatism will suffice for the cure of a severe case. The worthy and benevolent Mr. Thomas Slater, a man who is an ornament to the Astronomical Society, of which scientific body he has become a fellow, from the notice his scientific attainments attracted, when he constructed his splendid telescope in the New Road, was one day in want of the services of a plumber. He sent for a man who lived in his immediate neighbourhood. The man arrived, and wished to see the nature of the work required of him. He was, however, too ill to undertake it, for he had for six weeks been laid up with gout in one of his feet, and, eager to attend to business, had hobbled on crutches to No. 4, Somers Place. Mr. Slater, pitying him in such a plight, persuaded him to sit down in the kitchen. He prepared hot water for the poor sufferer, and washed his feet. Then he proceeded to mesmerise him. In less than two hours he could walk well, threw aside his crutches, and was able to complete his work. I need not add that Slater is a good optician, but he is no doctor, or the chances are that his mind would not have been so enlarged as to perform so noble a work of Christian humility and of real spiritual charity.

It was about the year 1835 or '36 that my late friend, the Lord Beresford, accosted me on the pavement fronting his own house in Cavendish Square. He was anxious to know if I had ever asserted that it was in my power to cure a little boy, nine years of age, who lived in the next house to him, of a blindness resulting from amaurosis, which had been treated for three months by Dr. Farre, Mr. Lawrence, and Mr. Travers. I replied it was quite true, but was not a little surprised that the topic of a conversation at a dinner-table in Fitzroy Square should have reached his lordship's ears. While we were in conversation the father of the boy came up, and his lordship informing him of the topic which occupied us, this gentleman angrily upbraided me for conduct he described as unbecoming the character of a physician, his temper allowed him to use the words charlatan and quack. I smiled, and bid my interlocutors good morning, observing to my irascible friend that I should be glad to restore eyesight to his boy in ten minutes, whenever, in his cooler moments, he should think proper to bring him to my house. Not two hours afterwards, he called upon me with his son, and desired me to explain the principle of the treatment I proposed to adopt, for as far as he was concerned he had no faith in my crotchets. Examining the boy's mouth, I observed, "Here is a malformed jaw, it follows the elongated type of the dolphin rather than the rounded mouth of the porpoise; there is not room for the development of the four first molar teeth, which should have been cut at the age of six years and a half. If your son will sit down opposite the light, I will make a free crucial incision in the gum of each of these teeth, and I predict that this young gentleman will be able to see directly." When the operation had been performed, the boy seizing his father's hand in great delight, called out, "Papa, I see you."

When they were quitting my house, I observed, it is not impossible that blindness may again occur, and if such a misfortune should arrive, it will be owing to a slow development of the jaws, and it will necessary to remove by extraction the four, apparently healthy, first permanent double teeth. The occasion did arise, for the boy, having returned to school, was one day in his arithmetic class suddenly taken with blindness. Bursting into tears, he requested that the master would send him immediately to London, where, having undergone the severe operation of the extraction of four teeth, he was again able to see.

It may be asked, what relation has this case to the subject of Animal Magnetism? All disease results from abnormal currents of vital force, and the removal of the cause of the inverse current effected the immediate cure. There is, then, little doubt that a philosophical view of the facts relating to the operations of the animal economy, rather than the envious imputation of indulging in wild crotchets, was my motive in forming the conclusion to which I had arrived. The father, on being questioned at various dinner-tables, as to the truth of his son's recovery of eyesight so rapidly, invariably replied that the credit of the cure should be divided between the three gentlemen who had for three months so anxiously attended the boy, and had salivated and setonized their patient, and myself, for it was impossible to say which of the expedients adopted had effected the cure. Nobody became the wiser by the adoption of my new principles.

Animal Magnetism does not more successfully rivet the attention of medical men when applied by passes. One morning in January, about the 20th., now nearly three years ago, a physician called upon me, requiring me to visit his wife; he drove me to his house, and informed me that on the Christmas day previously, a family party had assembled, and his wife sitting at the head of the table, while drinking a glass of champagne, was suddenly seized with such agonising pains all over her abdomen as to render it necessary to remove her immediately to her bed-room. She was undressed, and put into her bed. She could not bear the weight of the bed-clothes, she could not bear the light touch of a finger on the abdomen. The case was so urgent that Sir Thomas Watson's advice was had recourse to, as well as that of twelve or fourteen other eminent medical men. I found that the sufferings of the patient had not yet been alleviated; I proposed to mesmerize her immediately, requesting, as a favour, that no more medicine should be administered during my attendance. Mr. Capern was invited to assist me during the day, and Miss Cooper sat up with her during the night, to make passes when required; drowsiness was soon induced, and some refreshing sleep followed. The husband, impatient to have the opportunity of saying that medicine might have had some share in the case, administered a compound draught of senna. This ill-judged interference brought back all the agonies she had previously suffered. On my visit the next morning, I was disheartened at the breach of faith towards me, and proposed to abandon the charge of the case; however, upon the promise that my directions should be implicitly obeyed, I resumed the mesmeric treatment, which, being continued steadily, I had the satisfaction of seeing the lady in the drawing-room, in about a week, teaching her children, for she had completely recovered from her illness.

My friend and old colleague at the Middlesex Hospital, has been kind enough to compliment me upon this case, of gout in the peritoneum, but I do not learn that it has stimulated any one to inquire into, or to practise the important art of Animal Magnetism. Medical men will not give up their darling black doses, or their calomelizations, until they be forced by public opinion to run in more rational grooves.

This case having been stated to several sceptical persons who believe it wisdom to doubt what they cannot comprehend, I have requested my friend, Mr. Alfred Lidington to give me some cases of his own that are well authenticated, and the truth of which may be relied upon as far more accurate than the habitual statements of unbelievers. I beg to subjoin them.

Case 1. A child six years old was seized with severe cold and remittant fever, and soon the disease increased with great heat in the head, tongue coated and very dry, thirst very great, and copious vomiting, first of food, afterwards of sour liquid; he could not keep anything on his stomach, no medicine or food whatever, and for more than three days nothing could be made to pass through him. I was called in to mesmerise him, and in about two minutes I mesmerised him to sleep, and placing my hands upon his stomach, in less than five minutes he awoke and was relieved, and the following day he was up, and only had the usual feelings of excessive weakness; each day he obtained strength very quickly by the means of mesmerism, and in a week after was quite recovered; whenever this child is attacked with illness he will never take physic, but always sends for me to mesmerise him to sleep, and I have never failed to make him well, and very often he has declared he is well after I have mesmerised him for a few minutes.

I have many times excited the different phrenological organs of the brain of this child, and he has answered to each one most correctly, for instance when I mesmerised the organ of tune, he has declared to me that he can hear beautiful music, and so with veneration, he has felt irresistably impelled to pray, and speak of God and Heaven.

I have more often operated upon little children than adults, purposely to convince people of the truth of your science, for surely children so young, and many of whom I have never seen before, could not be guilty of any deception.

Case 2. I was sent for to mesmerise a child, three years old, who was suffering from paralysis of the hip downwards on one side, which his parent informed me came on while cutting his teeth, and finding it a very bad case, I was accompanied by Mrs. St. Aubin, the well known clairvoyante, whose marvellous powers have been published in the number for the month of January 1848, Vol. V, of the Zoist, and having put her into the deep sleep, from which in a few minutes she passed into the sleep waking condition, she then discerned the malady with her usual correctness, and then mesmerised the child, who almost immediately fell asleep, and when he awoke he walked across the room, exclaiming to his mamma, "Oh, mamma, my leg is well! see, I can walk on it," and every time I mesmerised it afterwards, he always endeavoured to convince every one that his leg was well.

Case 3. In the autumn of 1865, a very interesting case came under my notice, I was called in to see a lady, aged seventy, I found her very dangerously ill with bronchitis, and it being her third attack she feared she would never recover on account of the previous ones being so severe. I requested her to attend to the usual diet in such cases, and more especially about the atmosphere of the room. I then mesmerised her throat and chest for ten minutes and left her, telling her that she would have a good night's rest, and would find herself better in the morning, I called early the next day, and she informed me that she had slept all night and never coughed once, it being the first night's rest she had had since she was taken ill, she felt herself much better in consequence, and her appetite began to improve at once. I continued my visits each day, and in three days she recovered surprisingly, and was able to sit up. On the fifth day she left her room, and in a fortnight after was perfectly recovered, and felt better than she had done for many years past; she now advocates the marvellous powers of mesmerism, and recommends it to all her friends; I omitted to visit her on one occasion, and singular to relate, she did not sleep that night, and consequently felt more wearied the following day.

About twenty years ago, this lady had two tumours, one on each eyelid, of two years' formation, she underwent the operation of having them cut out by Mr. Morgan of Guy's Hospital, and he informed her that she would be likely to have many more; in fifteen years after she had another formed on her right eyelid, much larger than the former two, and she said she would never go through the operation of having it cut out, but would try mesmerism; she came to me, and after it had been growing five years was entirely removed by mesmeric passes in a fortnight.

The lady is still living, aud would be pleased to testify to these facts to any one who would feel interested in so important a subject.

Case 4. A young person of strumous temperament, with a diseased knee, was brought to me; suppuration had taken place, and nothing would allay the inflammation, she had been to several doctors, and they advised her to go to the hospital. She then went to St. Thomas' Hospital, and saw Dr. Croft, who told her that the knee was in great danger, and she would very likely lose her limb, unless proper precautions were adopted, he ordered her a blister and absolute rest. Finding it did not get better but worse, she came to me, and after I had twice mesmerised it, she was able to walk up stairs, using it without pain, which she had not been able to do hitherto. In a fortnight she was able to walk from London Bridge to Brixton, and is now quite recovered.

I could add many similar cases that have come under my personal experience. I have always been very successful, and more especially with maladies connected with a disturbance of the nervous system, I have very frequently cured them without the induction of sleep, and in some instances the cures have been instantaneous.

My mesmeric influence has been so powerful that I have habitually mesmerised a patient at any chance moment of the day; from a distant room, or from many miles distant, this I have done many times, and in going or sending to the room invariably find the patient in a mesmeric sleep, and very often in this wonderful sleep they have revealed to me things of the future which I have taken note of, and found them to come true. On one occasion I was with Mrs. St. Aubin, there were five other persons present, and I placed some blank note paper upon the table with a lead pencil over it, when we all witnessed a very marvellous phenomenon. A beautiful blue light was seen, by those present, to encircle the room, and it then rested on the paper, which had been placed upon the table. The light having disappeared, we examined the paper and thereon was written " God loves you all," which was done with such remarkable dexterity that no mortal hand could have written it in the time, it having been done in the space of about one second, and no hand was near the paper; on other occasions I have often witnessed in the presence of this most excellent and good medium, the chairs and table to move about without hands, and sounds have been produced in every room of the house, and on many occasions the thoughts of every one present have been revealed to them, through the magnetism of Mrs. St. Aubin; on one occasion two little children were present, and their feet and clothes were pulled and yet not a hand was near them. Sounds have often been produced on a hat, book, or a table, at a distance of five or six feet from her, and yet she has not been able to hear these sounds herself, although they have been heard very loudly by every one present.

These phenomena are facts, and for this reason, they have been seen and heard again and again, by myself and always four or five others present.

CHAPTER XII.

PHENOMENA OF ANIMAL MAGNETISM.

WHETHER correctly or not, the Author of these chapters supposes, and all suppositions or hypotheses result from imagination, this may possibly be a phantom of his imagination, that his defence of Animal Magnetism, undertaken as a defence of a holy truth, contains many ideas that are forcible from their novelty. All that is true may not be new, but, accompanied by God's blessing, all here that is new is true. Without this aid, our labour would cease; for if we are not servants of God we are nothing. If we do not carry the holy standard of truth, we are not sturdy and unflinching warriors in the cause of Animal Magnetism.

We have endeavoured to show that Mr. Rutter's instrument, the Magnetoscope, has introduced a new era in science. In the dim haze of vanity which envelops the minds of so many men of the present age of science, the influence of cliques is so powerful, that truth requires patronage for its protection. If the patronage be not forthcoming, the chucklers among the cliques, banded together to arrest and to obstruct the force of truth, glorify themselves as rulers of God's will. Without our patronage they cry, Woe betide the truth! Dr. Elliotson and his friend M. Chenevix both averred that Animal Magnetism was true. After an experience of more than a quarter of a century, I re-affirm their propositions, and I add Mr. Rutter's Magnetoscope is true, and whoever is destined to develop the future facts that instrument shall furnish, will become a powerful agent in extending God's mercies to mankind.

The endeavour has been made to show that the phenomena of Animal Magnetism have led to the explanation of the hitherto occult mysteries of sleep and wakefulness, and through the agencies which direct and govern them to the laws regulating health and disease. These interesting topics naturally lead us to enquire into the facts that have favoured the inferences at which we have arrived, otherwise it might be enquired, where are the warrants for the theories that have been propounded? These, as far as we have proceeded, we have endeavoured to demonstrate.

It is clear that a steel magnet appears to emit an influence which men who have paid attention to the subject for more than two centuries, acknowledge to be a force attractive as well as repulsive, according to circumstances. That this influence has two poles of opposite characters. The Baron von Reichenbach has traced the same relations to exist in crystals, besides discovering other properties relative to light in these symmetrically or geometrically arranged substances. Mr. Rutter has irrefragably established the existence of these opponent poles, not only in all crystals, but in all geometric and organic arrangements in nature, even up to man. Thus, the fact that has emerged from the light of his genius, is especially, that God having created man in his own image, has endowed him with magnetic life and a magnetic soul. In truth, the whole of nature is regulated by magnetic forces. Man lives, breathes, thinks, dreams, sees, hears, feels, tastes, has the sense of odours, and the enjoyment of consciousness, because he is essentially, next to the angels of heaven, a part of God's own great magnetic force. Mr. Rutter's modesty and religious humility may not allow him to assent that he has, by God's gift of genius, been enabled to establish all this; but nobody can deny that, by the grace of God's blessing, he was enabled to invent the Magnetoscope,

That instrument will be found to be my ample warrant for the assertions I have made.

Man, then, being a Magnet of a high order, is enveloped in a magnetic sphere of force. It would be idle to doubt that that spheres of force must necessarily place him in relation with all the objects and all the beings that approach it in the relation of attraction or that of repulsion. Being, by the Magnetoscope, proved to be a magnet, composed of numerous smaller magnets, there can be no question as to the polarities of the various parts of his body, and consequently of the existence of magneto-electric currents in that body. It does not rest with us, the investigators of Animal Magnetism in England, to prove all the facts which have been observed on the continent of Europe, as well as in America, Africa, and Asia, connected with the magnetic phenomena of the human being, and of those of the lower animals, and of the vegetable kingdom. We have fearlessly pursued our labours. The French and Germans had, however, preceded us, and never wanting in high purpose, and in industry as well as ingenuity, they have collected a large store of knowledge, and have produced numerous works of great value on Animal Magnetism. They have the merit of having proved the truth. We, their successors, can claim credit only for following in their steps, and of gleaning a few grains after the rich harvest they had reaped in the fields which open to our view such a wide prospect of future good.

Well might the sordid obstructors of truth place before their own eyes such examples as the physicians of France have set them. I was called to a wealthy merchant in Gloucester Place, whom I had known for years. I found he had been attended by a doctor who could do nothing for him, but talk, write prescriptions, and grasp fees, unless it was to tell him that he had a hopeless disease of his liver. If that patient had been advised by a physician who had had experience of the blessings which attend upon the earnest and the energetic treatment by

Animal Magnetism he might have been cured. When the ignorant doctor was informed that mesmerism was proposed, he warned the family against the treatment, as it would induce insanity. If this poor sordid obstructor of the truth, would apply to Mr. Fradelle, at the Mesmeric Infimary, No. 36, Weymouth Street, he would learn that patients have been daily mesmerised without such a consequence, and that Mr. Fradelle has had many years experience as a mesmeriser. I have had the experience of more than a quarter of a century, and, except from cross mesmerism or gross mismanagement, such a case could not occur. Years roll on. Men exist in the profession, I have taught myself to consider as the most exalted, the most laborious, the most self-denying of all professions, who give themselves up to envy, hatred, malice, and all uncharitableness, who regard such a bold falsehood as that which the wretched creature hazarded in Gloucester Square as a venial offence. In charity to him, I suppose him to be devoid of the sense of moral obligation. However that may be, the truth of Animal Magnetism shall proceed.

To teach the world to value the useful and the holy truths connected with Animal Magnetism, that is the object of our chapters, irrespective of the obstacles we may meet with in sordid money-clutchers, or men who forget what they have read respecting their noble prototypes, the silversmiths of Ephesus.

It is a remarkable fact, to be denied only by those pig-headed men who are careless with respect to the truth, that man can be made to sleep by several artificial magnetic expedients that may be employed for the purpose. All these, however various they may appear in their modes, may be resolved into the originating power of the human will. Every person, by a sufficient number of repetitions of the processes by which this power is brought to bear on that person, may be induced to sleep. Every person, then, said to be not sensitive, and so insusceptible of mesmeric influence, may be made to become sensitive, and so susceptible of being made somnolent, drowsy, and, finally, of being induced to sleep. It is really a process of magnetic induction. Some persons are far more susceptible than others. My friend, Mr. Henry Stafford Thompson, is possessed of a most unusual power of will, by which he has been enabled to effect a vast number of most marvellous cures by Animal Magnetism. It may not be generally known that very susceptible persons, either male or female, may be put to sleep, and then made to exhibit the attractive influence of the mesmeriser's passes upon the surface of a table, of a door, of a wall, of the back of a chair, of a walking stick, of a book, or any other object that does not too speedily conduct away the attractive force left by the operator's hand, or by his silent will. Several noblemen and gentlemen of good family were wont to visit me daily to make themselves masters in the knowledge of the phenomena of Animal Magnetism, produced by the attractive and repulsive forces, when I lived in Wimpole Street. I had, at that time, some young women attending, who, when put to sleep by tractive passes, were made to place the palms of their right and left hands to each other, when it became almost impossible to separate them. Some were more adherent than others, but it was almost ridiculous to find a strong man pulling at each arm of a young female whose hands were stuck together by their palms, quite unable to pull them asunder. Yet this was the fact.

One might ask how this phenomenon differs from that of an adhesion of two heavy pieces of iron together, influenced by forces arriving from opposite poles of a powerful battery; the phenomenon, in the one case, is from metallic magnetism, in the other, it is from human magnetism. Mr. Faraday, the autocrat, chooses that the one shall be "science," and the other kicked into space, until Satan shall protect it under his wing. One may say to Mr. Faraday that God's laws do not wait upon the pleasure of the cliques of the

Royal Society. Universal magnetism will not be kicked from pillar to post. The higher grades will assert their privileges, and the coarser forces shall not prevail over the powers of magnetic mind. We shall now offer an example of the force of the human will in producing an adhesion between a human being and a chair. Many a time have I made a pass over a comfortable dining-room chair with my hand, and have placed a susceptible young man, or young woman, upon that chair, and the adhesion has been so strong, that the person could not move from the chair. I have performed the same experiment upon a cane-bottomed chair, and upon a good, strong, elm-wood kitchen chair, with exactly the same results. Fancy a young woman who has been mesmerised, and has been awakened, and then coaxed to sit down upon a chair which I have previously, without her knowledge, rubbed a few times with the points of my fingers, having adhered so firmly to the chair that she has been quite unable to move away from it. This has been a very common and a very oft repeated experiment. Now, let us ask, how has the adhesion, in such cases, been loosened? Simply, by cross passes of the hand, or by blowing in the direction of the adhesion. If I blew gently, the force was dispersed. But the human will can effect the same result, as well as a few more.

In September 1847, Dr. Elliotson requested me to throw upon paper, any notes and observations I might have made on the force of the human will, I wrote him a hasty letter which was inserted in the 19th, or October number of the Zoist. From the preliminary observations he makes, I take the following remarkable passage, "Cornelius Agrippa ab Nettesheym asserts that a man naturally and without any miracle, unassisted by the Holy Spirit or any other, may convey his thoughts, in the twinkling of an eye, to another at any distance:" *et ego id facere novi, et sæpius feci. Novit etiam fecitque Abbas Trithenius.*

A professor of philosophy at Padua, Petrus Pomponatius, born in 1462, had contended, before Van Helmont, for the power of the imagination or will of one person to send forth an influence upon another, and enumerated the conditions of the exercise of this power, in nearly the terms of modern magnetism. He, too, surpassed all, for he point blank declares, that inanimate matter may obey this influence. *Cum hominis animæ voluntas et maxime imaginativa fuerint vehementes venti et reliqua materialia sunt nata obedire eis.* Page 664.

At a certain stage of our knowledge relating to the facts of Animal Magnetism, only about twenty years ago, this phenomenon of a powerful force being displayed or manifested by a human being, simply by the agency of his will, was regarded as a very extraordinary, and by some as an incredible affair.

My friend, Mr. H. S. Thompson had repeatedly performed feats by the efforts of his silent will, which were then considered marvellous. He begins a paper, which he hastily drew up at the request of Dr. Elliotson, by an allusion to the case of a young Frenchwoman, who was then being exhibited in London, by a M. Laurent and his daughter. She was remarkable in her somnambulism for the very exquisitely graceful *poses plastiques*, which she assumed when under the will of her mistress, Mlle. Herminie Laurent. She had been a cook in M. Laurent's family, and the daughter having repeatedly mesmerised her, she became a somnambule; and during her somnambulism, Mlle. Herminie finding her docile under the influence of her will, cultivated, by great patience, and after many repetitions, the faculty she displayed of imitating the *poses* of many of the finest statues of antique sculpture, and while in these *poses*, her mistress was wont occasionally, by the silent communication of suggestion, to oblige her to enact in the most natural manner, pathetic scenes. They called the exhibition *Oneiromancie*, perhaps little dreaming then that they were affording a philosophic illustration of

Andrew Baxter's thoughts on the phenomena of dreams, promulgated by him in the year 1733.

When Dr. Elliotson fearlessly pointed out to the world that there were facts in the works of Cornelius Agrippa well worthy of the deep consideration of mankind, no doubt the poor headed among the fellows of the College of Physicians, sneered at him for having had recourse to the statements to be found in those works. Nothing can be more remarkable than that the facts so frequently developed by the will of my friend, Mr. H. S. Thompson should have been noticed by Agrippa, between three and four centuries ago, and that they formed the subjects of profound study and thought for the gifted mind of the great Van Helmont, who not only pursued the phenomena of Animal Magnetism as a philosopher should pursue them, but elucidated their truth with wondrous acumen.

The phenomena of the human will are worthy of the attention of all philosophic minds.

Mr. H. S. Thompson, in the paper I have already alluded to, remarks on the action of the will or the power of communicating by the will alone, ideas and thoughts to persons, in a *mesmeric state*. He found that he had the same power with persons wide awake, or in a non-mesmeric state—on such as had never been mesmerised, in fact on persons whom the Baron von Reichenbach would designate as sensitive. Connected with the subject of will, Mr. Thompson speculates on the phenomenon of thought-reading. My friend would possess a more solid stay for his reflections, if he would study St. Paul's first epistle to the Corinthians, where in the 12th verse of the 13th chapter we read, "For now we see through a glass darkly: but then face to face, now, I know in part: but then shall I know even as also I am known." This intimation of thought-reading in a future state of existence is simply an illustration of clairvoyance. Concealment of our thoughts will not avail us. Every idea will be read by all who choose to read it. I once had under my care the very interesting daughter of a clergyman who could, at a distance of nearly two hundred miles, read her father's thoughts. She was insane, and our obstinately ignorant doctors who have the charge of the insane, do not believe that many of their lunatic patients are clairvoyante, and very easily cured by persevering in the practice of Animal Magnetism. My poor insane patient watched her father, when on Saturdays he was engaged in framing his sermons. Tillotson was often put under contribution by the reverend gentleman, and his daughter could cite long passages of eloquence from Tillotson, which she said her father was then incorporating with his own composition.

The lights that are thrown on the philosophy of the human mind, by the phenomena of Animal Magnetism, are of a very interesting nature. They may be scoffed at and ridiculed, and not believed, when actually witnessed by foolish, silly, vain sciolists; but these truths will find their way through space, and will enlighten those who come after us.

Community of taste and of feeling evinced by clairvoyant persons, in somnambulism, are classed by Mr. Thompson with the power of thought-reading. In both, these phenomena are traceable to that sympathy which results from the high species of Magnetism residing in organic arrangements of matter, they may certainly be considered analogous, but I should decidedly place the manifestations of thought in a higher grade than those of sensation; but the subject is very deep, and Mr. Thompson may be right.

This gentleman has often induced not only trains of thought in some persons, but corresponding actions, nor has it been necessary for him to be in the same room with some highly sensitive persons. He being in London, has produced the inclination to move, in persons in the county of York. Several cases of the influence of his will are related by him, which many persons thinking themselves wondrous philosophers would not believe. I credit the statements not only because I have witnessed them, but because I have done the same things by the power of my own silent will. Dr. Elliotson placed my letter to him immediately after Mr. Thompson's paper. I refer my readers to the October number of the Zoist, 1847. It is in the fifth volume of that valuable collection of facts.

The phenomena attendant upon the exercise of the will, are like all the phenomena of Nature, illustrative of attraction and repulsion. I have repeatedly forced persons to approach me, and have repelled them from me by the simple effort of silent will.

The strong adhesion of the hand to a surface that has been mesmerised by passes, or by the will, is by no means a rare phenomenon. Often has it been exhibited at Dr. Elliotson's house, and hundreds of times with my patients. A hazel stick in most cases requires no process of magnetising. I have seen many persons so sensitive that they would instantly grasp a smooth hazel stick, and fall asleep in less than a minute, and continue asleep several hours. In most cases of magnetic adhesion of the hand to the surface of wood and most other substances, simply blowing upon the hand, or a few dispersive passes will dissolve the influence. But in the case of the hazel, some wedge-shaped piece of iron or zinc is required for the purpose. This, introduced gently between the hand and the object to which it adheres, generally at once loosens the attachment, but I have seen cases in which the tendency to adhere has been not so quickly overcome,—idiosyncrasies contending with the usual law in these cases, time, gentleness and patience in continuing the means have at last succeeded. Repulsive agencies being always kept in view, when we have to conquer an attractive force.

With some patients, the attractive power of certain metals in producing deep sleep has been found to be very remarkable. As a general rule, these stand in the same position as they do in the list of metals in relation to their grades of polarity assigned to them by the electricians. Gold and platinum applied to the nape of the neck in sensitive subjects are found to favour sleep in the shortest time, and other metals more slowly. Of course, the more easily oxydable metals are so repulsive as to produce only a cold sensation. I have witnessed in some extremely sensitive subjects, very disagreeable effects from copper. I have witnessed some remarkable results in experiments with solutions of oxydes of metals, through which currents from a voltaic series of Smee's arrangement of a battery of six cells of seven inches by three and a-half have been passing. The patients were selected for extreme sensitiveness, having been remarkable for the great facility with which sleep was induced. Some of them could not gaze at me for thirty seconds without falling into a deep sleep. I had procured a strong glass basin, perforated at opposite points for platinum wire, riveted to platinum strips about an inch wide, which served as electrodes. The first experiments were made with a saturated solution of nitrate of potass. The patient was requested to place the fingers of both hands in the solution. This produced little effect, except in each case a sense of coolness in the mouth. The number of cells was gradually increased, and at last, when fourteen cells were used, each of the seven cases slept deeply, in times varying from four minutes to twelve. I have tried to induce sleep in a great number of persons with this same apparatus without success, but in some few, more remarkably sensitive than their neighbours, sleep has come on in from half an hour to an hour.

Having replenished the basin with distilled water, twenty minims of a solution of sulphate of mercury were added to it. Jane Love put her two forefingers into it while the electric current was passing. She was instantly asleep, and quite rigid. I caught her as she was falling, and placed her horizontally on the ground on a hearth-rug, where she slept for two hours on her back, but when she was at last awakened, it was found that she had a mercurial sore mouth, which, in three days, was effectually cured by the deep rigid sleep obtained daily by plunging her hands in the saturated solution of nitrate of potass, while a current of electricity was passing through it. I deepened the sleep in this case, and rendered her rigid by the application of a disc of pure gold to the nape of the neck. When she shewed a tendency to become relaxed, I re-applied the gold, and thus kept up the refreshing and curative rigidity for three hours.

A similar experiment was tried on Mrs. H., a highly sensitive lady, with the same results, except that the rigid sleep kept up four hours daily cured her in two days. Some little time afterwards I tried this lady with twenty minims of a solution of sulphate of copper, in the basin of distilled water, under exactly the same circumstances as with the sulphate of mercury. She slept instantly, and had immediately very offensive, and almost putrid breath, with great tenderness of the mouth and gums when she awoke. The electric current through a saturated solution of nitrate of potass, was had recourse to, and her hands plunged in the basin, were retained there for an hour, while she was again asleep. The cancerous odour was gone, and I awoke her to ask her questions; when she told me that her mouth was delightfully cool, but she felt languid and faint. I made her sleep rigidly for three hours, and she was awakened well. The next day, however, she had a severe head-ache, for which I made her take a long rigid sleep, and she awoke refreshed and quite well. All these experiments are not only illustrative of the fact that tonic effects from Magnetism are curative of disease, but that disease itself is produced by the cathodic current carrying copper influence in one instance, and mercurial influence in two instances into the human system. I use Mr. Faraday's own language, and I show the importance of his doctrine of reversals of polarity, in showing that his cathode had reversed a current to produce a repulsive agency carrying poison to produce disease; which disease is cured by the direct current productive of deep sleep.

The influence of other metals in producing sleep is in some cases very remarkable. I have stated in one of my notes to the English edition of the Researches of the Baron von Reichenbach, (p. 12,) that I have, in certain individuals, twelve in number, produced the rigid or tonic spasm, by touching the nape of the neck with pure gold, or with platina, or with rhodium, or with nickel, or with cobalt, or with antimony, or with bismuth. In every case but in one, I was always able to dissipate the spasm and awaken the patient by means of iron, applied to the nape of the neck. In that one I held a newly cast disc of cobalt, about two yards off, directed to the nape of the neck, without the patient's knowledge. She fell forward insensible upon Mrs. Charles Lushington, who was talking to her. She was not only insensible, but quite rigid. The pulse was, for a time, quite imperceptible. A current from a single coil electro-dynamic apparatus, which happened to be in action, was passed from the pit of the stomach to the nape of the neck. Colour gradually returned to her cheek, and the pulse and breathing removed our alarm. Dr. Thomas Mayo, late President of the College of Physicians, was present. The poor girl slept on that occasion fifty-seven hours. I state the fact without hesitation, because she slept in an attic room in a house I then occupied myself. A fortnight afterwards I was induced to repeat the experiment, and she slept forty-seven hours. If I endeavoured to awaken her by any mesmeric means, I found her idiotic, and I restored her to the deep sleep which always had cured her, and on this occasion she awoke spontaneously, vastly refreshed and improved in vigour. The case of this young woman was very illustrative of the phenomena of Animal Magnetism and it shall be given.

Mary Ann Douglas was a slight built small person, twenty-one years of age, of a highly nervous temperament. She was passing the railings of the Middlesex Hospital, when she fell down in an epileptic fit, foaming at the mouth, and quite insensible. She was taken up and brought into the hospital on the 16th March, 1845.

It was concluded that her epilepsy was dependent upon a severe form of hysteria, and she was treated for three months with iron and valerian, and the usual routine of remedies in such cases. Her epileptic fits remaining as severe and frequent as ever, she was transferred to my care on the 16th of August, 1845, when I mesmerised her. She fell into a heavy sleep on the second trial, and during two months more she exhibited no phenomena more remarkable than a common deep unconscious sleep. But she never had a return of her fits. Her case, as the development of phenomena proceeded, became deeply interesting. Her temperament was highly nervous. When the pyramidal termination of a rock crystal was pointed to the pit of her stomach, her eyes became heavy, and after a few drowsy winks, she fell fast asleep. If, now, the flat end, or it may be called the butt end or opposite pole were held to the pit of the stomach, she soon became wide awake. She was placed standing outside the dining-room door, at the distance of three feet, while the door was held ajar, so that she could not see the person who held the crystal. When the pyramidal end was held towards her, she approached the door till her forehead touched it, and she was found to be fast asleep. Then, the flat or butt end of the crystal was turned towards her, and first her head was repelled, for her forehead receded from the door, and then she stood wide awake, tottering a little, and apparently somewhat astonished. Stupidities may not have the capacity to feel interested in nature's developments. Such experiments as these to a philosophical mind are of deep interest; for they illustrate the fact that these forces can, and do permeate thick obstacles of wood, penetrate them, and exhibit most strikingly the opponent powers of attraction and repulsion; placing beyond a doubt, that attraction promotes sleep, and that repulsion is the cause of wakefulness.

But Mary Ann Douglas's case afforded many more illustrations of the phenomena of Animal Magnetism. If, when she was wide awake, a piece of zinc, of the size and shape of a penny piece, were placed at the nape of her neck, and two small bits of copper, each much smaller, were placed upon her toes, or on the shoes covering them, she would immediately sleep deeply. The pieces of metal allowed to remain there ten minutes were sure to produce rigidity. Continued awhile longer, the body would become curved backwards, as in Opisthotonos. All this curious state of system was changed in less than three minutes, by changing the positions of the zinc and copper discs; by reversing the polarities.

Mary Ann Douglas was not only a marvellous subject for the exhibition of most exquisitely delicate phenomena in Animal Magnetism, but though the child of a washer-woman, was really in conscientious feeling, and in refined affection, a most exemplary young woman. I knew her many years after she became my patient at the Middlesex Hospital. Her disposition was quiet and amiable. She was characterised by strong good sense, and had been strictly and morally brought up by a very worthy mother, who was a poor widow, and it fell to my lot to attend this woman in her last illness, when she was in charge of an empty house in the Abbey Road, far gone in consumption, hardly able to stand. I found her beyond all hope of cure, comforted by her affectionate daughter, who supported her by her hard earned pittances. This was a picture of poverty, but it was a touching scene of benevolence and amiable self-abnegation and sense of filial duty. The mother lingered long, and at last died. I called after she was dead, and found that daughter in bed clasping the cold body in her arms. She had been piously hoping for six hours that her own warmth would resuscitate that body, which was dearer to her than her own life.

Mary Ann became very ill in consequence of her excessive anxieties and exertions, but rigid sleep was an effectual restorative. A gold sovereign, the gold chain of my watch, or a small disc of platinum held to the nape of her neck was sufficient to put her to sleep rigidly, and this sleep with the rigid condition of her voluntary muscles was marvellously efficacious in restoring her to health and strength. She never left my house after undergoing a great variety of mesmeric trials, without feeling, and really being much better and stronger than when she entered it. I have put her to sleep, and have whispered to her that when she awoke she would have a leg of mutton hanging to her nose. Then by dispersing passes, I have wakened her. The hallucination remained after she became awake. She would stand, with her head forward, beseeching me to remove it. " Do take it off, Sir, pray do?" " Take what off?" " This heavy leg of mutton hanging to my nose." If one pretended to cut it off, it was of no avail. She persisted that it was still there. The cure was to make her sleep again soundly for five minutes, and when she again awoke, she remembered nothing of her hallucination. By the same process I have impressed, at different times, a great variety of hallucinations upon her mind, which have invariably been dispelled by the induction of deep mesmeric sleep. Dr. Elliotson first shewed me these important experiments upon some of his patients. Have the doctors, who make so much money by mad-houses, ever thought of curing their patients who may be afflicted with hallucinations, by inducing the mesmeric sleep upon them? O dear no! Dr. Winslow knows it would not pay. Psychology is the word! Glory to the great Sir Benjamin Brodie. He wrote a book on Psychological Enquiries. Vive la bagatelle! Teach us no Animal Magnetism; teach us no wonderful modes of cure. We do not choose to believe them; we read your tales; but who believes them? Thus the audacity of our opponents has prevailed against our indomitable industry. Dr. Elliotson has endeavoured to draw the attention of the public and of the medical profession to a full knowledge of great truths in his valuable collection of facts in the Zoist. I have wrought hard too to instruct mankind in facts connected with their best interests. The medical world point to the ignorant public and desire only to keep them ignorant; while they know of cures performed by the agency of Animal Magnetism, and do their best to smother all knowledge of them. Their sordid tactics will not succeed. Truth marches. Truth is Illumination from Heaven.

Do the medical world believe that the Demigods they worshipped, Mr. Wakley and Sir Benjamin Brodie, *(par nobile fratrum,)* succeeded, because for a while they retarded the knowledge we have endeavoured to diffuse.

Mary Ann's case offered illustrations of many more phenomena. She was a somnambule of a high order, for when fast asleep, she could describe events occurring in another room. She has described a scene of a suffering person in bed on the second floor, and the events occurring in that bed-room with perfect accuracy. Then she has turned her attention to the next room; in which she described a seat that had a long cushion with pillows and chintz covers upon it. "That seat," she said, "is a box lined with pink calico, and contains eight beautiful bonnets," and then she described each bonnet to the astonishment of those who came to witness the wonder. Her statements were completely verified. Often has she in the waking condition, but the expression wide or broad awake would be very incorrect, for the process requires a condition we are apt to call dreamy, been subjected to all the trials of electro-biology, improperly so called. Hallucinations, witnessed even in madhouses, are dreams, which those ignorant, but would-be thought wise and learned in their trade, for they are but trades' unionists banded together to ignore the higher learning of their craft, would do well to know something about. Mary Ann, when awake, could easily by the mere exercise of my will, be put into the dreamy condition, and then be made to believe that the same glass of water was, at one time a glass of lemonade, and five minutes after a glass of senna tea, and five minutes after, a glass of strong brandy and water; which would have the effect of making her very giddy, and then of making her reel and tumble about like one thoroughly tipsy. To remove these hallucinations completely from the mind, I have felt the necessity of inducing the deep sleep, which is the effectual cure.

To endeavour to inculcate the results of the experience of Dr. Elliotson and of myself on medical madhouse keepers, would be like spouting to baboons, only that the latter gentlemen would afford more amusement from their gestures.

In illustrating the various phenomena of Animal Magnetism, it has become expedient to recount the facts as they actually occurred in the most remarkable cases. I have had some individuals, who, in the earlier stages have exhibited effects more rapidly from their greater impressionability; but upon the whole, the wide range of phenomena in Mary Ann Douglas's case, constituted her, being as she was a most respectable, a most truthful, and guileless person, the most reliable case for investigation. Many facts first observed in other cases, I have hesitated to note as facts, until I had tested them by repetitions with her. One very extraordinary mesmeric fact exhibited afterwards in numerous cases, related to the fixing by my will, or by a pass with my fingers, a line of light, or as some of the somnambules would have it, of a bluish fire, on the floor or the carpet of the room. I tried this first with Jane Love, who asked what I did that for, as "it was quite impossible for her to step over it." She fancied she would be burned to a cinder, if she attempted it. "Nonsense," I said, and endeavoured to force her over the imaginary line. She screamed, resisting me, but fell upon the line completely rigid. Another young woman present said that Jane would never waken until I removed that bar of fire. Jane remained insensibly rigid until I made dispersive passes across the length of the imaginary line, when she resumed the state of somnambulism. Mary Ann Douglas was always alarmed when I placed the imaginary line near her, and she would gather up her clothes to avoid the evil consequences of it. The results, if she were forced to come upon it, were precisely the same as with Jane Love. In several other cases, I have witnessed exactly the same phenomena. They have especially occurred in those cases wherein the susceptibility to the effects of the rock crystal was most remarkable. Harriet P., a highly impressionable somnambule, who is mentioned in my note on the hazel rod, at page 100, of my edition of Baron von Reichenbach's Researches, as the person who, in a condition of Somnambulism, at Buntingford, in Hertfordshire, discovered, by holding a hazel rod in her hand, a spring of excellent clear water; she was so delicately susceptible of the influence of rock crystals, that one morning at my house she was put to sleep by the presentation to her face of the pyramidal pole, and wakened by that of the opposite pole, twenty times in four minutes. Mrs. H. was always easily made to sleep by the pyramidal pole of a rock crystal, and if, in her gentle, placid sleep, the rock crystal were placed near her, she would stretch forth her hands, and finding the pyramidal pole, she would, by her gestures, and the expression of her countenance, evince delight. But if the crystal, even at the distance of a foot from her, were turned about, so as to present the repulsive pole, she expressed a disgust and horror at it, and if the unpleasant direction were persevered in, if she happened to be standing, instead of wakening, as is usual with most persons, she would fall flat upon her back, perfectly insensible. The repulsive pole, in her case, did not waken, but produced the retrogression of the body, a fall backwards, and insensibility. Persons uninformed concerning the peculiarities of highly sensitive individuals, should not officiously intrude their experiments. I have seen this amiable lady three times submitted unwittingly to this alarming experiment. It is true that she, on each occasion, recovered completely on the application of the attractive pole of the crystal, and expressed herself refreshed and strengthened beyond measure.

In all these cases of extreme sensitiveness, a remarkable fact to be always remembered, and well pondered on, is Cross Mesmerism. This is a very curious magnetic phenomenon. I have illustrated many of the laws regulating Animal Magnetism by the case of poor Mary-Ann Douglas. Latterly, she never resisted the influence of a friendly will, confiding in the honest morality of her surroundings, but when I first made experiments with the will, she always tried to be stubborn in her resistance to the behests she received. Sometimes she would say, "I know what you are willing me to do, but I do not like to do it." "Tell me what it is, my poor child?" "You want me, Sir, to take a bunch of keys out of your right waistcoat pocket, I could come and do it, but it seems like picking a pocket, and I wont do it." If I persisted in willing her to do it, she would burst into tears, and tell me she had an awful headache! she was strictly and deeply conscientious. If our crass Stupidities could estimate the beautiful facts relating to the human mind, to which they might be introduced through the facts of Animal Magnetism, they might become enlightened, but God has not ordained for them the capacity of intellect which would enable them to feel the charms of such a study. Here and there may be found a mind capable of estimating the conscientiousness and simplicity of a creature educated in the depths of poverty, cheerfully devoting all her best energies and affectionate solicitude to the hard duties of a toilsome life to supply the wants of a sick mother, to be her nurse at night, and a washerwoman and servant-of-all-work in the day. This is grand prayer, for it is practically that holy, that sober, righteous, and godly life that really is spent to the glory of God.

The poor girl who exhibited such a touching trait of conscientiousness was asleep. She was a somnambulist.

When I told her I was the last man who would, in any way, make her swerve from any of the paths of strict honesty and rectitude, and explained to her my motive in willing her to take an object with her finger and thumb, out of my waistcoat pocket, she was reassured, and as I had, in my own mind, changed the object she was to select, she quickly came up to me, and took a buck-horn-handled penknife from my pocket, the object I had willed in place of the keys. Her first resistance gave her a headache, which, with all my dispersive passes, was not removed. Finding so many influences of spectators in the room, I sacrificed the experiments to prevent Mary Ann being cross mesmerised. The real meaning of which expression is, that if the subject under the influence of one mesmeriser receives the influence of one or more other individuals, in the same room, who may be more or less distasteful to the mesmerised person, that person becomes cross mesmerised. I have witnessed some dire effects from Cross Mesmerism. A stupid lot of surgeons, half eager to know rather more than they do, and half led away by the brutal wickedness of the late Mr. Wakley, tried their hands successively on a poor girl who was very sensitive. They succeeded in making her idiotic for a time, but she luckily fell under the care of a man who understood the value of the deep sleep, and adopted the treatment at once. After a continued sleep of twenty-four hours, the patient was better. A repetition of the deep sleep cured the patient. Cross Mesmerism, where the mesmerisers are persons of low, vicious, and malignant feelings, I have known to bring on an attack of suicidal mania, which has finally recovered after a sleep of eight days. These phenomena seldom occur, very fortuuately, except in cases of great sensitiveness, generally aggravated by very injudicious and clumsy treatment.

CHAPTER XIII.

PHENOMENA OF ANIMAL MAGNETISM—JANE MURRELL.

PURSUING the consideration of "the phenomena of Animal Magnetism," and following the evolution of facts according to the range in which my experience has led me, I arrive, after the thoughts suggested by the very beautiful case of Mary Ann Douglas, at the phenomena presented by the case of Jane Murrell, which were, in some respects, analogous to those we have just been considering, but, it will be perceived, offering points of deep interest, not usually met with in such cases.

Jane Murrell was a servant in the family of two of the most simple-hearted, sweet-minded, and truly pious ladies I ever had the happiness to know. They were two sisters, artists, who lived not far from Orme Square, in the Bayswater Road, and I verily believe if the world could be searched for specimens of virtue, intelligence, and taste, it would be extremely difficult to find the equal of these accomplished ladies; but they were more, they were most charitable, kind, and benevolent. Poor Jane Murrell, when I was first called upon to attend her, was quite insane, and had, in one of her wild fits, taken it into her head to seize her two mistresses, and to shut them up in the coal-cellar of the house, where they resided. Her force was overpowering, and when I arrived at the house I learned that all resistance to her muscular strength was an idleness. She was in a state of great exaltation. By the force of my will, and of coaxing, she gave up the key of the cellar, but soon showed symptoms of awful mania. We succeeded in getting her to bed, but when she got there, suicidal spirits prompted her, and if nothing else was allowed her to accomplish her dire purpose, she would lick the rough wall of the room, hoping that the blood issuing from her tongue should kill her. I proceeded with passes and my own strength of will to mesmerise her. In about an hour, she was conquered, and slept soundly. From that time, she became quite manageable. She had occasional outbreaks of temper, but was, upon the whole, very amenable. This creature had, in early life, been subject to epileptic fits, and these had been accompanied by aberrations of intellect. She was a hard-working young woman, and, before she became the servant of the ladies with whom I found her, she had been in the service of their sister, a highly respectable and much-respected married lady at Brighton. Jane's antecedents were all of a nature to produce a favourable impression on all who had occasion to inquire into them. If I were to remark on her head phrenologically, I should very decidedly have predicated a preponderance of the benevolent feelings, with strong attachment to her friends, but with a great tendency towards the sly and cunning propensities. She had great sagacity, great wit and love of fun; but her strong affections appeared to control all these powerful incentives to the irregulation of her moral ties. Circumstances finally determined her ingratitude to her indulgent benefactresses. She married my servant, and the couple, from having been well conducted, adopted sly and cunning courses, and behaved very ill.

These preliminary statements are made only for the purpose of showing that the extraordinary benevolence and charitable feelings of the two amiable ladies could not, in this world, protect them from the consequences of ingratitude. If the whole tale were told, how glorious would appear the conduct of those two ladies! They seek not their own glory, but the glory of our Father, who rules in Heaven.

One of these charming women was kind enough, at my request, to keep a diary of observations upon the case of their servant, and from this diary, as well as from some notes of my own, I am enabled to give the following interesting statement.*

On the 28th August, 1848, Mrs. S. first came to mesmerise my sister M., and during a week visited her twice a day. Our servant Jane Murrell, a simple but most estimable country girl, who, from personal attachment, preferred our service to that of any stranger, even with much higher wages, which fact I mention to prove her indifference to money and the impossibility of bribing her to act a part, had never heard of mesmerism; but when Mrs. S. first came to our house, expressed the greatest dislike to her and said, "she must be a witch," and when she found that this suspected person was to take up her abode in the house, only the influence of her great attachment to my afflicted sister and the sight of the blissful sleep and freedom from pain produced by the mesmeriser overcame her dislike to sleeping under the same roof as Mrs. S.

At length, however, Jane even consented to be mesmerised by Mrs. S., in less than five minutes she slept soundly, snored, and on being questioned replied, "she liked mesmerism very much, was very, very comfortable," answers which so amused my sister, she burst into so loud a laugh that it awoke Jane.

The next evening there was a violent thunder storm, and as Jane was always so terrified by lightning as to lose her self-possession, we called her into our room and tried in vain to soothe her. Mrs. S. said, "Jane, let me mesmerise you, a little sleep would tranquillize you," she consented, and was soon asleep, but not tranquil, crying out, "How dreadful! how high the flames are! how near the house too!" Her mesmeriser wounded her hand with a pin, and she was quite insensible to pain, but

* Mrs. Skelly, a most energetic mesmeriser, a pupil of the celebrated and benevolent negro's friend, Mr. Garrison

when Mrs. S. pricked herself, Jane cried "Oh! oh! what are you doing?" Experiments were then made to try her clairvoyante power, and without the means of knowing by her natural eyesight—for her eyes were fast closed—she named correctly several articles held over her head, or pressed to the pit of her stomach. A few days after this Jane was complaining of weariness and head-ache, and wishing to be put to sleep I offered to try my influence, and with as few passes as Mrs. S. had used, I succeeded. I had promised not to question her, but to let her sleep quietly, but she was no sooner asleep than after some uneasy exclamations, her countenance took an expression which was perfectly heavenly, and she cried out, "What joy!" "What blessedness!" "Ah, you do not see what I see!" "Such glorious happiness!" Then she seemed to perceive in the sky the spirits of beloved friends and relations who had died. My little niece coming into the room at this time, (touched Jane's hand in the way we usually did when we wished her to hear us,) and asked if she saw her mamma there, "Yes," she replied, "but she is higher than I am, and sleeping with angels round her." She saw other forms she knew, and at length burst into tears, and a passion of prayer, over the despairing spirit of Captain ——, who had lived a most evil life in the part of Sussex she had come from, and who entreated her to help him into that blessed company. Her idea seemed to be that this was not the final rest, for whilst she spoke to this despairing spirit she solemnly reminded him that there would be a day of judgment.

A few days later my nephew, who, though much interested in Jane, could not believe without testing her power in his own way. After hearing her relate an imaginary journey to her own home by the Brighton Railway, with my sister E., he asked her to accompany him on a short trip; "with great pleasure Sir," she said, and immediately commenced the description of their progress to the Eastern Counties Railway Terminus, which, with Epping Forest, the old Church at Chingford, and many other things remarkable there, (where he had resided,) and upon all of which he intently fixed his thoughts, she minutely described, though she had never set her eyes on any of them. My nephew held her hand during the description. Mrs. S. who mesmerised her on this occasion, wishing to prove the truth of phrenology, touched various organs of the head, and displayed their characteristics in the most exact and decided manner. Thus the girl became an object of continual interest to us, and we mesmerised her by turns, and when in the sleep, promiscuously took her hand that she might hear us speak, &c., (though this was wrong, as we afterwards found, for none but the mesmeriser should touch the patient,) and truly wondrous were the results, for often when asleep, the walls, &c. of our house were made transparent to her, and distance annihilated, and she described persons and events in her own home, though forty miles distant, as minutely and distinctly as though she was actually on the spot.

Some pieces of rock crystal sent by Dr. Ashburner to be tried on Mrs. M. (in whom it was difficult to produce sleep,) were, by her own desire, tried upon Jane, with great effect; only looking at the pointed end instantly sent her to sleep, whilst the pressure upon her chest of the blunt end as instantly awoke her. When asleep, she was usually good tempered, and more than naturally intellectual, and if disposed to work would do so, so vigorously as to do in half-an-hour what she or any other very active servant could not have done in two, when awake. She even washed during half a day, and only ceased when there was no more work to be done. No one could prevent these fits of industry, but Jane always suffered from them. Dr. Ashburner wished her to try to mesmerise Mrs. M., and she succeeded in doing so, she had but to look at her and Mrs. M. fell asleep, but Jane continued the passes till she slept also, and then she would call our attention by whisperingly saying, "Come, look at her, how happy she looks, &c." "but oh!" she added, "what a cruel sight to my eyes! You only see the surface, but I can look within, &c."

One luckless day she playfully defied my niece to put her to sleep, who somewhat unwillingly merely pointed her fingers at Jane's eyes, and the effect was instantaneous. Soon Jane saw Dr. Ashburner's carriage coming, and begged to be awakened quickly, that she might open the door to him; my niece tried hastily to do so, but failing, called to me, I tried also in vain. Whilst in this dilemma, Dr. A. arrived, and strongly expressing his disapprobation of her being thus trifled with, succeeded, after some effort, in awakening her. When she awoke, she was confused by the doctor's stern manner when he said "Now, Jane, I shall be seriously angry with you if you go to sleep again before twenty-four hours." "Yes, Sir," said Jane, and went down into the kitchen, but fell asleep on the way, and all the doctor's efforts were unable to wake her. She showed unusual temper, saying "she would go home; none should prevent her going immediately, for she had money." Dr. A. affected to doubt this, and swiftly running to her box she returned displaying two sovereigns, these Dr. A. persuaded her to lend him. She told him he had sent her to sleep by being angry with her, and she should not wake till five o'clock the next day, and she did not. Dr. A. after soothing her, and telling us not to thwart her, left the house. Soon after, Jane having thrown herself on her bed, was seized with terrific fits, tearing her hair, beating her head and chest in the most violent way, and raving at the hateful straight jacket which she declared we had sent for.* Dr. A. kindly left his dinner to re-visit her, and to him Jane again and again appealed to save her from the jacket. By pressing the organs of benevolence and of attachment, and by mesmeric passes, Dr. A. at last overcame her malady, and having instructed my sister E. how to mesmerise Jane, and to manage her during

* In Jane's early days she had been so treated, when from distress of mind, she laboured under fits of temporary insanity.

the fits, a strange result was soon apparent, for in her frequent long sleeps, she sympathized completely in all my sister's thoughts and feelings, and even fell ill when E. suffered from an unlucky chill. One evening Jane came from the kitchen fast asleep, and asked leave to sit with us whilst we took tea. Dr. A. soon called, and whilst he was talking to Mrs. M., poor Jane showed signs of a coming fit. Dr. A. made passes from head to foot, and extending widely her arms, made passes along them also, to our great wonder we saw her fixed immovably in that position, and only able to move her head and speak. Of her tongue she made free use, scolding and threatening the doctor. Dr. A. wound his gold chain round her wrists and greatly increased the rigidity; she cried out that it burnt her, and she scolded still more. Thus she remained till the fit ceased, and the rigidity was gradually dispersed by contact with iron.

These experiments led to Jane's going daily to Dr. A's. house, to try the effect of a galvanised bath. She had only to dip the tips of her fingers in, and sleep came before she could withdraw them. We usually found her so, and rigid also, when we paid our visits; but one morning she contrived to protect one hand from the influence, by holding in it a small key, and, having thus released herself, we found her, though with fast shut eyes, very busy finishing an apron which she was working with wonderful neatness.

From the time that Dr. A. became her daily mesmeriser, she could bear none but him, and the touch of any *other person burned her so*, she cried out with the pain. (This is no exaggeration, for I have known the touch of another person to raise a blister on her skin.) Our only means of communicating with her was by writing, which she, through her closed eyelids, read. Sometimes she would say, "Let me see you speak, and I shall see the words come from your mouth." But this mode of communicating thought, so familiar to the deaf and dumb, was to her so difficult she seldom practised it, except when Mrs. M. spoke, when she would give the most patient attention. On going to bed at night, she always became rigid at the precise time Dr. A. told her he would make her so, only desiring her to turn her face to Grosvenor Street, and think of him. Jane had a great objection to this cold rigidity, and would protest over and over again she would not submit, but she could never resist the power of Dr. A's. will, though acting two miles off. Once she tried to defeat the doctor's purpose by placing a number of knives in her bed, but we removed them, and the scheme failed. On a second attempt, she was more successful, and, sitting up in bed, she made, without light, during the night, a bonnet-cap, and mended her clothes. But Jane's condition, bad as it was, became worse, for, during a prolonged mesmeric sleep, she lost the sight of both eyes, and became delirious. At this time, my sister's bad health had so increased we dared not tell her of Jane's condition.

Jane had intervals of reason, and in one of them she predicted she would awake on the following Sunday, Nov. 15, at eight o'clock, A.M. On the night of Saturday, my sister was so alarmingly ill that I never left her till four o'clock on Sunday morning, more than once we thought her dead, but she rallied, and finally slept; then we counted the hours for Jane's awakening, which happened exactly as foretold. Dr. A., calling at two o'clock, and seeing Jane's sad state, mesmerised her eyes, till, to our intense relief and astonishment, he restored her sight once more; but with this, her delirium returned, and she implored Dr. A. to take her from the horrid wretches she lived with, and let her live in his galvanic bath-room, where she promised she would sleep, and be quite rigid, and give him no trouble till she awoke. Dr. A. kindly consented to this, and she remained in his house, till, in much improved health, she left to visit her friends in the country., (Nov. 22, 1848.)

On the journey down she was much disturbed, and fell

into a short mesmeric sleep, on discovering that a white kitten was in the carriage with her. On the way from Hayward's Heath Station, she told her brother, who had met her, that she should go into a mesmeric sleep at eleven o'clock; so, on arriving at home, his first words to his mother were, "Is Jane's bed ready? she must go to bed immediately." "Not till she has supped, I hope," replied her mother, but Jane had already fallen asleep on the bed, whilst begging them not to touch her lest they should burn her. The mother's dismay and grief may be well imagined, and she passed the whole of the night in watching and lamenting over her beloved child, whom she fancied dead. At seven o'clock next morning, Jane's rigour relaxed, and she opened her eyes, to the joy and relief of her mother. The next night the same rigidity and sleep occurred, but in the morning, the sleep did not pass off with the rigidity, and Jane, in a sleep-walking state, went mentally the round of her neighbours and acquaintances, amusing herself with examining themselves and their houses, &c., thus displaying to her startled family her wonderful clairvoyante power. My poor sister died a week after Jane's return to her friends, and I believe she *knew* of our sad loss *at the moment it occurred.*

At my request, Jane went to Brighton, to stay with my niece at that place, and whilst there, a remarkable proof of her prophetic power was given, for one day, my sister E. was told by Jane she had received a letter from an old friend in America, who had written to take a last leave of her, and the letter had been brought by one who said the writer was no more. E. told her she was sorry, but not the least surprised, for Jane had herself predicted this death in one of her mesmeric sleeps. On the 15th of January she came to London. February 8th, Jane went to see Mrs. M. and my nieces. One of them, who had been to Brighton the preceding day, told her of the murder of Mr. Griffith, a brewer residing there, and they read the account of the murder in her presence. Jane had often seen the unfortunate man. On her return home, she soon as usual fell asleep, and then related to me all the particulars she had heard of the murder, described the personal appearance of Mr. G., and told me that she knew the spot well, where the body was found, and the house to which they had removed it. She continued to dwell on the subject, shuddering with horror, and becoming every minute more and more disturbed, lamenting over him, his wife, and his miserable murderers. "Now," she said, "I must go over and look at the body," then the change which came over her face was such as no acting could have produced, and we sat gazing at her, as silent, and almost as awe-struck, as herself. At last, starting up, she said she could bear it no longer, she would go to bed and try to forget it, but when there, she continued as distressed as before, sobbing over the murdered and the murderers. "Miserable men," she cried, "you have escaped for the present, but what will your thoughts be when you lay your heads on your pillows?" Then she fancied herself laid beside the body, begging the pistols and knife might not be put near her, and with difficulty we kept her in bed. My conviction is, had Dr. A. been present to employ the influence he possesses, Jane could have been led by him to tell all that justice could require, but she would never have known a moment's peace afterwards. If some less tender-hearted clairvoyante were employed in such a discovery, would not the fact, known, make the hand of murder falter? At length she slept composedly, and became rigid.

February 9th, Jane awoke not in the least occupied about the murder, she fell ill, and I proposed to take her to Dr. Ashburner.

10th. My sister-in-law had come to stay with me for a short visit, she did not know of Jane's fits having returned, or of her sleeping; but as I knew we might soon expect her to have them again, and as we wanted more assistance I this morning told our trouble to Mrs. S., and asked if she felt strength and courage to assist us. She showed much alarm, and though very sorry, declared she could be of no use. Whilst we breakfasted, Jane fell asleep. Mrs. S. now grew very uneasy about the return of Jane's fits, and although I assured her that Jane had never deceived us, and had promised us one week's respite, and that she might safely remain till the evening after the next, she took her leave. I then prepared to send for my niece E., or for other assistance. Jane asked where Betsey was going, I told her without disguise. She then kindly and earnestly assured me I need not send for any one, that her fits would return no more. "A week ago," she said, "I could not see so far, but now you may be sure I shall have them no more, and so I have told the doctor." In addition to this most welcome news, she told me she should only sleep a few days longer, and after Monday Dr. A. would not be able to put her to sleep, and then she would only want a little country change to make her quite well and fit for work, though she should never recover the sight of her right eye.

11th. Jane's clairvoyant power was shown in a most remarkable way, for after returning from Dr. A's today she said, "I have travelled a great way to-day, I have been home twice to see how Betsey got on with the cooking, and as you were not at home, I went to the chapel and found you there, then I went to Miss S— at —— (mentioning a place one hundred and eighty miles away from us), and looked at the letter you wrote yesterday. You think I did not? Well there was not much in it, but this you said in it, 'it is indeed a great blessing;'" and this was quite true, I *had used those very words* in the letter I had written and sent to my friend one hundred and eighty miles off. After this there would be little interest in recording many other similar examples of her wonderful powers.

13th, 14th. Jane had no sleep at Dr. Ashburner's and complained much to me of the pain and horror she felt when he obliged her to look long at his fingers pointed at her eyes, said they seemed to grow so frightfully large and something came curling about their points, and she dared not look fixedly at his eyes because she had been reading "The Seeress of Prevorst," and feared to see the reflection of more than one pair of eyes in his. She had no natural sleep either of these nights and became very tired and nervous.

15th. Saw a white cat again and was much discomposed, but slept a little at Dr. Ashburner's.

16th, 17th, 18th. Nothing particular occurred.

19th. Jane went to her aunt's in the country.

21st. Received a letter from Jane, saying she had been much upset again by a white cat, &c.

27th. Jane having returned from the country went with me most unwillingly to Dr. Ashburner's, but when there slept well; we returned about three o'clock to my sister's, where when asleep she told me, nothing should induce her to go to the doctor's on the morrow; I remonstrated long with her and asked her, "Are you then resolved to be blind and mad? Are you quite indifferent to all the anxiety we feel for you?"

At length her wild spirit subsided, and bursting into tears it was long before I could soothe her. When calm she rested on the sofa, and asked me to promise not to tell her aunt how she should be to-morrow. I asked, how she would be, ill? she said "No." "Shall you have a fit or be delirious?" "No, you have not guessed one thing." My little niece then said, "Blind?" "Yes," but she could not tell when; or whether she should ever recover, and in less than five minutes she became as blind as she had been twice before, and as before, when she awoke to a sense of her misfortune, her sincere piety taught her to calm her despair and submit herself with patience to God's will.

28th. She had scarcely any sleep last night, her blindness continuing the same, in the morning and during the day. She slept frequently and asked whether it was yet night. Dr. Ashburner came when she was awake and, against her wish, put her to sleep.

March 1st. Jane still quite blind, and very meek and submissive. I walked with her to Dr. Ashburner's, when there, and asleep, she told him she should recover the sight of her right eye at twelve o'clock, and the other she would perhaps tell him about to-morrow. After her return home she slept long, but twice complained of falling from a house; at length she awoke and asked if she had had a fit, or a fall, for she felt dreadfully shaken—at half-past ten she went to bed and was soon in a mesmeric sleep from which she did not wake that night. When the clock of the neighbouring chapel struck twelve, my niece and I were near her, she lay perfectly quiet till a few minutes after, showing no sign of having regained her sight, but soon with a joyful smile, she cried, "Thank God!" We echoed her exclamation. Then she added, "your clock is five minutes faster than that of Hanover Square."

March 2. This day Jane rose at seven, she had been rigid all night, and ran into my room crying out with delight, "I can see, I can see!" I congratulated her and sent her back to bed.

3rd. On falling asleep this morning, Jane told us she had been vexed and much annoyed by a few hasty words respecting her which my sister E. had written whilst she was blind, that she could not have thought Miss S. would have said such a thing of her. These few words, expressive of strong vexation at her unwillingness to go to Dr. Ashburner's, and written without thinking of Jane's wonderful powers, I had burned almost before I had fully read them, congratulating myself in silence that the poor girl's clairvoyance was suspended during her blindness, but now she told us that she could see more than she did before her blinduess, and though whilst she was blind she could not read anything that was written or know much that was done, *all came to her after she recovered her sight.* Of this she gave me many proofs, telling the contents of my own and my sister's letters when they concerned her, and several things which had been done. At this time she felt a strange irritability, and had taken a great dislike to some ladies whom she daily saw at the doctor's, one of whom she said she should like to beat. She told us Dr. Ashburner wished her particularly to be at his house early on the morning of the 5th, because a gentleman would be there who did not rightly believe in the power of the crystals, and had made some misstatement on the subject in a lecture to H.R.H. Prince Albert. She said she never told lies, or deceived, and she could not bear to be suspected, therefore should this gentleman display any scepticism she should certainly fly at and shake him, and then run behind the doctor for protection, who perhaps with Dr. Elliotson would help her to beat him; but if the obnoxious lady was there she certainly would not go in, however much the doctor might wish it, for she could not be in the house with her. She slept her usual rigid sleep on this and the previous night (these remarks she made sleeping; but awake, her dislikes and desire to beat those she disliked remained the same).

4th. Having slept her usual rigid sleep from 11 to 7, Jane prepared to go to the doctor's, but before going, gave us some wonderful *proofs of the power of reading our thoughts.* I had arranged that, if able to keep awake, Jane should accompany me to church, but I went elsewhere; on arriving home, I found Jane had been brought there by the doctor, she having been much burned by a young lady whom Jane, by the power of her will, had thrown into sleep, and who, in that sleep, had clasped Jane's head in her hands. Jane appeared to suffer much pain from the burns, but felt no resentment towards the young lady, exclaiming at intervals "Oh, how pretty she was!" "Oh, what a sweet creature!"

March 5th. The doctor having arranged that the ladies obnoxious to Jane should not be there, she went to his house and I awaited her return with some anxiety; when she came she fell asleep, and told us Dr. F. had burned her head very much, but he did not intend it, that when she had fallen on the floor into the dead sleep, which the crystals always produced, she believed he had tried to raise her head to place a cushion under it, that he had put her to sleep with something in sealed papers which he held inside his hat. She thought after all that he did not believe, but she had not shaken him, for Dr. Ashburner would not let her, Dr. F. had a friend with him, and Dr. Elliotson was not there.

7th. This day Jane talked much of my late dear sister, saying, how much she regretted not being able to see her once more before her death, or her interment. "I have seen her since," said she, "I often go and look at her, but that is not the same. She asked me when we parted to come and look at her once more, but she told me I should never see her again alive." Then poor Jane, earnestly, as she often does, expressed her strong desire to die, to be with Mrs. M. and the blessed, for now she had no fear of dying since *she had seen how happy they were.*

8th. Her irritability has been again excited, and in her desire to possess a hazel stick, with which Dr. Ashburner had tantalized her, she had lost her temper, had shaken the doctor, and shown other signs of temper. This stick, or any hazel stick, I have since seen attract Jane irresistably, when inclined in the least, or pointed up, but when reversed or held point downwards, she has no desire to touch it. If Jane succeeded in grasping it, she dropped instantly down in sleep, but a touch with a crystal upon the stick made her hold relax. As she returned home in an omnibus, a gentleman unluckily had such a stick, and laid it across his knees, Jane instantly tried to seize it, and a struggle ensued, which ended in her being put out of the omnibus. She felt shame and sorrow in relating this.

9th. Jane suffered much in the head this morning. I walked with her to Dr. Ashburner's, and left her there, promising to return at half-past 7 o'clock, which I did, and found her more interested in endeavouring to get possession of a hazel stick, than in the impending blessing of restored sight. At five minutes before eight she looked at the clock on the chimney-piece and said, "Doctor, your clock is three minutes too fast, but I go by this time." At the moment the hand pointed to the hour she awoke, and putting her hand hastily to the eye, exclaimed, "Oh! what's the matter with my eye? I can see!" The expressions of wonder and gratitude were delightful to listen to, because they were truly heartfelt. Three gentlemen, Sir W. N., his son, and Mr. R. were witnesses of this cure.

10th. Dr. Ashburner called in the evening, he made marks on the floor with his hand, and these Jane could not pass without becoming insensible.

13th. Had no mesmeric sleep last night; has gone to the doctor.

14th. Slept only a little at the doctor's, and with much exertion on his part.

15th. Very ill, and occasionally delirious. Dr. Ashburner called, put her to sleep, and gave her mesmerised water to drink, which did her much good.

18th. A clairvoyante lady gave Jane a prescription, and assured her if she took it, and followed her advice, she would be cured, impiously adding that the Almighty himself could not have prescribed so well for her as she had done. The lady has ordered her to take wine and meat, both of which disagree with her, and since many months Dr. Ashburner has allowed her to drink nothing but milk and water. I felt very anxious, but was preparing to get the medicine when she fell asleep, and then told me she would take what the lady had prescribed, it would do her harm, but she would take it. I asked her if she knew what would be better for her, she said, "Yes, but I would not give it her." I said I would, for I felt confident that Dr. Ashburner would have more reliance in her opinion than upon any other person's. The remedy which I gave her was very simple, and I felt by no means hopeful of a good result, but in a few minutes she awoke, and asked me what had happened to make her feel so well. She did not sleep that night, but Dr. Ashburner was surprised to see her so well.

21st. Jane would scour her kitchen floor, though I begged her not. She came up irritable, and threatened to put my niece and me to sleep, I put my arms round my niece to shew I would protect her. Jane then said she would not hurt her, but me she would shake or put to sleep, and the doctor considered it would do me good. I gave my niece a piece of gold to protect herself. Jane confined her arms while she shook it out of her pocket, and then taking it up in the tongs ran out of the room and hid it, then returning, she fixed me in a corner of the room and tried for some time to make me sleep, telling me she knew I would not burn her, but I resisted the influence, and she at length desisted. Soon after, without awakening, she apologized for the liberty she had taken. But when later, Dr. Ashburner called, her wild mood returned, and displeased at something I said, she, in spite of his commands, to which she usually submits, carried my niece down and shut her in the coal-hole, and insisted on returning for me. Dr. Ashburner put on the appearance of great anger, and scolded her, she became delirious, and it was long before he could calm her. Jane has asked me and my niece to remember that she must be mesmerised on the 15th May, and the 8th June, or she would be seriously ill, and possibly lose her life, but we must speak to her on the subject when awake.

21st. Jane has again replied to my thought, which was a wish to know how I had lost some money. She said, "I will not tell you, I had it not, M. did not take it, but I never will say who did." Dr. Ashburner calling, I mentioned this, and he endeavoured, by the power of his will, to compel her to declare what she knew, but Jane, though much distressed, resisted his efforts. Dr. Ashburner did not speak whilst exercising his power. At length, at my entreaty, he desisted, for I could not bear to worry the poor girl, who is honesty itself. Jane laid the contents of her pockets on the table, and admitted she knew who the thief was, but Dr. Ashburner said we might tear her heart-strings out, and yet she would be silent. When he was gone Jane burst into tears, and with difficulty I calmed her.

23rd. I wanted to send part of a gold locket to a jeweller's in Regent Street, it was merely the inner ring with the glass, into which I had fitted a miniature picture. Jane offered to take it, I objected, as I feared the gold would stiffen her. She said she did not think it would, but she would try, I then put it into her hand without effect. But a gold ring made her rigid, and her hand closed so forcibly on it, I could only release it by forcing the knob of the iron poker into her hand. When the ring fell she recovered; but she took the locket very safely.

24th. Jane went to the doctor's, and slept in the kitchen. Just as we were going to bed, my sister arrived. Jane was overjoyed, but had scarcely spoken to her when she fell asleep, and, to my sister's amazement, began to count the gold and notes, which, for safety, had been put in the front of her dress. Jane then taking her purse, which lay on the table, asked why there was a shilling short, and said that four shillings and sixpence must have been given to the cab-driver, instead of three shillings and sixpence; "Indeed I am sure of it," she said, "for I see what money he has now in his pocket, and he had but one shilling and a few halfpence before." She told my sister she knew she was coming, for she had seen her drive from B—— to the terminus, with an elderly gentleman and lady. All this was quite true. Jane said she had often seen my sister at B—— Hall, the seat of Lord G., and noticed the visitors, &c. She described all the particulars of a ball given at D——n whilst my sister was visiting there.

28th. Jane returned, accompanied by Dr. Ashburner's housemaid. Major Buckley had accidentally cross-mesmerised her by making passes over the doctor's face in trying to act on another patient. I ought to have said Jane said, on the 26th, that M. Louis B. would dine with the doctor that day. Dr. Ashburner was much astonished at this, because he had no intention of inviting him; but nevertheless a particular conversation led to M. Louis B.'s dining there by invitation.

April 11. Jane went to the theatre with Dr. Ashburner's servants, slept at his house, and returned home on the afternoon of the 13th. She was delighted, and said how kind the doctor had been in paying for the whole party, and that Mrs. Ashburner, who had returned from the country that day, had kindly consented to dine earlier that the footman might be at liberty.

My sister tried to persuade Jane that it would be a meritorious act to try and read through one of the envelopes sealed by Mr. B., and gain the reward offered for doing so; she need not keep the money for herself, since she could not reconcile her conscience in doing so, but give it to the Mesmeric Hospital. But nothing can reconcile her to the idea of being called a clairvoyante, a term equivalent she thinks her parents would say to a witch. "No," she said, "if I cannot get money honestly by working, I will never get it by *them* hypocritical ways." For, awake, Jane will give no credit to anything she is told respecting her wonderful powers when sleeping.

Soon after this she fell asleep, and, in high good-humour, related much of what she had seen at the theatre, making amusing comments. But, in the incidents of one piece, she found so near a resemblance to the simple, but sorrowful history of her own short life, that we were moved to tears as she recounted them, and, in a low and touching voice, drew the comparison, sighing over her blighted prospects of happiness, which her high principle and firmness of character had made her renounce. She spoke, as she had often done lately, of her conviction that she should drown herself. We both entreated her to remember her duty to God, her parents, and herself, and not to commit so great a sin. She replied, If God allowed her to drown herself, it would not be a sin, because she should not be in her right mind if she did so. At present, she said she saw nothing else before her, but if her health improved, that might make a change in her views. I should mention that at the play, Jane was with difficulty prevented rushing on the stage to seize a hazel stick, which one of the actors carried.

Jane, asleep, has become very learned in the science of phrenology, on one occasion, she sat and examined my niece's and my head, from the further side of the room, and then began to scrutinise my sister's. I moved my head, because I sat between them. "Why do you move?" said Jane, "do you think I cannot look through you?" And she made her remarks on E's. head, really looking at it through mine. She tells us she can look through any opaque body, a thick wall is more transparent to her than glass to us, and night and darkness exist not for her when she is asleep. She now takes a quick survey of the heads of any strangers who may be brought in the way of her observation, and her opinion of them is at once formed and fixed.

On the morning of the 21st., I went with Jane to Dr. Ashburner's. Some ladies were there, and one of them showed surprise and doubt of Jane's power and animated sleep-waking. This lady asked a few trivial questions. Jane's answers were short but truthful, and she told her she was just such another as M. Louis B., and him she did not much like. I could easily perceive Jane gave her the most earnest and penetrating attention. "I should like to know if she has anything to say to me," remarked the lady. "No," said Jane, "but I may *think* what I like." Asleep she has no respect for mere rank.

Dr. Ashburner offered me a piece of blue glass to take

home, but Jane assured me if I took it she would not return home, for if I were to put her to sleep by its influence, she knew she would not wake again. I therefore left the glass, though anxious to have it, for I knew Dr. Ashburner made her sleep very deeply, by allowing rays of light to fall on her eyes through such stained glass, and that morning she had been powerfully influenced by the blue glass of a stained window. Jane returned in the evening, in high spirits, and told us she had seen a great many persons at the doctor's. She arranged our tea, and got her work; inquiring after my sister, I told her she was out, and probably would call for her at the doctor's. When she knocked, Jane ran to let her in. My sister sat down, and began to tell how she had called at Dr. Ashburner's, and he wished her to bring the glass with her. Jane contemplated her wildly, asking if she had it with her. "Yes." "I felt you had," she said, and fell on the floor in the most alarming state of rigid sleep, her mouth fast closed, her head drawn back, and her hands clenched on her chest. She groaned, shrieked, as she had never done before, and whilst her voice strove for a passage through her clenched teeth, her cheeks, brow, and neck were purple, and her swollen veins seemed bursting, while every breath she drew made an alarming gurgling in her throat. My sister drove with all haste to Dr. Ashburner's. The doctor came, and under his passes her arms gradually relaxed, the other terrible symptoms gradually disappeared, and she burst into tears. When a little recovered, she refused to be mesmerised to sleep, said she had now two influences on her, my sister's through the glass, and the doctor's, and that she felt she should have fits; contrary to her wish, Dr. Ashburner mesmerised her to sleep, and *she had* two fits, but afterwards slept till morning.

22nd. She awoke much better than we could have hoped, though much worse than she had been, and, by the doctor's persevering attention, she gradually recovered from the effects of the blue glass; but for several nights after this, she scarcely dared to attempt to sleep, because of a feeling of suffocation which could only be eased by sitting up in bed. Jane had predicted that on the 22nd. the blue glass would lose its power over her, and, on entering Dr. Ashburner's hall, this was found to be the case, for the glass of the window had no longer any effect.

25th. Jane has suffered from pain in the ancle, this is entirely through sympathy with Dr. Ashburner, who has hurt his.

May 1st. To-day, and for many days, the diary is prætermitted, because no new features have occurred in Jane's case. One day, however, towards the end of the month, my sister complained of drowsiness, and asked Jane what she could do to get rid of the uncomfortable feeling. Jane laughed, and said E. did not know always what happened to her, for the other night, when there was a storm, and E. sat working, and seemed disinclined to go to bed, Jane, thinking if Miss E's. candle was out she herself could possibly sleep, had willed her to go to bed and sleep, and had laughed to herself at seeing how quickly my sister was acted on. This was true, and I remember that it appeared to me my sister could not go to bed quickly enough, though, some minutes before going she had said she really must finish her work first.

May 4th. Jane went early to the doctor's, at about half-past seven I felt as if compelled to lie down to sleep, protesting Jane must be "willing" me to do so. I slept soundly for an hour. On her return in the evening, Jane told me that Dr. Ashburner had been absent all day, and that whilst sitting in the kitchen quietly with the cook, she put herself to sleep, had fixed her thoughts earnestly on me, and had intently willed I should sleep. This she told me whilst sleeping, and soon went to bed begging I would go up stairs, because such a little thing as my niece could not keep the spectres (by whom she has been haunted of late,) away. I sat on the side of the bed, and she held my dress fast, pretending she would mesmerise me, suddenly she felt in my pocket a note, on which I had a memorandum about her. "I shall go and look in the cabinet," said Jane, still holding me fast. She then mentally examined the cabinet, and told me the book* had slipped between the wall and the cabinet, and starting up, grasping me tightly, almost screamed "You have been again writing to the doctor about me." She was easily diverted by my sister drawing her attention to another subject, and soon after her sleep deepened and we left her.

6th. Jane went in good time to the doctor's, and after sleeping some time, recovered the sight of her right eye, but on her return she seemed low and flurried, and being disturbed by some conversation which took place about my late sister, had a long fit of delirium, from which she recovered with alas, the right eye dark again. I asked her if brushing and combing her hair would prevent her waking from the sleep she then had,—she said "yes," and I therefore did so for an hour to her great comfort† and advantage, for she fell into a deeper sleep, which lasted till seven the next morning.

8th. Jane has been actively busy all day washing, and in high spirits, but at noon fell asleep for an hour. She said she would have some roast mutton before she woke, and so lose no time; on awaking, she had resumed her work, when the doctor called and forbade her continuing, she obeyed, but vexedly. He put her to sleep soon, and after he had gone, my sister and niece, who admire and esteem Jane much, said, "don't disturb her, her thoughts are far, far away!" but certain uneasy sensations about my head made me suspect her thoughts were near and

* I had lost an account book during the day.

† Jane has often found great relief to her sufferings from having her hair brushed and combed. We feared doing so at first, thinking it would cross mesmerise her, but she says we take her influence by first handling her hair only, and at length may even touch her head, without danger, and the sleep thus produced is delightful.

fixed on me, and that the mischievous girl was "willing" me, my head grew more uneasy, and my face hotter every moment, and at length without any reason, I burst into a laugh. Thus she had made me obey her will, and I was conscious she had ceased trying her power on me. She afterwards acknowledged she had played this trick, and said, "You had better put that down." Jane is very anxious about the approaching 15th May, and also about her intended visit to her family at Whitsuntide, she fears she shall be ill, and that her mother will oppose her return; she says, should she be ill in the country, and obliged to remain there, and any of her friends attempt to mesmerise her, they would only do it once, for they would handle her head and drive her delirious, and then she would go and drown herself. I have begged her, if possible, to tell me why water is so attractive to her, (for it is so now, and has been so for some time). She says, "Still water has no power over me, it is only running water," but she either does not know the cause or will not tell it.

Since writing the preceding papers, circumstances have prevented my making the usual notes respecting Jane, but certain it is that she continued to display at will always more and more wonderful phenomena. The evening before the 15th May, though she appeared in no respect less well than usual; she wept and lamented in sleep over the approaching fearful morrow, entreating us to keep the house quiet, and warning and cautioning us. She went to bed without awaking, and said she should sleep comfortably till five o'clock, when she should wake very ill. Soon after that hour I awoke, and on speaking to her found she had predicted truly, she had entirely lost all power of moving herself, except in one foot, and when eight o'clock struck, was quite blind. In the faintest whisper she asked to have her head turned on the pillow, or her limbs moved occasionally, and thus with the patience and resignation which we had always found so touchingly in her, under the heaviest afflictions, she lay till two o'clock, when Dr. Ashburner called, and having put her to sleep he breathed through his closed hands on her eye, she, faintly smiling, said, "Do that again," and on his repeating the experiment she said, "Now I can see with one eye," but the other remained dark. Dr. Ashburner made passes over her, and in a short time she raised her arms over her head, turned in her bed, and then became quite rigid, in which state she remained for some hours, from this fixed state she recovered, sat up in bed, and the same evening walked about the house.

On the following day, Jane went as usual to Dr. Ashburner's, but she felt a weakness in her right side.

On the 29th, Jane went to pay a short visit to her friends in the country, but before going she promised to resist the influence of the water and the hazel trees, also to return at latest on the 7th June, she had gained considerable strength, and could pass the Serpentine without much trouble. On the evening of the 7th she had not returned at ten o'clock, and we had nearly given her up, and were in great trouble on her account, when she suddenly appeared, she had stopped on her road home to see Dr. Ashburner's servants. Her own family had placed every impediment in the way of her leaving them, but she had not been able to break a promise given, she knew not why awake, but she said she had run away from home, and persuaded a friend of the family to drive her to the station. She went cheerfully to bed, but when asleep, told us she should be dreadfully ill at five o'clock on the following morning; and she awoke at five o'clock, but her reason slept, never had we seen her in so fearful a state of delirium, or her mind disturbed with such fearful ideas. But again a happy state was wrought, and her reason restored by Dr. Ashburner, who having (with that unwearied kindness for which he is so distinguished) kindly called in, put her into a mesmeric sleep, from which she did not awake till 11 o'clock the following morning, at Dr. Ashburner's house. I had walked with her there at her own request, whilst she still slept, and when she awoke in his library she could not comprehend that she had been home, and was very unwilling to be left there. Since that time, after many alarming fluctuations of the state of her health, at length a very happy change begins to appear; she is much stronger, is no longer to be influenced either by the crystals or hazel wands, and on the 15th inst., at three o'clock, p.m., the long lost sight of her right eye again returned, she suffered most excruciating pains in the head at the time, but they subsided, and she continues to this day blest with her eyesight, and daily increasing strength.

Thus I have endeavoured to give a faint outline of the part which has passed under my own observation of the wonders of this extraordinary girl; and I have written it in compliance with Dr. Ashburner's wish, and in gratitude for his kind attention to the dear sister whom he could not save, and to this hapless victim of ignorant presumption whose cure appears at last to be happily effected. Many events of a much more astounding nature than any I have related I have withheld, partly because Jane, always averse to publicity by writing, has asked me to let them pass unnoticed, and partly because one does not like to be thought mad.

How I have sometimes wondered at the extreme prejudice displayed by those who, in their determined scepticism on this one subject, have laboured to prove to my sister and myself that we are the dupes of an imposture carried on for nearly eleven months, without the possibility of the assistance of a confederate, by a simple and inexperienced young person. My opinion is, they refuse to believe in these powers, which they call "useless and superhuman," simply because the Almighty has not deigned to inform them expressly why he bestowed such powers. Jane, who always proudly refused to gain money by clairvoyance, now that her health and eyesight appear so happily restored, looks forward with honest pleasure to being soon able to resume the duties of a servant, and we participate in that feeling, for no servant ever served us as well as she. M.A.S.

CHAPTER XIV.

PHENOMENA OF ANIMAL MAGNETISM.—SPIRITUALISM.—CLAIRVOYANCE.—PREVISION.—HUMAN WILL, &c.

We have arrived at a serious question. What is it? Physiologists can make nothing of it. Who are the philosophers who shall pretend to answer it? The question is this: Can these phenomena be accounted for on the principles adopted by our modern sceptics—those of positive materialism? I was a modern sceptic myself. I admired Hume; I thought the wit of Voltaire transcendent; I had no idea that Gibbon deserved one tenth of the abuse he got. In those days I should have ridiculed the pretensions of Andrew Baxter in setting himself up to controvert doctrines which were so fashionable in his day, and which have since swayed the minds of those who call themselves psychological philosophers. We live in more advanced times. I have already said, in essays I gave to my friend, Mr. Morrell, that we were children of light, and that being so, we were bound to find our future ways illuminated. The question does not relate to physiology alone. We have to consider the subject as belonging to our studies in Animal Magnetism. Now, we think that men ought to consider well before they repudiate a subject which may turn out to be very holy. Those who obstinately adhere to their religious prejudices are to be pitied. Those who act on the principle of Christian charity are to be respected. We have to show that our subject is too vast to allow us to fear the foolish opposition which may be offered us by any sect. The idea of necromancy is upon a par with the sacrilege of taking dead bodies out of their coffins for purposes of science. Necromancy is just body-snatching. The term is silly enough, but the people who make use of it are more silly. What possible connection can there be in the idea of consulting the witch of Endor, and in reading the Epistles of St. Paul? If the one be very wicked, the other is acknowledged to be very praiseworthy. Whoever got any harm from the Epistles? It is possible they might not understand them, and yet he who runs may read them clearly enough. You may be assured that John the Evangelist was no fool, and he was one of the writers upon whom we lean when we wish to obtain evidence of the holy nature of Animal Magnetism. We are not now inclined to go into polemics, but we are determined to show that Animal Magnetism received a sanction from John, who was himself a healing medium. He says, in the Epistle to the Hebrews, that he had laid hands on some of the disciples, who were inclined to become oblivious. We are not sure that these disciples were not possessed, for why should he mesmerise them, and bring them out of their troubles, if that operation of laying on of hands was not intended to convey into them an influence quite incompatible with unholy possession. This subject has occupied the minds of several philosophic persons. The learned Dr. Hugh Farmer wrote a book full of sound reasoning on the subject, and though he did not comprehend the nature of the influence he spoke of, or rather wrote about, it is evident that no other force but that of human magnetism could be intended. In those days, people wisely assumed that the Almighty chose his own methods of accomplishing his ends. He was able to make men believe that unseen spirits could, and did, visit them; and there was no question about the character of the spirits. The Epistles point out the modes of ascertaining their characters. He, however, who will not believe in the ministration of angels, has no occasion to rely on the teaching of the Epistles. We do not contend for the Book of

Revelation, St. John would be no better authority there than in the Epistle to the Hebrews. He is put on the shelf, as an authority, the moment he tries to inculcate the doctrine of a world of spirits, and the possibility of their having any influence on the denizens of this world. How shocking all this would read to one who really believed what he professes ignorantly to believe. Not one man in a hundred believes that which he professes to believe. This is a sad state of things. For we can hardly tell who is a real Christian. Profession does not qualify a man. He may profess what he likes, but unless he can refute Dr. Farmer's book, he has no right to believe that he is *au fait* at the doctrines of Christianity.

We are not to worry people into a belief of what is true. But we may take the liberty of rebuking those who profess much knowledge, and yet do not know the very elements of the Christian faith. We are sure that a large majority of persons educated in the doctrines of the Church do not believe in a world of spirits. They may have some vague idea of the existence of certain angels, but they have no idea that the Apostles urged them to communicate with the world of unseen spirits. They think a man must be quite heterodox who believes it worth his while to withstand ignorant public opinion, and assert his faith in that which the Apostles undeniably taught. For myself, I am little troubled about the opinions the world may entertain of me. I know that as long as I make the Gospels and Epistles my rule of life, I need care very little for the opinions of worldly men. I have done with this matter, for I shall convince no one. Conviction does not arise from facts alone, it arises from the influence exerted upon the mind through the agency of holy spirits. This may sound strange to those who do not read their New Testament aright, but it is nevertheless the fact. All truths are holy, and they can reach us only through holy influences. How remarkable a view is this of all truths. Is it intended to insinuate that no one can tell the truth without a holy influence? Quite so. He who tells the truth is not to be despised, however unimportant that truth may be considered. It may be the beginning of a habit of telling the truth. If so, it is the beginning of a holy life. For habit, once acquired, is not easy to shake off. But how as to the Holy Spirit prompting the telling of that truth? Those who do not choose to believe in any forces, except those that are believed in by the listeners to Mr. Faraday, will not believe me when I say that the atmosphere is full of forces. We know that light is a force, heat is a force, magnetism is a force. These, by the way, are pretty nearly all Mr. Faraday's forces, and they are so precious to him that he would make a grand conservation of force. How he would accomplish it, his audience may be better able to tell than I can. This I know, that our forces are many of them quite unseen, and yet, under the title of motives, they influence the actions of men. Who that has read anything of the psychological literature of the last hundred and fifty years, has been so blind or deaf as not to have seen books or heard talk of the doctrine of Free Will.

No man can be ignorant of the many battles that have been fought on free will and necessity. One set of men contending for the absolute necessity of a man's thinking according to surrounding circumstances; another fighting for absolute free will, irrespective of all motives. Yet they both fought foolishly; for there is no necessity for doing what you do not like, and there is no absolute free will, for all men acquire habits. A bad habit straightens a man's will. He cannot quit the groove in which his bad habit has lodged him; but if he have acquired those good habits which lead him to a communion, with holy thoughts, he has ample free will, since he wills always to do right. How is this clearly made out? It is not necessary to define free will further than to say that it consists in freedom of action. A man is free to act wisely, but everybody has an interest in wishing to constrain a man who is about to act foolishly. He may not be constrained, but nevertheless, no man can act foolishly, and be at the same time a wise man. A fool is sure to be watched and mistrusted by those who know him, and is not this a bondage. We often speak of bondage as if it were only bodily, but mental bondage is far more melancholy. It is a bondage to vicious habits, and no man can have a free will who has a single vice. It weighs down his mind, and prevents the free current of thought. He can think only in one way, and that is in the direction of his vice; he is sold to an evil spirit, who keeps his thoughts intent on that vice. In time, he acquires another, and that other familiarises him with a third; he is at last an object of great pity, for he has lost all control over himself; he has taken to habits he cannot overcome, and is the victim of evil spirits. It is the language that is now made use of, it is not the idea that shocks my readers. If one could consent to leave out the word spirit, my readers would say, "we heard he was mad, but the man speaks common sense enough." Suppose the man is able to prove that all this common sense is no common sense at all, without the doctrine of a world of spirits! Would you guarantee that he talks nonsense, by good heavy bail? Not you! You know too well this is not likely; you must try if the nonsense I shall write can be read with patience; you may be assured that your patience will not be much tried. It is proposed to shew you, in the first place, that the task we have been engaged in, however dry some parts may have been, is not so dull now. Our enquiry was how far Animal Magnetism was allowable? We hear you observe it was on the doctrine of a world of spirits. It was indeed truly, as you say; but we must be allowed to begin at the beginning. When we have shewn that Animal Magnetism, practised as we have shewn you how to practise it, is the source of all health, we shall proceed to shew that it is so, because no health can reach us, except through the agency of our lungs. We breathe health, and the breath of life means that we inhale God's fostering care. We are not sorry to be obliged to bring this idea before you, because it is something unusual for you to reflect upon. It is not a new truth, but it is a truth. You must be content to take it for one. We are anxious to show how it is connected with Animal Magnetism; we know that the breath of life is compounded of oxygen, and a force that has some intimate relation to light. That force is called actinism, and is an electric force. It is hardly necessary to tell my accomplished medical readers that it is the agent in the chemical changes which proceed in the photographic camera. There are many reasons for believing that photography has been a God send, and yet few think of thanking God for the revelation. His bounties are all revelations. Who could have thought of such a profane idea? This is the mistake many religious persons fall into. Why profane? All God's bounties are revealed to us. But you must confine the word to one revelation. Then why not to the book of Revelation? People are led away by a word. Every act of God is a revelation to man. It strikes either his frame or his intellect; it is either a blessing or a curse, as it happens to be received. No revelation of God's will can be so holy as the balm of comfort he sent through Christ! But we are not to conclude that His bounties ceased at the Resurrection of our Saviour. We are creatures living daily upon his bounties, and each of these is revealed to us in a different way.

Are we prepared now to understand that Animal Magnetism is a revelation of God's bounty? That it is owing to it we receive communications from the world of spirits. We are past all recovery when we admit this one great fact, and it is quite impossible to withstand the evidence in its favour. We proceed to shew what this evidence is; we are content to believe in the influence exerted by one person upon another; we think we know how this proceeds; we believe in a magnetic fluid or force; we are bound to say that it is quite likely; but we do not reflect that a magnetic fluid or force from a human being is quite a different affair from the fluid or force from a steel magnet; oh! is it so really? We were not prepared for that. But we went over the grounds for believing that every form of matter was moulded into its own peculiar shape by a force destined to control it, and to make it obey its own law. So when we arrive at human magnetism, we find reason to believe it to be a higher description of force than any belonging to minerals or vegetables, or even to animals lower than man himself. We are now prepared for an acknowledgment that human magnetism is the highest description of magnetic force. But, in truth, all forces obey God's will; not one emanates from any matter without being permitted by a law. That law is the rule by which that force is guided. Animal, or let us now speak of Human Magnetism, is controlled by a law which allows man to dream. When he dreams, his soul quits his body and wanders on its travels. It may be said this is very wild; it cannot possibly be proved. It was proved more than a hundred years ago by a Scotch tutor, whose name was Andrew Baxter. He was a man of no common mind; his book is a powerfully reasoned one. We do not now propose to go into all his arguments; but we may mention one fact developed since his time, which is an irrefragable proof of our assertion. It is that in sleep we are in death, and if the soul did not return, we should never come to life again. This is simply an assertion. How can it be proved? We can say, that no one whose soul had ever left his body was found to come to life again. But that is not enough. Then we say, that the man who is content to believe in a future state of existence has no other proof of the fact than this one of the soul leaving the body; but we must have proof of this too. You will not be content with all the proofs I offer. I will try another. Christ said, "We are children of Light." Did he say this without a meaning, or was he in earnest? He must have meant that we were emanations from light, and this is the truth. Having brought you thus far, we are not inclined to let you go. The time is precious that is devoted to these subjects. We are obliged to condense our arguments. Light emanates from the sun. The sun is the great source of light, and of magnetism as well. We are aware that for all purposes of rest we sleep in the dark; but we had light all day. Have we not inhaled that light? Has it not impinged upon our eyes, upon our skin, and penetrated through all the pores of our bodies? Is there any limit to the light finding its way into our systems? We believe that each ray carries with it a portion of God's will, for he lives everywhere, and his omnipresence necessitates the acquisition by the light of a portion of his law, in order that it may regulate the functions of the magneto-electric currents of the body. We find, then, reason to believe that light is an universal agent, and not a simple agent for illuminating all it touches, but one for distributing ideas from the brain, under the direction of the will. What is the will? The observations I made some years ago led me to the conclusion that the will was a motive force. Those who are curious to see the facts on this subject may be referred to certain papers, published under the direction of Dr. Elliotson, in No. 19, Vol. 5, of the Zoist; October, 1847, and to a note in the first part, page 29 to 32 of the Baron von Reichenbach's Researches, Baillière, 1851.

It is not necessary to detail minutely the various kinds of spiritual manifestations, as proofs of the gift that has been vouchsafed to us of communicating with the spirit world. The admirable book by T. Brevior on "The Two Worlds" may be studied with advantage on this part of our theme. Mrs. de Morgan's book, "From Matter to Spirit," is also a valuable collection of facts, and deserves deep attention. Mr. Wilkinson's book on "Spirit Drawings" introduces us to another kind of Manifestation.

We may proceed to recount some of the phenomena which have been observed at circles assembled round tables, at which manifestations of the presence of unseen intelligences were undeniable. This is a subject which will admit of no absurd cavils, and conclusions of scientific *savants*.

From seven to nine persons of sensitive constitutions being assembled on seats round a table, with their hands placed gently on its top, at the edges; in the course of half an hour, sometimes longer, sounds resembling the ticking of a clock may be heard on the surface of the table. In most cases, if an alphabet be now taken up by one of the circle, and the letters called out by that person slowly, the sounds will cease until the intelligent agent makes a sign by the small sound on the table. The letter which has thus been arrived at by a clear indication is written down. The alphabet is recommenced, and, by a similar process, an intelligible word is spelled out. Long communications have thus been made to relatives and friends of persons who have departed this world. According to my experience, most of these communications have been of cheerful and happy import, some have contained advice exceedingly cogent.

All the considerations in which we have hitherto indulged, respecting the kinds of spiritual manifestations, form but a part of the wide range of our subject. In examining somnambulism apart from the ordinary phenomena of sleep and wakefulness, we cannot but be struck with the subject of prevision. This I have witnessed so frequently during the last quarter of a century, as to leave no doubt in my mind that it is a high spiritual fact.

CHAPTER XV.

PHENOMENA OF ANIMAL MAGNETISM; SOMNAMBULISM, PREVISION AND CLAIRVOYANCE.

It is not to be denied, that facts in abundance have been adduced to warrant the students of Animal Magnetism to claim for the science they have been investigating a rank of no trifling importance. Man, philosophically considered, offers to the deep thinker a vast range for observation. The anatomist has disclosed to us extensive ideas on the structure of his frame. The physiologist, aided by the labours of the histologist and of the chemist, has not been remiss in storing up numerous facts, which have elucidated the obscurities that long defied our efforts to advance the boundaries of medical knowledge. The electrician has done his best to aid the development of our knowledge on the physics of physiology. But, except here and there, he has not very widely scanned the horizon of his ken. Faraday and Matteucci have given us the most important facts we possess in animal electricity, the facts relating to the direction of the currents from the brain in the gymnotus, and in the torpedo, to the caudal extremity. Some confusion was present in the mind of the Italian philosopher when he made his experiments on the nervous currents in the frog, for he appears to have mistaken an inverse current for what he calls "the current proper of the frog." It is a curious fact that stolid prejudices should so completely have obfuscated the intellectual powers of these men as to oblige them to ignore the multitude of facts connected with the higher classes of magnetism and electricity, those belonging to the organic or structural developments of matter. It is still more strange that a man like Mr. Faraday should be in such haste to declare himself in opposition to a subject which he so superficially looked into, and committed himself upon, without due examination. Of course, I allude to that portion of Magnetism which has been called table-turning, on the forces of which he has so egregiously blundered.

Many considerations stimulated me to proceed with the publication of facts I have witnessed on this subject, the most important of which rest on their close connection with Animal Magnetism, and which must lead us incontrovertibly to the spiritual agencies or forces, correlative to those involved in the agencies which obtain the name of physical forces. One may hint at the possibility of all forces being properly named physical, or the forces, innumerable though they so appear, as of universal nature. God is not to be limited, except by his own power, and the largest-minded naturalist who may attempt to classify his forces, must yield himself to the conditions required by the series of Mr. Babbage's wondrous calculating machine.

Our readers will apprehend the respect we feel for the science of mathematics. Accustomed in our earlier years to the mental exercise of geometry and algebra, we have never divested ourselves of the correlations of physics and mathematics. Now, called upon to investigate the laws regulating the psychology of Animal Magnetism, we find ourselves most at a loss to connect the abstruse phenomena of clairvoyance and somnambulism with mathematical science. That the connection will be established, there can be no doubt. How this will come to pass we need not be anxious to determine. Who could have foreseen the great discoveries of Gall and Spurzheim? The works of both these philosophers are full of important facts, and of observations rich in benefits to humanity. Who could have anticipated the glorious discovery of Mr. Rutter's Magnetoscope? Who could have planned the idea that a zealous, poverty-stricken physician should have emerged from obscnrity, to link the vast discoveries of Gall, so full of truth as they are, so closely allied with Animal Magnetism, to the grand science of mathematics? Who could have thought of these things beforehand? No one but a being gifted with the prophetic spirit of prevision. But is there in nature any human being gifted with such a faculty?

Who, of late years, has not read of the prophecy of Cazotte? It appeared in print in a well-known work, in La Harpe's "Œuvres Choisies et Posthumes," 4 vols., in 8vo., Paris, 1806; Tome I, p. 60. This prediction is taken from page 65 of Deleuze's "Mémoire sur la Faculté de Prévision," which has been translated for me by the kindness of my friend, Monsieur Adolphe Didier, the celebrated medical clairvoyant. It is as follows:

"It seems to me as though it was only yesterday; it was, however, in the beginning of the year 1788. We were dining with a brother academician, who was a nobleman of distinction, and a man of superior understanding. The company was numerous, and of all sorts: courtiers, lawyers, literary men, academicians, &c., &c. We had, as was then the custom, had a sumptuous repast. At the dessert, the wines of Malvoisie and Constance went round, and raised the general merriment to that pitch which did not, in those times, always keep within the bounds of discretion. Society, however, was then in that state when anything that caused a laugh might be said. Chamfort had read to us some of his tales, full of impiety and libertinism, and all the aristocratic ladies present had heard him without being under the necessity of making use of their fans, which circumstance gave rise to endless jokes. A passage from Voltaire's 'Pucelle' was recited by one; by another some philosophic verses from Diderot.

"The conversation then became more serious. Every one was loud in his admiration of the revolution effected by Voltaire, and admitted that he was entitled to the highest pinnacle of glory. He has given, they exclaimed, a character to his age, and has caused his works to be read as well in the ante-chamber as in the drawing-room. One of the guests then informed us, convulsed with laughter, that his hair-dresser had said to him that morning, while in the act of powdering him, 'You see, Sir, though I am only a very common sort of fellow, I have not any more religion than any one else.' They arrived at the conclusion that the revolution would be soon accomplished, and that it was high time superstition and fanaticism should give way to philosophy; from this they were led to calculate the probable time of the event happening, and which of the present company would live to see the age of reason. The older ones amongst us regretted that they could not flatter themselves with such hopes, whilst the young men, on the contrary, rejoiced that they might safely do so. The Academy was then highly complimented for having prepared the great work, for having been the chief place, the centre, the grand mover in liberty of thought.

"There was only one amongst us who did not seem to take any part in our joyous conversation, he even allowed, in a very quiet manner, a few jokes to escape from his lips on our wondrous enthusiasm. This was Cazotte, an amiable and original character, but who was unfortunately infatuated with the reveries of the sect called the *Illuminati*. He at length began to speak, and in a very serious tone said,

"'Gentlemen, be satisfied, for you shall all see this great, this sublime revolution, which you desire so much. You know that I am a prophet; yes, I tell you again you shall see it.'

"They answered him with the well-known chorus: 'Faut pas être grand sorcier pour ça.'

"'So be it,' continued Cazotte, 'but perhaps it may be necessary to be something more than a conjuror for that which I have yet to tell you. Do you know what will happen at this revolution, what will happen to all of you who are here? what will be its immediate result, its undoubted effect, its solemn, its serious consequence?'

"'Ah, let us hear what that is,' said Condorcet, with his sullen air and foolish laugh, 'a philosopher is not sorry to meet with a prophet.'

"'You, Monsieur de Condorcet, you will die on the floor of a dungeon; you will, in order to escape the hands of the executioner, die of poison administered by your own hand, of poison which the happiness of those times will compel you always to carry about you.'

"At first this excited great astonishment, but on our recollecting that our dear friend Cazotte was accustomed to dream while wide awake, our laughter increased to a roar.

"'Monsieur Cazotte,' continued Condorcet, 'this story is not so pleasant a one as your '*Diable Amoureux*.'* But what pray has put this dungeon, this poison, and this execution into your head? What, in the name of common sense, has all this to do with philosophy and the age of reason?'

"'It will be exactly as I tell you, it will be in the name of philosophy, of humanity, of liberty; it will be during the reign of reason that you will end your days. And it may be well styled the reign of reason, for reason will then have temples dedicated to her—indeed, there will be then no other temples throughout all France.'

"'Truly,' said Chamfort, with a sarcastic laugh, 'you will not be one of the priests of those times.'

"'I hope so; but you, Monsieur Chamfort, you will be one of them, and worthy of the office will you be, you will gash your veins with a razor; you will not, however, die from your wounds until several months

* The name of a Novel by Cazotte.

afterwards.' We looked at one another, and continued laughing. 'You, Monsieur Vicq d'Azir, you will not open your veins yourself, but in order to be sure of the result, you will, while suffering from a fit of the gout, cause them to be opened six times on the same day, and you will die in the night. As for you, Monsieur Bailly, the scaffold; you, Monsieur Malherbes, the scaffold.'

"'Ah, heavens be praised,' said Koucher, 'he bears ill-will only to the members of the Academy, he has just executed some of them in a most cruel manner, but as for me, thank God!—'

"'You too,' interrupted Cazotte, 'you will die on the scaffold.'

"'Oh, he has laid a wager,' we all exclaimed, 'he has sworn to exterminate every one of us.'

"'No, it is not I who have sworn it.'

"'But we shall then be subdued by the Turks and the Tartars?'

"'No, it will be as I have told you, you will be then governed solely by philosophy, solely by reason. Those who are to treat you thus will all be philosophers, will have continually on their lips the same sentences as you have been uttering for this hour past; will repeat, will recite, like you, Diderot and the 'Pucelle.''

"The company whispered to one another, 'You see he is deranged,' for his look was still grave.

"'Don't you perceive,' said some one else, 'that he is only joking? and you know that he always seasons his jokes with a touch of the wonderful.'

"'That may be,' answered Chamfort, 'but his wonderful is not now very pleasant, it relates to the gibbet. And when is all this to happen, Master Cazotte?'

"'Six years will not have flown by before all I have just said will be accomplished.'

"'Behold many miracles' (and it was now I who spoke) 'you have not put me down for anything.'

"'You will be at least as extraordinary a miracle as any of the others,' answered our prophet, 'for you will be then a Christian.' Loud exclamations followed this assertion.

"'Ah,' cried Chamfort, 'I am no longer afraid, if we are not to perish until La Harpe becomes a Christian, we may look upon ourselves as immortal.'

"'Whatever may happen,' then said the Duchesse de Grammont, 'it is very fortunate for us women that we have nothing to do with revolution. When I say that we have nothing to do with it, I do not mean by that that we do not meddle with it somewhat; but it is not usual to make us in any way responsible, and then our sex—'

"'Your sex, Madame, will not this time protect you, and it will be in vain that you will not interfere, you will be treated like the men, there will not be any difference made betwixt you.'

"'But what is that you are telling us, Monsieur Cazotte? You are surely preaching the end of the world?'

"'I cannot say; but what I know is this, you, Madame, you will be led to the scaffold, with many other grand ladies, in the executioner's cart, and with your hands tied behind your back.'

"'Ah! I hope that in that case I shall have at least a mourning carriage.'

"'No, Madam; and still grander ladies than you will be carried, like you, in the cart, and having, like you, their hands tied behind their backs.'

"'Grander ladies! What, the princesses of the blood royal?'

"'Grander ladies still.'

"The assembly appeared much shocked, and the sensation appeared to be communicated to the master of the house. It now seemed to us that the joke was being carried rather too far. The Duchesse de Grammont, in order to disperse the gathering gloom, made no allusion to this answer, but said, in the lightest possible tone of voice, 'You will see that he will not allow me even a father-confessor.'

" 'No, Madam, you will not have one, nor you, nor any one else. The last sufferer who will be allowed a father-confessor will be—'

" He paused for a moment.

" 'Well then, who is to be the happy mortal that will be allowed this privilege?'

" 'It is the only privilege that he will then retain, he will be the King of France.'

" Upon hearing this, the master of the house rose suddenly, and so did we all. He went over to Monsieur Cazotte, and said to him, in a deep and solemn tone,

" 'My dear Monsieur Cazotte, this gloomy pleasantry of yours has lasted long enough, you carry it too far, even to the pitch of compromising the company in which you are.'

" Cazotte made no answer, and was preparing to withdraw, when the Duchess de Grammont, who was still anxious to banish our seriousness, and restore our previous good-humour, said,

" 'Master Prophet, you, who are telling us our fortunes, say nothing about your own.'

" He remained for some time silent, with his eyes cast down.

" 'Madam, have you read in Josephus the Siege of Jerusalem?'

" 'Oh, to be sure I have. Who has not read that? But I will imagine that I have not read it.'

" 'Well, Madam, during the siege, a man went seven days in succession round the ramparts, in the sight of the besiegers, crying incessantly, in an ominous voice, 'Woe to Jerusalem!' and on the seventh day he cried, 'Woe to Jerusalem, and woe to me!' and at that instant an enormous stone, which was cast by one of the machines of the enemy, struck him, and crushed him on the spot.'

" Cazotte having given this answer, made his bow, and withdrew.

" Another prediction which Cazotte made, which appeared at least fully as extraordinary as the above, was that which Cazotte made on entering his house the day his daughter had succeeded in snatching him from the monsters who were dragging him to the scaffold. Instead of participating in the joy of his family, as they gathered around him, he announced to them that in three days he would be again arrested, and that he should then be obliged to submit to his fate. As he foretold, so it came to pass, he perished on the 25th of September, 1792, at the age of seventy-two years."

Called upon therefore to consider the whole of our subject philosophically, can we hesitate when we reflect upon the numerous facts of somnambulism which have been presented to our notice, to declare that prevision is not only a phenomenon which may occasionally occur, but which, during the last twenty-five years, has repeatedly been observed by myself. I believe it is always a spiritual fact. In declaring this belief, I hope I cut away all connection between our subject and all artificial magic and divination, or the nonsense of necromancy. We have now to discuss only the phenomena occurring in the due course of nature, and falling out before us, by God's permission, in the ordinary course of the events resulting from his physical laws. There are certain persons we meet with in our intercourse with the world, who are far more susceptible in their nervous constitutions to impressions from external influences than others. One shakes the hand of a man or a woman who is of a cold and repulsive temperament, and there is no mistake about the nature of the grasp. A Frenchman would at once conclude that there was not any heart in that *accueil*. The fact is that the person was not of a sensitive temperament. The magnetic influence determining a cordial reception was not present, and with certain individuals never is present. Their dispositions towards the majority of their fellow-beings is decidedly repulsive. They lose much of the enjoyments which those of a more sensitive and sympathetic disposition obtain. They are the sympathetic and the sensitive, who are generally the most attractive members of society. They are those, who being submitted to the condition of mesmeric sleep, exhibit the higher phenomena of our subject. The more delicate temperaments are found in the lower stations of life as well as in the higher. From their sensitive temperament they often become the victims of disorder of the mucous membranes, consequent upon a want of knowledge that they should avoid acid fruits, beer, cyder, and rhubarb tarts, and numerous other articles of food that tend to produce nettle rash, or other disordered states of skin. I had a cook who was a woman of a highly sensitive nervous temperament, very liable, from errors in diet, to *urticaria*, and to other cutaneous eruptions. Magnesia, sulphur, soda, potass, and the usual routine of remedies had been by several medical advisers prescribed for her. I lived in North Audley Street when she was in my service, where I had a large room, in which I had daily a number of poor people mesmerised for various maladies. My cook was very eager to try whether a course of mesmerism would remove her tendency to cutaneous disease. A few minutes sufficed to induce deep sleep, and in a few days she was a good specimen of somnambulism; walking about in her sleep, talking with other somnambulists, of whom I had seven or eight, occasionally, as her companions. One day she became very serious, and predicted an illness, which would affect her on a certain day and at a certain hour. I went to my table, on which was a desk, and noted the time for the accomplishment of the prediction. It was nine days distant. In order to try whether breaking up the associations connected with the somnambulism might not influence the prediction, the practice of mesmerism was suspended; but on the day and hour designated by this woman in her sleep, and not remembered or alluded to in her waking state, she was seized with headache, accompanied with inflamed eyes, and feverish heat of skin, thirst, and accelerated pulse. I mesmerised her to sleep, and she prescribed for herself the best treatment that could be adopted; a diet of barley-water and effervescing draft of lemon juice, and carbonate of potass, under which treatment she recovered in a few days. A month afterwards, in somnambulism, she again predicted an access of a similar attack. At the stated time she was taken ill, as before. I had been suddenly called down to Bath, and was not witness to the facts, but Dr. Elliotson, cognisant of the circumstances of the prevision, was kind enough to attend the case for me, and under his care the woman got well.*

Prevision regarded only as clairvoyance must be deemed very unsatisfactory. Deleuze tells us, "that the essential characteristic of somnambulism is the development of new senses, of new faculties, among which the faculty of prevision occupies the first rank. We cannot realize such a faculty, but we can, as in those born blind, *recognise* the existence of a sense *different* from touch, by the assistance of which we perceive the forms of bodies we cannot reach, by which we distinguish them by their colour; though their forms may be perfectly alike. For a blind person, though he may recognise the existence of a faculty he does himself possess, can no more comprehend how we can distinguish objects a hundred yards

* It has been spiritually communicated to me that I have been without knowing it myself, a clairvoyant medium.

When I lived at Calcutta, one of my most intimate friends was T. D. well known for his extensive acquirements for his conversational powers, and for many charming social habits. He practised in the Supreme Court as a barrister; he was of course a welcome guest at my dinner table, at Garden Reach, just opposite to my friend Dr. Wallich's house, in the Botanic Garden. T. D. generally passed the night on these occasions at my residence, and my immediate neighbours being on very friendly terms with us, we passed many a pleasant evening in lively discussions. T. D. and I then established an intimacy that was renewed in England.

His religious opinions were often the subject of debate between us, and as I earnestly pressed upon him considerations not to his taste, a certain amount of coolness intervened. Why, or how I was strongly impelled to write to him, about a fortnight before his death, I know not; I stated to him that he was not long for this world, and pointed out to him the day on which he would die. This was *prevision*, for my friend, T. D. departed this life on that very day.

off, than how it may be possible to perceive a future event. Time and space are, for him, two obstacles of the same order. Let us act as the blind do, let us assure ourselves of the reality of phenomena by their results, let us observe somnambules as the blind observe us; we shall then assure ourselves that the human soul is gifted with the faculty of prevision, that this faculty which, in the natural ordinary state is not exercised, develops itself more or less in certain circumstances, and that it can give us notions, which are complete strangers to those we owe to our other faculties. Do not let ourselves suppose that this prevision results from communication with spirits, or intelligences, for beside the fact that nothing proves the reality of this communication, we only remove the difficulty further from us; this prevision not being more easily explained by spirits other than ourselves, than it is by supposing the faculty belonging to the human soul."

Andrew Baxter, who was the opponent of David Hume, as well as of Bishop Berkeley, wrote an admirable work on the Nature of the Human Soul, and in the course of his argument, he found it necessary to examine the phenomenon of dreaming. The facts upon which he relies to prove the immaterial nature of the soul are those on which I desire to lean in my endeavour to explain the phenomena of somnambulism, of clairvoyance, and of prevision.

Baxter shews that dreams "are *involuntarily intruded upon the soul.* It hears, sees, and feels objects at that time; not as it would itself, but such as they are made to appear to it, and is just as passive in receiving these impressions as it would be in receiving the like impressions from real external objects, by means of the senses when broad awake; shewing as much backwardness to them and suffering as much from them; awaking sometimes with trembling, sweating, and crying; and as much fatigued by night with such visions, as by labour and toil by day." After having drawn a vivid picture of various "*disagreeable, shocking, and unnatural circumstances,*" he makes the pertinent remark that "undoubtedly in these, and such other instances, it is absurd to say *the soul would lay a plot to frighten itself*, and then be foolishly in *real terror* with its own designs." It is not necessary to refute the plea of double consciousness, or of the soul being two distinct entities, "each ignorant of the other's consciousness and designs." The hypothesis of Deleuze, and of later writers on a *faculty* possessed by a sleeping person in prevision, applying to clairvoyance equally, is quite untenable, as necessarily implying some kind of mechanism; and the supposition of any mechanism in the body productive of dreams, or of any analogous phenomena, has been ably demolished by Baxter.

The learned word *mechanism* is the *cheval de bataille* of the atheists; as if any mechanical arrangement could possibly enable us to create thoughts, and their consequences. The result of Baxter's cogent reasoning leads to the conclusion that the phenomenon of dreaming "must be produced by a *living intelligent cause.*"

The date of Baxter's work must be borne in mind. The above is quoted from the third edition, 1745. The first having appeared in 1730. The arguments were not then brought forward in favour of any spiritual facts connected with Animal Mgnetism, although the subject of sleep and of dreaming must occupy an important place in any philosophical inquiry into phenomena, which it is impossible to disconnect with our vast subject.

It is clear, then, that we are at variance in our attempts to explain the facts connected with dreaming, with Deleuze, and those writers who assume, or imply, that the mechanism of the body can account for the production of dreams or of the faculty of thought under any of its conditions. Let us inquire into the modern facts relating to the influence of the will in the production of dreams in the waking state. We know well, from the results of a great number of experiments made by the investigators of the facts in Animal Magnetism, that we are able by their exercise of the faculty of will, to make persons whom we have put to sleep by any of the various known means of inducing that condition, dream, and, in their dreams, perform acts to which they are urged by the silent power of our will. There is no doubt as to this fact. I have demonstrated it not once or twice, but hundreds of times. The Zoist, which must immortalise the name of that shamefully-treated benefactor of his race, Dr. Elliotson, is a great storehouse of facts, many of them illustrating the truths now insisted upon. If, then, so many persons in a state of sleep, artificially induced, can be made to dream by the exercise of the will, and be influenced, in the condition of sleep, to perform numerous acts prompted by spirits in the flesh, is it to be considered a matter of wonder that a cogent reasoner, of more than a century ago, should have clearly shown that dreaming could take place only by the ideas intruded upon the dreamer, by the will of living intelligent agents or spirits.

But we spirits in the flesh have made persons dream while they have been broad, or wide, awake. Who has not heard of electro-biology, the absurd compound word to indicate the phenomena of obedience to the will, in a condition in which the person influenced is not asleep, but is in that dreamy state between sleep and wakefulness. Some people are gifted with a power of will the influence of which is so clear and gentle, as not to produce the slightest tendency to sleep in the person willed to perform a series of actions that appear not to be under his own control. My friend, Mr. Henry Stafford Thompson, well known in the county of York, is an example of a person gifted with this clear power, or as I have called it, in a note to the English edition I published of Baron von Reichenbach's Researches, *motive force* of the human will. A very remarkable instance of the exertion of this power by my friend occurred some years ago, one morning, at my house, in Grosvenor Street. Ten persons were present, and Mr. Thompson called to know if I had any interesting phenomena for his observation. Two of the party were very sensitive and impressionable. The one, a lady who had often been made to sleep by will, the other, a well-known musician, a man of acknowledged genius and of great reputation, who had been three or four times made to sleep heavily and deeply by my will. It is a curious fact connected with the influence of the human will, when exerted upon sensitive persons, that the characters of the influence vary in different individuals. Mine has been described by sensitives as heavy, obtunding, and crushing, like a broad wheel of a waggon, while Mr. H. S. Thompson's has been compared to the quick passage of a mail-coach wheel.

Mr. Thompson looked for a minute in the direction of Mrs. H., who was seated on one of the ten chairs placed against the wall of my dining-room. She fell asleep. Then he looked at Mr. J. B. C., who was seated on a chair in the recess of a window. The worthy gentleman thought that I was mesmerising him by my will, and exclaimed, "Pray do not lift me up, Dr. Ashburner. I know you must be mesmerising me, for I feel myself going into the air." The next minute he was asleep. He was allowed to sleep for a few minutes. Then my friend, Mr. Thompson, willed him to become awake, while he was fixed to his chair, with the absence of all power to move himself from the position he occupied. He said it was strange he could not get up. "I am positively fixed to this chair, I cannot move myself from it." In another minute he said, "It is very strange, there were several persons in this room, I see now but Mr. Thompson." Then suddenly he saw Mrs. H., who had walked in her sleep from her chair round the chair on which Mr. Thompson sat, at the bottom of the dining-room table, to the chair occupied by Mr. J. B. C., but remaining at a distance from it, sufficient to be out of the reach of his hands, when they were stretched out towards her. Mr. H. S. Thompson then commenced a set of experiments, by playing, with his will, on the phrenological organs of my susceptible and sensitive man of genius. People are little aware that the sensitive temperament is one of the conditions necessary to constitute the true man of genius. Our friend was instantly influenced. He apostrophised the sleeping lady, who stood like a statue before him, in all the grace which accompanies the higher development of somnambulism. "Thou angel of light and loveliness! Come, let me clasp thy fair and lovely hand! Let me imprint an ardent kiss on that fair and delicate hand! Beautiful woman, come to my embrace!" Then, as if struck by remorse, for Mr. Thompson had directed his will to organs productive of conscientious feelings, he said, "O wretch that I am! What would my wife say if she were here? My dear, my darling wife!" and he held his hands pressed against his temples in an agony of remorse. The next minute found him again impenitent. "What care I for my wife. Thou art all to me! Oh darling angel! best loved of my soul!" In vain he endeavoured to reach the lady, for he was fixed to his chair. Mr. Thompson, to shew his power over the organs of this sensitive brain, caused the whole scene to be re-enacted, and then released the automaton lady from her sleep, and the man of genius from his hallucinations. Will any man of thought venture to say that the phenomena which have now been detailed, are not worthy of deep study? Here is a man, endowed or gifted with an extraordinary power of clear will, able to influence the phrenological organs of his neighbour, so as to excite his amorous feelings, and cause him to suspend his sense of moral duty; and the next minute lead him, by exerting spiritual forces on the organs of conscientiousness, to produce remorse and melancholy, and the philosophic world of science is to gape and wonder, or to deny boldly and flatly truths as sacred as any that God has ever poured out to man's cognisance.

The tale I have told is full of instruction; it tells us

that man can make his neighbour *dream* at his will, for the whole was a *dream* forced on the soul of the man of musical genius, and in his *dream* the victim of my wilful friend was obliged to obey the impulses which governed him; or as Baxter would say, "which were *intruded on his sensory.*" All these facts are so many evidences of the existence of an immaterial force which we call soul, or really the principle of consciousness, which, when not properly educated, or habituated to self-control, runs riot, and soon becomes the victim of evil influences. Is Animal Magnetism then, which teaches us the philosophy we have here displayed, to be spurned by men calling themselves scientific. Are we, who have cultivated this deep, this extensive knowledge to succumb to prigs, whom we know to be grossly ignorant on the subject of which they profess to be adepts.

Will sciolists for ever rule their fellow-men? Is society to be for ever blind to the light of such truths as have been now brought before them? Is pig-headed obstinacy for ever to overwhelm the will of men gifted with common sense?

The scene I have described as it occurred in my dining room in 65, Grosvenor Street, is replete with the substance that ought to excite deep thought. It is not only illustrative of the philosophy of dreaming, it illustrates the motives of human conduct. For man in his waking state, unconsciously to himself is often dreaming. We are not bound to be always afraid of probing our own meaning. When we say we dream, if we have convinced ourselves that we have reasoned rightly, we mean that we are influenced by forces that impinge upon us from sources external to ourselves. These forces, be they *intelligent* or *brute,* impel us to act, and often unconsciously to ourselves, form the impellent motives to our actions.

How can we know that such forces operate upon us, and produce the motives to our actions? It may be said that, however plausible may be the train of reasoning here, chaining our facts together, there is no certainty of the existence of forces, intelligent or brute, of whose presence we have no ocular demonstration. Precisely the same objection may be adduced against the air of the atmosphere we breathe. We do not see the air around us; we cannot touch it; we are unable to grasp it. It is not that men are simply unable, from stupidity, to be convinced by the force of reasoning, but their minds have been prepared by habit for rejecting truths, however cogently reasoned out. There are, however, repulsive agencies or forces, preventing their minds from adopting sound conclusions.

A wondrous wise man, of an excellent opinion of himself, beseeches and teazes me to accompany him, in order that he may witness the phenomena in the presence of the celebrated medium, Mrs. Marshall. He sits down at the table, he hears sounds like gentle rappings or tickings, resembling the noises of a watch. He takes up an alphabet and slowly and deliberately touches each letter with the point of a dry pen, or perhaps with the point of a porcupine quill. At a certain letter the raps or ticks arrest his attention; he puts the letter with the aid of a pencil, upon paper; he begins the alphabet again; the raps arrest his attention at a different letter, which he adds to the first he had noted with his pencil, and so, by the repetition of the process, the full name of his mother was spelled out. How can one be sure that a spirit prompted this intelligent result? Ask mentally, not speaking aloud, for the precise date of your mother's death, or the day when her spirit quitted her earthly body. The answer was accurately indicated by the alphabet process. "Are you happy?" was asked, and the reply was, "Very happy." This eager man had become satisfied of the possibility of the facts he had witnessed. He got up, and the table moved towards him, impelled by some unseen force, and followed him as he moved from one part of the room to another. As he left the house he appeared vexed and irascible, simply remarking "*cui bono!*" Such a man is a specimen of a very common order of intellect, and may have given rise to the old adage of the impossibility of making a silk purse out of a sow's ear.

I did not press my good friend, the Baron von Reichenbach, to give any opinion on this subject, well knowing the reticence of men of high scientific character on the facts they may have witnessed of marvellous spiritual manifestations. During his visit to London in 1863, the Countess Powlett was kind enough to invite the Baron one evening to her house in Hanover Square, to assist at a séance, where Mrs. Marshall and her niece were present. I accompanied my friend to the countess's house, where we found a limited number of distinguished persons assembled. Having seated ourselves round a large table, supported on a solid strong pillar on four claws, the Baron, having heard that the phenomena resulted from certain tricks performed by the agency of the tendons of the peronæi muscles of the legs, requested that the shoes of the two mediums should be taken off, to which they readily consented, then when an abundance of rapping and ticking sounds were heard, he got under the table to hold the mediums' legs, and soon satisfied himself of the absurdity of his suspicions. Then he took a card with a printed alphabet on it; the word spelled out was the name "*Rieka*," the pet name by which he, and he alone was wont to call his wife; the next name spelled by alphabetic indication was that of his daughter "*Ottone*;" and afterwards the name of the Baron's brother-in-law, with whom a short colloquy in German ensued. After some preliminary manifestations, the table was moved from its place by an unseen force, and the Baron wishing to satisfy himself as to the amount of power exerted by the unseen agencies, again got under the table and grasped the four claws "like grim death," as a distinguished lady present observed; at that moment the table moved rapidly with a force irresistable, sweeping the Baron bodily, with no small velocity, round a very large room. When it returned to its place, the table showed no signs of unwonted animation. The Baron felt a little astonished! but Mr. Faraday, a man of cool mind, would have concluded that the table was moved by unconscious muscular force, and he might have relied upon the velocity of the motion for a corroboration of his theory.

Having entered upon forces spiritual, we may at once say that it is not proposed to define a very specific line of demarcation between the higher forces regulating the combinations of matter. We are not able to make those severe boundaries between heat and electricity, which the profound correlators of force imagine themselves able to establish. Until we become better acquainted with the mathematical value of the forces governing the minuter shapes of matter, we shall have no occasion to trouble ourselves with the confusions of works on the correlation of force.

When we reason from very strong and very numerous facts, all strictly belonging to the domain of nature, it cannot be denied that our opponents are bound to produce facts equally cogent and equally numerous if they propose to overthrow our propositions.

It has been proposed to show that man often dreams in his waking state, simply because, in that state he is often under the influence of spiritual forces. We have seen that Dr. Leger could measure the forces of the separate phrenological organs of man's brain, in virtue of the magnetic emanations proceeding from those organs through himself to the pendulum of his magnetoscope, which was an ingenious modification of the original instrument invented by Mr. Rutter. The same organs that can emit magnetic influences are capable of receiving magnetic influences. Hence we may comprehend how our fingers send magnetic influences through the skull and its coverings to the various organs of the brain. Who that has witnessed repeatedly the facts in phreno-mesmerism, which Dr. Elliotson and I have, at our respective houses, shown some hundreds of times, could doubt the fact that a sensitive somnambule could be influenced by the Animal Magnetism emanating from our fingers,—themselves separate Animal Magnets,—to sing or to joke, or to pray fervently, or to sing hymns, or to sing martial or comic songs, or to feel and exhibit pride or selfishness, or attachment, affection, hatred, or combativeness, or the lower feeling of amativeness, or their negations, as often as we mesmerised and de-mesmerised the organs we wished to stimulate, or to render quiescent.

That the sensitive human being can dream in sleep, in that sleep-waking state we call somnambulism, and even in the state of wide-awake consciousness, if experience and fair logical inference are to avail us in our researches, there can remain no doubt. I have a remarkable tale to tell of an instance of clairvoyant prevision in a gipsy, whom I saw at Roehampton more than twenty years ago. There was a garden party in celebration of a birthday. I met there the two Misses Byrne, of Cabinteely. The elder one asked me to accompany her to a field in the grounds, abutting on the Barnes common road, where some gipsies had assembled. One of their women began to banter me immediately on my depressed state of spirits. "That thousand pounds is the cause of your fretting," she said, with a smile, and Miss Byrne asked me if there was any truth in what she said, and I replied there might be some truth; but how she came by the knowledge, I could not guess. The woman then desired me to cross her hand with a piece of silver. She desired half-a-crown. Accordingly, I did as she desired. Then she told me not to fret any more, "Providence has been kind to you on several occasions, I may assure you that between this day and this day fortnight, the thousand pounds will arrive, and your mind will be at ease." The day was Tuesday, on the Tuesday after, at six in the morning, I was roused by the ringing of my night-bell. I got up, expecting a call to summon me to a patient. Instead of that, I found, on a good horse, a man, who, asking my name, delivered into my hand a long, official envelope. On opening it, I found a thousand pounds in bank notes. The stranger who brought it hurried off as fast as the horse could gallop. To this day, I am not aware of the person who was my benefactor, nor do I know how the secret, which I thought was my own, could have transpired.

These mysterious events form part of the philosophy of Spiritualism. They fall out into this world in their due course, by the ordinances of Providence, or of the will of God, a high magnetic force, which regulates the angels or ministering spirits in their moral government of the world. Spirits intervened then, in the clairvoyant prevision of that gipsy, who was made to foretell the arrival of the thousand pounds.

Taking spiritual impressions on the sensorium as the cause of dreams, we proceed naturally to a description of clairvoyance which I first published in the 20th Number of the Zoist for April, 1848, 6th vol., p. 96. Subsequently, I republished this paper, with the addition of eighteen pages of important matter, in the form of a pamphlet, entitled "Facts in Clairvoyance, with observations on Mesmerism, and application to the philosophy of medicine, and to the cure of disease."

If the cases described in the pages of the paper and pamphlet alluded to, were the only examples of clairvoyantes reading out the printed words of mottoes, enclosed in the scooped out shells of walnuts and of hazel nuts, which had been filled up with small comfits, in addition to the mottoes, and then closed, objectors might, even very frivolously, have opposed themselves to the facts. But in this case there was no loop hole of escape. Major Buckley, a man of a certain handsome income, had a carriage and his box at the opera. He was devoted to the investigation of somnambulism and of clairvoyance. People pretended to ridicule him, because he took young ladies to his box at the opera. It was a kind and good-natured thing to do. But it gave him abundant opportunities of trying experiments deeply philosophical.

I have repeatedly been with him when he was surrounded with a bevy of his charming young friends. I have been told that they humbugged him to the top of his bent. It is quite possible that a few of those who failed, might have boasted that they deceived him. These were the exceptions. I have myself witnessed the nutshell experiments too often to be turned aside by the bold assertions of determined obstructors of the truth. Major Buckley, between the acts of the opera, often instantly drew the attention of the sensitive ladies he had in the box to the passes he made with his two hands down his own face. One or two of the young ladies would say, "We perceive the blue cloud on your face." He generally had a bag containing the nut-shells in his pocket. When he produced them, he held one in his closed hand, and one of the young ladies not only read the printed motto, but told him the number of white, yellow, and pink comfits in the shell, which, upon being opened, was found to be quite accurate. On these occasions I have not trusted to the nutshells brought to the opera by the worthy Major, but have furnished myself with some purchased at M. Cœuret's, who was then a wholesale manufacturer of such articles, in Drury Lane. I must confess that the ladies in Major Buckley's box always read my nutshell mottoes with perfect accuracy. But I have, upon several occasions, met a very charming person at Major Buckley's lodgings, who is now married, who never failed to read accurately every motto in a nutshell, which I happened to present to her after the preliminary production of the blue cloud on Major Buckley's face. She never could see a blue cloud on my face, although I imitated Major Buckley in downward passes on my face. This lady is a cousin of a friend of mine, who sits at a board of directors, of which we are both members. I asked her to describe to me what she could perceive, when she read the mottoes which were enclosed in the nutshell then in my doubled up hand. She replied that on each occasion she saw a surface, like, the unfolded scroll we see at pantomimes, when something is announced, not according to the rule of silence that obtains at that kind of entertainment. The motto appears on the unfolded scroll in large letters. This fact is curiously corroborated by other spiritual facts.

The phenomenon of dreaming, then, is illustrated in these cases in the wide-awake state. The dreamer receives an accurate spiritual statement through impressions made upon her sensorium by invisible spirits. It is impossible to conceive that any impression made upon our brain by our own will of an accurate truth, of which we are quite ignorant, should be impressed upon our sensorium by ourselves. This would be too great an absurdity. These young ladies in their vivid waking dreams, were enabled to read mottoes in nutshells; those mottoes being folded up and buried among a number of very small sugar comfits, of different colours, which they were able to distinguish and to count the numbers of each colour.

Could any of *us,—spirits in the flesh*, be cognisant of the letters of the motto, or of the numbers of the comfits. Could we see through the walls of the nutshells? If not, it was not in our power to prompt the clairvoyantes to make their statements. It follows then, that a higher order of spirits was concerned in this affair; evidently a higher order, with faculties we do not possess.

The whole of Spiritualism is a marvellous subject. No one with a head on his shoulders, and an active and intelligent mind connected with that head, can doubt on this. When one hears of a man's doubts continuing on the existence of spirit rapping, as it is called, after he has witnessed the phenomena eight or ten times; for more than an hour each time; after he has had repeated proofs by repeated conversations, by the aid of the telegraphic alphabet in presence of a medium, with persons or minds with whom he was intimate in this world; what can be thought of the judging capacity of that person? I would much rather not be his judge, lest being one of the extensive family of my brethren in this world, I should be sorely tempted to call him a fool.

The following extract is from a letter which appeared in "The Reasoner," June 1st and 8th, 1853:

"My Dear Mr. Holyoake,—Allow me to make a few observations on your article concerning "Those Rapping Spirits." A logician should liberate himself from all forms of prejudice, and should possess the flexibility of mind that can adapt itself to all new developments of truth. You appear to me very susceptible of erroneous influences, from your desire to prejudge the question before you, on canons of dignity established in some school which few in these days of railroads and electric telegraphs would recognise. You give your opinion that "the rapping itself is an undignified mode of communication," forgetting that it is analogous to modes of communication established between the spirits of potentates, philosophers, and all high dignitaries who have occasion to use the telegraphs in various parts of the world. If you could pursue your train of reasoning fairly, you might in time establish the position that all reading and writing were most undignified, since they form modes of communication between *spirits*, or *souls*, or *minds*, or *intelligences*, by means of the letters of the alphabet. If we can for an instant allow that it is in the power of absent and departed minds to communicate with us, I do not perceive how the fact can be interfered with by establishing a want of dignity in the means used by those minds; but suppose that the means are through the agency of alphabetical symbols, how can you possibly argue on the question of dignity? Nay, is it not a poor prejudice to assume a want of dignity to such a process at the moment you are impugning the character of Mr. Hetherington for the want of that dignified politeness due from gentleman to gentleman, in attributing to him the possibility of an act which would be discreditable to even some of those rowdie cads of the press, who have been exercising their ingenuity in really *frivolous raps* at a subject far beyond their depth? I grieve to find your much nicer and wiser mind engaged in a train of thoughts biased by such inferior influences.

"These spirit manifestations are not to be cast aside by what any man may think of their dignity, if truth be at their foundation. In whatever form the ingenuity of vituperation may be couched, abuse will not avail for their destruction. I may be allowed to remark, that no man pretending to philosophy shall be able to define the limits of dignity, or of respectability, in the phenomena of nature. If facts be presented to us, we must take them as they come, and be content to examine them, and with a gentle and philosophically-humble frame of mind establish for them their due importance in the place they must occupy in our stock of knowledge. Certainly the Editor of "The Reasoner" ought not to reject them without thoroughly good and sufficient reason.

"It may be that your prejudices are too strong to allow you to receive the facts that should lead to the possibility of the existence of spiritual beings. I do not contend for immaterial essences, for my limited capacities allow me to conceive of the most highly refined essence to be only a form of material being. If I could categorize a thousand degrees of a thousand kinds of the essences of electric and magnetic fluids, I might denominate some of them as *soul*, or *mind*, or *spirit*, or *intelligence*. In the present state of our knowledge, we cannot definitely trace the exact nature of thoughts or ideas. But we know enough to be aware that there is a strong probability that an idea is a compound of light in some form with a subtle agent emanating from the brain—that crystalline and magnetic organ of ours which sheds, in each act of thinking, a specific fluid that unites with light and thus becomes capable of being seen, as it issues from the brain, by persons in a highly sensitive or clairvoyant state. It may be that you do not like to believe it possible that the Baron von Reichenbach tried at Castle Reisenberg, near Vienna, many experiments in the dark with magnets, and that he ascertained the fact of light emanating from their poles—visible to some individuals, but not to all—visible to many not previously capable of seeing the light, but rendered capable by remaining in the dark chamber during several hours. I can assure you that I have seen that magnetic light. Different sleep walkers have seen analogous light from the surface of the human body—more abundant, clear, and intense, from some parts than from others. And the Baron von Reichenbach has experimented upon numerous persons in his very dark chamber who—not in a state of somnambulism—have been able to see light emanating from all parts of the body, interior as well as exterior, and has thus established the fact of clairvoyant introvision, by a new process of experimenting upon living organisms in the depths of darkness. If you will take the trouble to read a long note on animal light, at page 438 of my edition of "Reichenbach's Researches," you will find numerous facts to lead to the conclusion that light emanates from those animal spheroidal structures, which render our bodies a congeries of crystallic molecules; and if inorganic crystallic spheroids emit odic or magnetoid light, we perceive why crystallic animal arrangements should in like manner be sources of light.

"The Baron von Reichenbach has, by numerous experiments, shown that *vegetable* organisms, placed in his dark chamber, emitted *odic light*, perceptible clearly not only to the delighted Endlicher, the renowned professor of botany, but to several other persons, establishing the great fact that all organic forms of crystals emit light as one of their constituent properties. You have probably looked over Haüy's plates in his books on crystallography and have observed that which has been copied, in woodcuts in so many systems of chemistry, that all geometrical forms of crystals are made up of innumerable tiny globules or spheres, each of which is an elementary crystal, and the aggregation of these, according to electric laws, produces the geometrical crystal, with all its magnetoid properties; and it is easily conceived why, if you break a crystal or a magnet into a thousand pieces, each piece is found to retain all its crystallic or magnetic properties. The philosophical Mr. Rutter, of Brighton, reflecting more deeply than some of his neighbours, constructed a magnetoscope, which for extreme delicacy is unsurpassed, and it enabled him to demonstrate these facts with the clearness and ingenuity that characterizes his fine mind. That plaything, consisting of a piece of silk thread, to the end of which is attached a small piece of sealing-wax, like a fuchsia bud, or a lucky sixpence, or a gold ring, by which people try the old vulgar experiment of striking the hour of the day, against the sides, in the interior of a glass tumbler (will you call it *an undignified pendulum*?), offered to Mr. Rutter a series of ideas, which, carried out, only as a philosophic and originating brain could carry them out, enabled him to trace laws reaching to the depths of physics, and to far clearer relations of mathematics to the philosophy of mind, than we have hitherto possessed; for, in the hands of Dr. Leger, Mr. Rutter's instrument has given us a basis *where*, on the statistics of phrenology, shall rest our unerring calculations on the forces directing our thoughts and actions. I need not now enumerate the many more facts to be found in Mr. Rutter's or Dr. Leger's unpretending works, for which we are indebted to the new magnetoscope, but I may remark that we now know of all vegetables and animals being possessed of magnetoid properties; that, in truth, each of these, and the component parts of each, are separate magnets. So an orange, an apple, a carrot, a walking-stick cut from a hedge, a human being, a limb of that human being, the bones of his arm, the bones of his fingers, all are magnets, having the polarities and mean points quite analogous to bar magnets and to crystals. Mr. Rutter's magnetoscope has established far more than all this, for he has, by its aid, been able to trace the existence of magnetic curves around so numerous a list of bodies, constituting magnetic or magnetoid spheres around each of them, that it is difficult to define the limit to which the law extends, obliging all the atoms of matter to range themselves under magnetic control, producing everywhere spheres or spheroids that have their axial and equatorial polarities, and obeying for ever the perpetual impulse to change. However infinite may be these evolutions, prompted by the law of change, every sphere, as well as every spheroidal atom, is productive of new evolutions of light. I do not know whether you will be so stern a philosopher as to reject all the exquisitely beautiful evidence offered by Mr. Rutter's magnetoscope, or conclude that such a man as the great Berzelius had become fatuous because he accepted the fact adduced to him by the Baron von Reichenbach—for this had been done by men pretending to be philosophers—but it will matter little, since truths will find their way. Eighty or ninety years of prejudiced opposition have not sufficed to crush mesmerism. It is a truth in mesmerism that many sensitives wide awake, and clairvoyants in magnetic sleep, have seen rays of light, made up of globules of a greyish or of a cærulean blue color, emanating from the eyes of individuals who have been exerting thought or will.

"I will not tire you by taking you over a wide field of natural history to tell you that luminous animals exist, and offer phenomena beautifully analogizing with the facts relating to the light emanating from the thinking man, from the excited ape or monkey, from the furious tiger at his prey, from the dog and cat which are our companions. If I brought you to the conclusion that *ideas* are shapes of light which emanate from our magnetic brain, you would be apt to say that the inferences are strained, for your *habits* of thinking require a closer catenation, and the forming of the chain by more numerous links. I am clear that *ideas* are not only shapes of light, but that each particle of those atoms constituting those shapes is subject to a magnetic law of attraction and repulsion; and to a magnetic law of curves combining to form spheres, with polarities obeying the directions of those which belong to this huge globe, the earth we inhabit, and to the tiny globules of ether that formed the subject of Sir Isaac Newton's calculations. If we acknowledge the tendencies of magnetic laws, and if we admit the facts that have been published by many experimenters in Mesmerism, from De Puységur, Tardy de Montravel, down to our own times, by the Baron von Reichenbach, by Mr. Rutter and Dr. Leger, we cannot escape from the inferences on the essence of *ideas* and their relations to magnetism. If we know that mental enlightenment is dependent upon the amount of *ideas* fabricated by our magnetic or cerebral machinery, is it so difficult to arrive at the assumption or hypothesis that the mainspring of that machinery is a magnetic intelligence or consciousness, connected with and stimulating our organism to the various actions it performs? a highly refined essence of magnetic fluid evolved in the electrical act of conception, developed with our unfolding processes of life, and released at death from its gross connection to realms of light, and "*a blessed and brilliant home*." You will tell me this is poetry, not logic, and that you will not allow me to tie such hypotheses together with my miscellaneous bundles of knowledge. You will deny that I can prove the fact of an intelligent existence separated from our corporeal frames, and dwelling in one of the seven spheres said to surround the globe we inhabit. Will you allow me the fact of the electric telegraph on this earth, and of the necessity of intelligences in existence at each end of that instrument? Some of you logicians are so severe that you hold the doctrine of the argument being precisely the same in relation to those not present and to those not in existence. If so with you, I need proceed no farther; but if I can satisfactorily establish the fact that communications have reached me by telegraphic signals, which could come from no other source than unseen *intelligences*, I am not bound to adduce you more evidence respecting the wires or channels of communication between those intelligences and myself, than the facts which are known and accepted by the students of mesmerism as the acknowledged phenomena of the will. See the fifth and some other volumes of the *Zoist*, and my notes to Reichenbach's "Researches," and you will perceive that there exists many facts on this subject. If it can be shown that unseen intelligences somewhere exist, and are in a position in which their magnetic powers can act at a distance, knowing the modes in which the magnetic forces of our minds act here, we can be at no loss to accord to them the faculty of making signals upon the wood of our tables, upon the glass of our windows, or upon the walls of our apartments. How do the wires of the electric telegraph convey the messages? Does not each molecule of wire receiving the electric impulse so arrange itself in relation to its neighbour as to allow the fluid to traverse the length of the wire in an astonishingly short time? Is not each spherule or globule of the metal so shoved into its proper relation with its neighbour spherule that a general consent may exist between the particles of metal so as to allow of the easy transmission of that particled fluid, usually called *force*, from one extremity to the other? Do you know the fact, that if by the wire which has conveyed a message from London to Paris, you wish to convey the fluid back again, there is an unwillingness in the spherules to obey the same impulse as readily as they did before? Why? Because you have to disturb the comfortable, axial, and equatorial relations that they had previously agreed to under the first influence. They are creatures of habit; and if you habituate them by repeated electric currents to one mode of opposition, and they become too lazy to like to assume any other. The wire that by one course of exercise likes to be straight, will, by repeated changes of currents, or *snubbings*, become curved, cockled, and brittle. What is a train of thought composed of currents of globules of highly-refined matter, but a magnetically arranged wire? What is the human will but an arrangement of globules, that have been seen, by highly sensitive persons and clairvoyants, to issue from certain organs of the brain, and, traversing its anterior parts and passing through the eyes, to impinge on a person or thing to which the individual exercising the will has directed it? Who likes to have his will thwarted or to be snubbed? Can we not explain now how the influence of habit is operative on the faculty of thought? To accuse unseen intelligences of want of dignity when they communicate with us by magnetic telegraph, is not proving the presence of that flexibility of mind which I know you to possess.

"The phenomena of thought-reading are so familiar to those who have studied mesmerism, that I shall not urge upon you the obvious explanation that is at hand, if the question be asked as to how the unseen intelligences read our thoughts and converse with us mentally? I may find it as difficult to satisfy you on these facts as on the *undignified* nature of the celestial telegraph. Some persons turn from all mesmerism with unmixed disgust, and speak of its resulting from Satanic agency; and it is not impossible but that you may feel so bothered with the thought of its marvels, as to consign it into the heart of your next edition of Father Pinamonti's Hell. Only remember, that the accounts which have hitherto reached us of life in the Spheres is diametrically opposed to the descriptions of the holy father, and in no respect more so than in the delicious odours inhaled by the virtuous spirit.

"I have a great difficulty to contend with in advocating the existence and powers of unseen intelligences, or spiritual beings, from the fact relating to the operations of the human will not being credited by great numbers of persons believing themselves to be philosophers. It may be that there is no possible mode of intercourse known to the inhabitants of the upper magnetic spheres than the employment of the will, a magnetic force or agent in the production of the raps, which appear to be when carefully listened to, vibrations or disturbances of magnetic relations between the molecules of the wood, glass, or other substance whence the sounds would seem to reverberate. We know well that the guiding the hand of a writing medium is quite analogous to many an experiment that has been made by my friend, Mr. Thompson, of Fairfield, and myself. If, in sitting in an omnibus, or in a railroad carriage, I have been able, by the force of my will, to make a person sleep, and for the purpose of establishing the existence of this power I have repeatedly made persons fall asleep in these vehicles—if sitting near or opposite to a passenger, I have induced that person to put a hand into mine, or to do other ridiculous things—and I have often done this—I have established an important truth. I have often, by the exertion of my will, obliged a person, who was two miles distant from me, to sleep instantly, and to continue asleep from eleven at night until seven in the morning, thus influencing a poor wretched victim of insanity, for her good, not only at the instant, but setting up a train of tonic forces in her nervous system which lasted eight hours. I have, by the force of will, obliged individuals to come to me from places at the distance of two miles, hastening over the ground at a quick pace. I know that Mr. Thompson has done the same thing, the distance being much greater; and he has influenced persons to sleep at two hundred miles. In Mr. Spurrell's little book on the rationale of Mesmerism, a fact is recorded of a person being willed to come from Norwich to London. I do not find it so hard to believe that the spirit of my father can, by his will, guide my hand to write sentences, the matter of which was not only not in my head a second before, but of which most often I cannot guess the purport, while my passive hand is guided in the formation of the letters. If you had become a writing medium, and had communicated as I have done with old friends long departed from this earth, you would perforce cease to disbelieve in the phenomena, and you would derive enjoyment from the knowledge that those who were your attached friends still live, to be developed into intelligences even more pure and refined than they were here. Who could have been a nobler or a finer character than the late Professor Macartney, of the University of Dublin? Large in intellect, he was necessarily free in thought. High in moral qualities, he was the most strictly honourable and conscientious man I ever intimately knew, and many a pupil and many a friend will vouch for his generosity and for the warmth of his attachment. His acquirement and his industry, such is the force of example, give a noble tone to the studies it was his lot to superintend. Is it not a source of exquisite enjoyment to be certain of being able to renew, even by the aid of the *undignified* telegraph, one's former affectionate intercourse with such a friend?

"I had the misfortune to lose my father fifty-five years ago. Although I was but a child, I have a vidid remembrance of him. By the natives of Bombay he was more than respected. He was venerated for his high talents, and for his great goodness. Is it nothing to feel that such an intelligence is able to make his ideas clear to his son? But you will ask, 'What proof have you of the identities of these persons?' This brings me to narrate to you the events of the first evening I spent in the presence of Mrs. Hayden.

"I had always regarded the class of phenomena relating to ghosts and spirits as matter too occult for the present state of our knowledge. I had not facts enough for any hypothesis but that which engaged for them a place among optical phantoms connected in some way with the poetical creations of our organs of ideality and wonder, and my hope and expectations always pointed to the direction of phrenology for the solution of all the difficulties connected with the subject. As to the rappings, I had witnessed enough to be awâre that those who were not deceiving others were deceiving themselves; and there really exist on our planet a number of persons who are subject to the double failing of character. Having been invited by a friend to his house in Manchester Square, in order to witness the spirit manifestations in the presence of Mrs. Hayden, my good friend can testify that I went expecting to witness the same class of transparent absurdities I had previously witnessed with other persons described to me as media. I went in any but a credulous frame of mind; and having, while a gentleman was receiving a long communication from his wife, whom he had lost under melancholy circumstances of childbed, some years ago, watched Mrs. Hayden most attentively, and with the severest scrutiny, I finally satisfied myself that the raps were not produced by her, for they indicated letters of the alphabet, which, written down in succession, constituted words forming a deeply interesting letter, couched in tender and touching words, respecting the boy to which that eloquent mother had given birth when she departed from this world. If Mrs. Hayden could have had any share in the production of that charming and elegant epistle, she must be a most marvellous woman, for during a good part of the time that the raps were indicating to the gentleman the letters of the communication, I was purposely engaging her in conversation. The gentleman would not himself point to the alphabet lest his mind should in any way interfere with the result; and therefore he requested the lady of the house to point to the letters for him, while her husband, seated at another part of the table, wrote down each letter indicated by the raps on a piece of paper.

"I was now kindly requested to take my turn at the table, and having successively placed myself in various chairs, in order that I might narrowly watch Mrs. Hayden in all her proceedings, I at last seated myself, relatively to her, in such a position as to feel convinced that I could not be deceived; and in fact, I was at last obliged to conclude that it was weakness or folly to suspect her of any fraud or trickery.

"There are some people who think themselves uncommonly clever and astute when they suspect their neighbours of fraud and delinquency. It may be wisdom to be not too soft and credulous, but depend upon it the statistics of the existence of roguery and knavery in society, and the relative proportions they bear to honesty, will not bear out the proposition that it is wiser to suspect every man to be a knave until you have proved him to be honest. The world may be bad enough in morals, but unless there were a great deal more of good than of evil in the human heart—I should say in the human brain—society would not hold together as it does. I know no man who has been hit so hard by the villany and knavery of his brethren as I have myself been; and yet, attributing much to the influence of surrounding circumstances operating upon the bad moral organisations it has been my misfortune to meet with in medical life, I should be sorry to come to the conclusion that my worst enemies were not to be far more pitied than blamed. As for Mrs. Hayden, I have so strong a conviction of her perfect honesty, that I marvel at any one who could deliberately accuse her of fraud.

"In order to obtain an experience of the phenomena in the fairest manner, I asked Mrs. Hayden to inform me whether it was requisite to think of one particular spirit with whom I wished to converse. 'Yes,' 'Well I am now thinking of one.' It was the spirit of my father whom I wished to enlighten me. No raps on the table. I had anticipated an immediate reply, but there was for awhile none. Mrs. Hayden asked if there was any spirit present who knows Dr. Ashburner? Immediately, close to my elbow, on the table there were two distinct successions of gentle rapping sounds. The next question was, 'Was the spirit he wished to converse with present?' 'No.' 'Was there any one present who would endeavour to bring it?' 'Yes.' 'Are the spirits who rap near Dr. Ashburner, friends of whom he is thinking?' 'No.' 'Will they give their names?' 'Yes.' These replies were signified by rappings to questions put, some audibly, some mentally. Mrs. Hayden suggested that I should take up the alphabet, which was printed on a card. I took the card into my hand and pointed at each individual letter with the end of a porcupine quill—my friend Mr. Hoyland, the gentleman of the house, kindly undertaking to put down on paper for me the letters distinguished by the raps. When I arrived at a letter which the spirit desired to indicate, a rapping took place; but at all the other letters there was a complete silence. In this manner I obtained the letters successively ANN HURRY, the name of one of the most beautiful and accomplished, as well as pious and excellent persons I had ever known. I had not seen her since 1812. She married two years after, and died in 1815. My father and most of the members of the family had been on terms of the greatest intimacy with several branches of the Hurry family, and I had, in youth and childhood, known Ann and her cousins, as companions and playfellows. By the aid of the telegraphic signals I have endeavoured to describe, I conversed for some time with the charming companion of my early years, I learned very interesting particulars relating to her happy abode in the spirit world. My curiosity had been excited by the different sounds produced by rappings that I heard close to those made by my friend Ann. I asked for the name of the spirit they represented. The name which came out by the letters indicated on the alphabet was ELIZABETH MAURICE, another companion of the childhood of myself and my brother and sister—another almost angelic being while on earth, but now with her cousin Ann, an inhabitant of the third sphere in Paradise. The authoress of the 'Invalid's Book,' and some other works testifying to a pure, gentle, and refined taste, conversed with me awhile; and at last a louder and more decided signal was made to me from the middle of the table. The name I obtained by the telegraphic raps was that of my father. I asked him to communicate to me the date on which he quitted this world for the spirit home, and the raps indicated '7th September, 1798.' I asked him where the event took place, and I obtained the answer 'At Bombay.' I asked his age at the time, with many other questions, the replies to which were all quite correct. I kept up mentally a long conversation with him on subjects deeply interesting, and it was productive of a communication from him, which I subjoin:—

"'My dear Son,—I am delighted to have this privilege of communicating with you, hoping to dispel some of those wrong impressions which now hover around you in regard to this spiritual being. Allow a spirit who inhabits one of the higher circles to decide for you on a most important subject, to try to remove from your mind the doubts which perplex you, and to establish in their stead a firm faith in the Creator of Heaven and earth. It is He who permits us to make these manifestations, through certain constituted persons, in order to impress mankind with the fact that the spirit shall live in a future state, in a more bright and blissful home. What proof can I give you of the truth of this? You have only to name it, and it shall be granted to you from your father, who has ever watched over you with the care of an angel. *Do not doubt* what I now say,

"'Your affectionate father,

"'WILLIAM ASHBURNER.'

"I am giving you a short narrative of the first part of my course of experience of the Spirit Manifestations. It is important not to be too diffuse. I am desirous of showing that if the subject be investigated in a calm and bold frame of mind, there is no danger of the bad tendencies which have been so fiercely deprecated. I may not be able to prove to you, and to such as yourself, that there is a sufficient amount of facts to satisfy you of the existence of intelligences absent from the immediate sphere of our own cognisances, but I have at all events been able to adduce to you a number of curious facts; and if these and more such be tied together in bundles, so placed as to affect the phrenological organs of a vast number of brains with the attractive force of agreeable conviction, many of the ideas advocated by the *Reasoner* will have a chance of being displaced and forced into the category of negative existence. To take up the impossibility of future existence, is to deny that we are beings of limited capacities, and to arrogate to ourselves the power of finality. No weakness is so ridiculous as that of fancying that we are arbiters of events—that our *will*, exercised by organs that soon shall rot, is to determine the fate of a holy truth. How ardently does the bigot fancy he is right. Sincerity may be his merit, if ignorance be the cause of pardon for a foolish sincerity. A new truth, a new event, which, established into a fact is a new light, makes the antecedent idea pale, and it vanishes before the force of new conviction. I cannot express to you the influence on my mind produced by the facts, rapped out by alphabetical signals, that my spirit friends Ann and Elizabeth, knew of their cousins Hannah and Isabella having called a few days before at my house at twelve o'clock, and that they knew I was going from Mr. Hoyland's house to 17, Palace Gardens, Kensington. They knew the persons I should see there; and on being asked if they were acquainted with any other persons residing in Palace Gardens, Ann replied to me that her cousin Henry Goodeve lived at No. 2—a house he had not long before purchased. If these be not facts demonstrative of a future state of existence, in which friends of former days are now cognisant of the events occurring here, I do not know what will be sufficient to force your mind to a conviction. But these are only a small part of the numerous proofs

I have had of the identity of persons with whom I had been acquainted years ago. I have in subsequent *séances* had many opportunities of holding intercourse with a score of other persons now in the upper magnetic regions of space surrounding this earth—intelligences, some of whom were friends here, and some of whom were individuals of whom I had been desired to learn facts that turned out to be marvellously true.

"Had I been inclined, I could have made an equally absurd affair of this serious inquiry, as some have succeeded in doing. My taste does not lean in that direction. When I am convinced that I have a good grip of a bold and sacred truth, it is not an easy matter to shake me from it. I have tested the fact of the spiritual manifestations most minutely and carefully, and I grieve for those who have concluded against it from a touchy disposition not to accept a truth simply because it does not originate from self, or on account of any other weak and personal consideration. It is easy to go to simpletons and say your neighbour is a credulous fool, and the simpletons believe it, because perhaps they have never seen a mesmerised somnambule under the influence of a magnetic impulse from the finger on the organ of self-esteem, obliged to utter the same class of words. You in your articles on 'Those Rapping Spirits' were influenced to trot in a groove on the point of *dignity*. Some *infallible* judges of dignity there are who cannot perceive in mankind any other high qualities but those of cunning and acquisitiveness. Man is a strange compound, and to the philosopher it is a curious subject of reflection how very trifling in themselves are the motives which make the wisest rush into the most foolish and illiberal courses. It is unnecessary, after the notices of the spirit manifestations in subsequent numbers of your periodical, to dilate on the deficiencies of philosophical taste that have characterised some of the would-be considered investigators of the subject. I may say that when I have been impelled by the lower feelings of our nature to feel desirous of attacking them, it has happened invariably of late that I have had affectionate warnings from the Spheres not to be guilty of the error of hurting unnecessarily the feelings of my friends. You will acknowledge that if the tendencies of Spiritualism are to make men more tender towards the feelings of their neighbours and more mindful of the obligation they owe to kindness and friendship, those tendencies cannot be very dangerous, or evil, or pernicious.

"With every good wish,

"I remain, my dear Mr. Holyoake,

"Yours truly,

"JOHN ASHBURNER.

"40, York Place, May 26, 1853."

I have myself so often witnessed Spiritual Manifestations that I could not, if I were inclined, put aside the evidences which have come before me. When Mr. Charles Foster was in London in 1863, he was often in my house, and numerous friends had opportunities of witnessing the phenomena which occurred in his presence. It is not necessary to enlarge this volume by a long list of names of witnesses. It may, however, be stated that many of them were persons of rank and of consideration in the higher walks of London society. They came, not many at the time, and thus had better opportunity for investigation. The second morning that Mr. Foster called upon me was about two weeks after his arrival in England. Accidentally, at the same time arrived at my door, Lady C. H. and her aunt, the wife of the Rev. A. E. I urged them to come in, and placed them on chairs at the sides of my dining table. Their names had not been mentioned. Mr. Foster having retired to the further extremity of the room, so as not to be able to see what the ladies wrote. I induced them each to write upon separate slips of paper six names of friends who had departed this world. These they folded into pellets, which were placed together. Mr. Foster coming back to the table, immediately picked up a pellet, and addressing himself to Mrs. A. E. "Alice," he said, which made the lady start, and ask how he knew her name. He replied, "your cousin, John Whitney, whose name you wrote on that little piece of paper, stands by your side, and desires me to say, that he often watches over you, and reads your thoughts, which are always pure and good. He is delighted at the tenderness and care which you exhibit in the education of your children." Then he turned towards me, and said, "Alice's uncle is smiling benignantly, as he is looking towards you. He says, you and he were very intimate friends." I said, "I should like to know the name of my friend," and Mr. Foster instantly replied "Gaven. His christian name will appear on my right arm."

The arm was bared and there appeared in red letters, fully one inch and a quarter long, the name William raised on the skin of his arm. Certainly, William Gaven was my dear old friend, and the uncle of the lady whose name is Alice. How, without yielding to the truth of the assertion of Mr. Foster, that he was a discerner of Spirits, the fact could be known to a complete stranger, who had all his life resided in the United States of America, and could know nothing, even of the names of the ladies whom I had brought into my dining-room from the street door, where I had accosted them, their names not having been known to my servants, is a phenomenon well calculated to puzzle the intellect of any one, not having faith in Spiritualism. Mr. Foster's arm retained, on the surface of his skin, the raised red letters for fully five minutes. I applied a powerful magnifying lens over them, and my two young friends and I watched them until they subsided and disappeared. It has been said that the skin was scratched by a pointed lead pencil, and I knew some persons who wrote on their arms, and succeeded in raising red letters; but the letters did not so quickly subside, and in some instances left sore scratches, marks or tokens of the want of common sense.

Mr. Foster next addressed himself to Lady C., whom he had never seen before in his life, until he met her in my dining-room. "Your mother," said he, "the Marchioness of —— stands by your side, and desires to give you her fond blessing and very affectionate love." He added, "Lady C., you wrote on a piece of paper I hold here the name of Miss Stuart. She stands by the side of your mother, and is beaming with delight at the sight of her pupil. She was your governess, and was much attached to you." He added, that charming handsome person, the Marchioness, "was a great friend of the doctor's. She is so pleased to find you all here. Her christened name is to appear on my arm." Mr. Foster drew up his sleeve, and there appeared in raised red letters, on the skin, the name "Barbara," which subsided and disappeared gradually, as the former name "William" had done. Here were cases in which it was quite impossible that the medium could have known any single fact relating to the families, or to the intimacies of any of the persons present. I had myself formed his acquaintance only two days, and the ladies had arrived from a part of the country with which he could not possibly be acquainted. It may be inquired very fairly, how it is proposed to connect such a narrative with any philosophical view of our mental functions? One need be at no loss for a reply, but it is more advisable at present to multiply our facts.

My father was, in his youth, addicted to the pursuit of knowledge, and besides physics and chemistry, although he never proposed to become a professional physician, he studied anatomy at the Borough Hospitals, and had the late Mr. Cline for his teacher, and Sir Astley Cooper for his fellow student. Mr. Foster had passed his life of twenty-four years in America. The son of a captain in a merchant ship, sailing from and to the port of Salem, in Massachusetts, he had never heard of Sir Astley Cooper. One evening, in my dining-room, a hand, as palpable as my own hand, appeared a little above the table, and soon rested gently upon the thumb and four fingers on the surface of it. Several persons were seated round the table. Mr. Foster, addressing me, said, "the person to whom that hand belongs is a friend of yours. He is a handsome man, with a portly presence, and is very much gratified to see you, and to renew his acquaintance with you. Before he mentions his name, he would like to know if you remember his calling your father his old friend, and yourself his young friend." I had forgotten it, but I remembered it the moment the name was mentioned: "he calls himself Sir Astley Cooper," said Mr. Foster, "and wishes me to tell you, that certain spirits have the power, by the force of will, of creating, from elements of organic matter in the atmosphere, facsimiles of the hands they possessed on earth." Shortly, the hand melted into air. Then Mr. Foster said: "two friends of yours desire to be remembered to you. They accompany Sir Astley Cooper, one was a military surgeon, and went to Canada. He was at Edinburgh your fellow student. He calls himself Bransby Cooper. The other was your intimate friend, George Young, who has communicated with you once before, since he left your sphere."

It would not be difficult to multiply facts relating to the Spiritual Manifestations of this very extraordinary medium. My friend, Sir William Topham, well known among all who have investigated Mesmeric phenomena, as the person who induced on Wombell, at the Wellow Hospital, that profound unconscious sleep, which enabled Mr. Squire Ward to amputate a most excruciatingly painful limb, above the knee, without the patient's knowledge, asked me to give him the opportunity of inquiring minutely into the phenomena, respecting which our friend Elliotson and I were so completely divided in opinion. Sir William, with the concurrence of Foster, fixed an early day for dinner. There were only the three of us at the dinner-table. The servant placed the soup tureen on the table. No sooner had I helped my friends to soup, than Sir William, who had preferred the seat with his back to the fire, requested permission to alter his mind, as the fire was too much for him. He went to the opposite side of the table, forgetting to take his napkin with him. Immediately, a hand, apparently as real as the hand of any one of us, appeared, and lifted the napkin into the air gently and gracefully, and then dropped it carefully on the table. Almost simultaneously, while we were still engaged over our soup, one side of the dining-table was lifted up, as our philosophic friend Mr. Faraday would conclude, by unseen and *unconscious muscular energy*, and the Moderator lamp did not fall from its place on the centre of the table. The decanters, salt-cellars, wine glasses, knives and forks, water carafes, tumblers, all remained as they were in their place; no soup was spilled, and Faraday's unconscious muscular force, or some correlative, or conserved agency prevented the slightest change among the correlative ratios of the table furniture, although the top sloped to very nearly an angle of 45 degrees. There was a wonderful conservation of my glass, china and lamp. The servant who was waiting upon us stared, lifting up both arms, exclaimed: "Law! well, I never!" and the next minute, he cried out, "Do, do look at the pictures!" which with their ten heavy frames had appeared to strive how far they could quit the wall, and endeavour to reach the dinner table.

The appearance of hands was by no means an unusual phenomenon. One evening, I witnessed the presence of nine hands floating over the dining table.

On one occasion, the Honourable Mrs. W. C. and her sister-in-law desired to try some experiments in my *dunker kammer*, a room the Baron von Reichenbach had taught me how to darken properly for experiments on the od force and the odic light emanating from living organised bodies. This room afforded opportunities for marvellous manifestations. When the light was excluded, the two ladies were seated on one side of a heavy rosewood occasional table with drawers, weighing at least seventy or eighty pounds; Mr. Foster and I were on chairs opposite to them. Suddenly a great alarm seized Mr. Foster; he grasped my right hand, and beseeched me not to quit my hold of him, for he said there was no knowing where the spirits might convey him. I held his hand, and he was floated in the air towards the ceiling. At one time, Mrs. W. C. felt a substance on her head, and putting up her hands, discovered a pair of boots above her head. At last Mr. Foster's aerial voyage ceased, and a new phenomenon presented itself. Some busts, as large as life, resting upon book cupboards seven feet high, were taken from their places. One was suddenly put upon Mrs. W. C.'s lap; others, on my obtaining a light, were found on the table. I removed these to a corner of the room, and put out the light. Then, the table was lifted into the air, and there remained for some seconds. Then, it gently descended into the place it had before occupied, with the difference that the top was turned downwards, and rested on the carpet. The ladies were the first to perceive that the brass castors were upwards.

One of these ladies had missed, on another occasion, her pocket-handkerchief, Mr. Foster told her she would find it in the conservatory behind the back drawing-room. It was behind a flower-pot. Mrs. W. C. went upstairs and found the handkerchief in the spot indicated. A similar event happened a second time. The question was, how the pocket-handkerchief could travel from the dining-room, all doors being shut, to the floor above, where it was deposited on a shelf in the conservatory. Mr. Faraday would aver that my facts were corroborative of his conservation of force.

In that back drawing-room stands a heavy Broadwood's semi-grand pianoforte. Mr. Foster, who is possessed of a fine voice, was accompanying himself while he sang. Both feet were on the pedals, when the pianoforte rose into the air, and was gracefully swung in the air from side to side, for at least five or six minutes. During this time, the castors were about at the heighth of a foot from the carpet.

Most sensitive persons are easily influenced to give way to appetite or passion by evil spirits. A Mr. Adams who lived for a while in London, with Mr. Foster, called me up at two o'clock in the morning, telling me that his friend was dangerously ill. He had returned late from a jolly party of young Americans. He had taken more than was good for him, and I found him lying on his back, snoring in insensible sleep. I prescribed a powerful dose of calomel and jalap, and returned to watch him in the bed-room. Suddenly, Mr. Adams and I being present, the bed-clothes were tightly rolled downwards as far as his groins. The shirt was then rolled tightly, like a cord, exposing to our view the skin of the chest and abdomen. Soon there appeared in large red letters raised on the surface, the word Development, which extended from the right groin to the left shoulder, dividing the surface into two triangular compartments. These were filled up with sprigs of flowers, resembling fleurs-de-lys. The phenomenon lasted nearly ten minutes, when the shirt and bed-clothes were unrolled gently and replaced as they were at first. Mr. Adams informed me, that on their voyage from America, during a severe storm which alarmed Mr. Foster, he said many spirits were surrounding him, and fearing mischief from them, thinking they would throw him into the sea, he threw himself on the floor of his cabin, when the same scene we had just witnessed occurred.

One evening, Mr. Charles Foster accompanied me to the house of my late friend Lord Arthur Lennox. Among those present were the Duke of Wellington, and the reputed editor of the *Times*, who appeared desirous of offering a test to Mr. Charles Foster; and the name of the spirit selected by Mr. D. was that of the *Times* Correspondent in China, whose name having been written by Mr. D. on a small slip of paper and properly folded so as to conceal it completely, Mr. Foster mentioned certain particulars of the death of Mr. Bowlby which were not only highly probable, but which gave the assurance that the name of the individual was correctly stated by Mr. Foster before the paper was unfolded. Various other similar phenomena occurred in the course of that evening. If the contemned, the much decried, although the truly important branch of human knowledge known as Mesmerism, or Animal Magnetism, has by degrees led us on to the acquisition of a power new to us, of communicating with our former beloved friends who now enjoy happiness under improved conditions, in other spheres of existence, are *we*, the cultivators of this knowledge, who have been always actuated by high and honourable motives, to quail before the sordid ignorance, the ignoble slander, the vile social persecutors of those who cater for the lowest prejudices of mankind?

Who that has heard of the facts relating to the turning, tilting, uplifting of tables, and other articles of furniture, who that has heard of the marvellous feats that have occurred in the presence of various American mediums, and especially in the presence of Mr. Home, at the house of Mr. Rymer at Ealing; at the Palace of the Tuileries, where a hand floated over the table, at which were seated the present Emperor, the Empress, one of her ladies-in-waiting, and Mr. Home, would doubt that some invisible agent was operative in the production of the phenomena? Without pretending to embark on a sea of doubts whether the facts be true or not, we may simply aver that we believe the hand which floated over the table at the Tuileries offered itself to be kissed by the personages present, and that it took up a pen, and signed the word Napoleon in the unmistakeable autograph of the Emperor Napoleon the First.

Why do we believe all this? Is it only because we have been told it? No, indeed. But we know, of our own knowledge, that things quite as wonderful have occurred in our own presence, by unseen agency. At this time of day, when we know of a book sent forth to the world by the Baron Guldenstubbe, reçounting phenomena that have been repeatedly witnessed by friends of our own; when we hear, on good testimony, of the remarkable phenomena occurring in the presence of the Baron d'Ourches in Paris, and of various of our personal friends in America, why should we doubt of the existence of invisible beings in the air, who are constantly engaged in works of good, or of evil? It is quite absurd to shut our eyes to facts. These either are facts, or they are not facts. If they be truths, no power on earth can put them aside. If they be not, they ought to have fallen into contempt, or oblivion, long ago. But they are daily occurring. What is to be done then? Let us be as obstinate in denying intelligent agency, as my good old friend, Sir David Brewster, or Lord Brougham, that will not mend the matter. We do not consent to bow to the dictum of either of these men. Why should we be guided in our conviction by their dogmatism, however eminent they may be in their respective positions? Lord Brougham's telling me that all dogs are wolves, would not make me believe them to be so, nor would my old friend, Sir David's assertion, that the last thing he would consent to conclude, should be, that the intelligent agents were unseen spirits, make me yield my common sense, that they could be no other.

This chapter might be lengthened indefinitely, for the tales that we could recount of the extraordinary phenomena which have occurred in presence of numerous persons who have assembled to witness them, are of a most marvellous nature; it is very ridiculous that numbers of these persons are ashamed to allow their names to be published because they fear to afford a testimony to truths so remarkable, we can only pity them, for they cannot be persons of the expanded intellect necessary to carry on the knowledge which the world requires for its improvement. We may be told that we are not men of science when we publish these truths, but we would rather sacrifice our reputation for science, than that for a sacred love of truth.

In conclusion, a desire to make more clear the relations of the somnambulist, the clairvoyant, the previsionist, and the prophet, we confess ourselves at a loss for any sharp definition. Each would appear to be but a hazy modification of the other. The first condition being a highly sensitive constitution. Many persons of this class require nothing but an extremely delicate magnetic influence from a neighbour to excite in them a clairvoyant state, and in the presence of such a state, that person may be able to read concealed words, or thoughts, or to predict future events from the power exerted by spiritual agencies, from the World of Spirits.

If any elucidation has been afforded by the facts in this chapter of the phenomena detailed in that Sacred Book, the New Testament, so that those who do not profess to believe, shall have reason to become animated with a desire to make deeper inquiry, and to be convinced that the physical laws of the universe are bound up with the grand moral and religious plan of the All-wise and All-beneficent being, by whose creative will we live, and move, and have our being, His servants who labour for His glory may remain amply satisfied.

CHAPTER XVI.

PHILOSOPHY OF RHEUMATISM.

Formerly, patients were wont to apply to their medical advisers on every slight occasion of disturbance in the condition of their health. It was not necessary to have any serious attack of illness, or to be the victim of some complicated disease. The doctor was consulted for catarrh, or for a tooth-ache; and as time passed, the physician, more observant, in consequence of the gravity of his calling, was better versed in symptoms, than in causes. Maladies were treated by the practitioner much in the same way as they are now by the Homœopath, that is to say, very little was done for them. The Homœopath, really the most illogical of reasoners, the least competent to judge on any matter of science, is now becoming the medical confidant in every family. It is not for want of knowledge that we are not homœopathic, but it is for having a kind of knowledge they do not know how to acquire. A homœopathic doctor is essentially a bad reasoner. He comes to a conclusion on a fanciful arrangement of his premises. He seeks for symptoms, and the symptoms he notes may lead him very wide of his true mark. Nothing can be more fallacious than the conclusion that a pain in a part is to be cured by a remedy calculated to increase that pain. The principle, if true, should be universal. Where can we find the case which is to be treated by an universal panacea? There is no absurdity too great for a homœopathic practitioner to digest.

All sequiturs and non-sequiturs are alike to him, and we have no occasion to refer him to the celebrated conversation, so often quoted, from Fielding's "Tom Jones," in which the serjeant twits Partridge for the supposed affront of calling him a non-sequitur. All homœopaths are little aware, that like the serjeant, they are busily employed in the mazes of a labyrinth of non-sequiturs. It is not our purpose to analyse the fallacies in which men of acknowledged talent delight to indulge. They are, themselves, not aware of the causes which produce in them the haziness of intellect, which ends in a conviction that an absurdity is a truth. We feel for them the same pity we must ever feel for men of weak and misguided intellect, although we must respect the courage which leads men to avow the conviction at which they have arrived. Some are given to one folly, some to another, but we will venture to predict that the Organon of Hahnemann will, in less than half a century, be as much scouted by the sound physicians of that time, as it has been by all the logical reasoners of the time in which we live. It is not enough that we express so decided an opinion on a work which is almost worshipped by the sciolists of our profession, we must go further, and attempt to shew how far the mischief of evil influences may reach. For this purpose we shall sketch a history of the opinions that have prevailed on the symptoms of the disease, which we know as Gout.

Many are the varieties of the forms in which gout makes its appearance. I have already pointed out the general symptoms by which its presence is marked. It is with no narrow view that one ought to be contented in tracing the causes of such a malady. I claim for for myself the merit of adopting implicitly the pathological view taken by Chomel, that in all cases, gout and rheumatism are one and the same disease, but I go further than Chomel, I embrace the views of Hippocrates, and assert that all rheumatic diseases are scions of the same stock as gout. Men of modern views will not admit my reasoning, when I point out the analogies of rheumatism, and many other diseases that are undoubtedly the offshoots of the same family. We are wedded, now-a-days, to the views of the school of Bichat. I was educated in the elements of medical knowledge, when the school of Bichat was in its zenith. I have no reason to decry all which was done by that great man. He taught me to throw aside all respect for authority, and to regard facts only as the sources of all our wealth in science. It is not necessary to dive into the subject of the changes induced by those maladies, which Hippocrates classed as rheumatic, in the different tissues of the body. Clear I am that one and all of the complaints which terminate in effusion are rightly placed in the category of rheumatism. I may be told that dropsy is an exception. No one who has studied dropsy, philosophically, can for a moment doubt that it naturally belongs to the family of rheumatism. This is not the place to discuss the subject, but whoever is willing to enter into the inquiry, will be soon convinced that all dropsies are the result of an overloaded venous system; or, in other words, of a catenation of those causes which emanate from Dr. Prout's secondary processes of destructive assimilation.

We are not now to discuss how far a dropsy is attended by rheumatic pain. We need not trouble ourselves on the subject of pain. We are merely tracing the causes of those symptoms which arise from errors in the secondary processes of destructive assimilation. We know that these are the diseases which may be fairly classed together as those terminating in effusion. Gout, then, is nothing more than rheumatism, attended by a rather more serious tendency to effusion. No one who has had much experience in gout, will deny that, in many cases, the danger, to the patient, is in proportion to the more or less rapid tendency to serous effusion. Whether it occurs suddenly in the theca of the tendon of Achilles, or in that of the Rectus sheath in front of the thigh, or in the pericardium is often a matter of mere accident. I have known the heart relieved in a severe case of gout by a copious sudden vomiting of a tremendous quantity of serous fluid from the stomach. The essential danger in gout is undoubtedly the same as that which occurs in dropsy. It is an effusion into a vital cavity. How, in the course of investigations into the diseases known as gouty and rheumatic, this chief danger has been overlooked is, to me, a matter of astonishment. We are apt to complain that men take narrow views of the subjects they investigate, but never has this observation been so truly a truism, as in the matter of gout and dropsy.

We are not now to consider the various ramifications of our subject. It will suffice for our purpose, to trace the more obvious causes of the effusions, which oblige us to place a number of diseases under the one category of rheumatism. There are *first*, the common catarrh, and its natural sequences; then *secondly*, the disease known as bronchitis, which, in fact, is catarrh of the bronchial surfaces; then, *thirdly*, the disease known as arthritic rheumatism, a malady varying so much in the degrees of its severity, as to constitute a great variety of nominal diseases, which are distinguished as scrofulous white swelling, morbus coxæ, podagra, acute rheumatism, rheumatic gout, chronic rheumatism of the joints, rheumatic affections of the muscular tissues, rheumatism of the theca of the sciatic nerve, lumbago, porticolis, and a host of other nominal distinctions, constituting a mass of confusion which is like the entanglement of a skein of thread. These are the results of narrow and circumscribed views of pathology.

We propose to widen our basis; and in order to be clearly understood, we desire to note the fact that not one of these diseases owes its origin to what is called inflammation. To be clear on this point, we may state that inflammation may terminate in gangrene, mortification, or the formation of pus, as in an abscess; but it never terminates in effusion. Those, who like the late Sir Benjamin Brodie, are given to dogmatise, may imagine that it is quite sufficient to state, authoritatively, that the fact is not so ; but it is quite impossible to yield to him, or them, on this point. Time will settle the dispute. Those who come after us will estimate our St. George's Surgeon at his proper value. By and bye, he will be known better for all that he has missed, than for all he has done.

Thus far we have sketched some ideas on the general propositions relating to the subject of Animal Magnetism. When we arrive at the special objects of the work we have in hand, it will be found that our views take a wider range than would be warranted, were we not obliged to explain views on physiological pathology, quite at variance with those accepted by the world of science. We do not affect to consider our own physicians and surgeons as the men constituting the world of science. They may be classed, for the most part, as the disciples of such journalists as the late important Mr. Wakley and his staff, who led our practitioners by the nose. Our views must be stated. When we undertook to produce a work on Rheumatism, Gout and cognate diseases, we were surprised at the mass of information gathered together on each special subject. We could not have believed that men had worked so long without arriving at some conclusion, that was, in some measure, philosophical. However remote from the common sense view of any subject men's minds may be, no one could anticipate, in persons holding the position of teachers, the absence of all power of logical abstraction. The books on Rheumatism and Gout are fearful examples of inconsecutive reasoning. One dogma, more especially put forward of late, is that of a poison in the blood of all rheumatic and gouty subjects. It obtains the name of Materies Morbi, and is a fanciful entity in the blood. How Dr. Todd, a Professor of Physiology, Dr. Garrod and Dr. Fuller, a physician of St. George's Hospital, could dare to risk their reputation on such absurd nonsense, is a question for themselves to answer. I am not their confessor, but the public have a right to much better reasons for the existence of a Materies Morbi than any they have put forth. We shall not go over the grounds of objection to the doctrine they choose to advocate. The time that would be occupied is too precious. To apply reason to matter of pure fancy and imagination is quite a work of supererogation. There is not in chemical or physiological principles the shadow of a foundation for the existence of a Materies Morbi in the blood in any case of rheumatism or gout. We will proceed, not to fanciful hypotheses, but to facts.

Rheumatism is a disease classed, by ancient authors, under the general category of catarrhal diseases. They had no special names for the complaints we designate as rheumatism or gout. Feeling sure that many ailments attended by defluxions of the nostrils were closely allied to those arising in certain states of atmosphere, and producing pains in various parts of the body, they naturally thought that the flow of which Shakespeare speaks, as *Rheum*, was the cause of the malady. The pain was the principal symptom, and the common people to this day associate the idea of rheumatism with pain, and that kind of defluxion we know as cold in the head. If our notable physicians were to tell their patients when they were called in to prescribe for a delicate lady of fashion, who had been obliged to use a dozen pocket-handkerchiefs, that she had got a severe catarrh, occasioned by a material poison in the blood; the probabilities are, that the lady would feel herself to be much wiser than the doctors; and yet this is the folly to which their books have committed them. They need not try to escape from this dilemma, by saying they did not write about cold in the head. If not, they ought to have done so, or to have acknowledged that they had no idea that one disease had any relation to the other. A research into the works of the ancients would have given them clearer views of what they were professing to write upon.

There are still some subjects, upon which it behoves us to remark, that do not appear to be within the immediate scope of our inquiries, but which must necessarily occupy our attention. These relate to the catarrhal diseases. No one who had not considered the subject of the common cold in the head, could imagine that this was the fertile source of so much mischief. We are apt to regard it as a trivial passing occurrence. We know that there exist an immense variety of degrees of severity in this, as well as in every other disease. We are willing to believe that cold is so great a trifle that we need not trouble ourselves, or our friends upon the matter. We are sure of our own approbation when we conceal our sufferings from others. But we little think that we bear about us the elements of a fatal disease. Where is the man, who has even a severe cold in his head, who would not be ashamed of the sympathy of his friends for his being obliged to suffer what he does. He knows well that his ailment is attended with severe headache, an inaptitude for business; a very inconvenient defluxion from the nostrils; a sense of roughness and pain in the root of the nose and in the throat; a sensation, I have elsewhere spoken of as physical grief. I was wrong then, for grief is always mental, and mental grief most frequently attends cold in the head. We are not now to discuss why this state of the mind should attend upon bodily disease; it is only here alluded to for the purpose of shewing that mental phenomena are present in all gouty diseases. We need not be alarmed at the decided tone of this assertion. We may be prepared for much more boldness than has yet been evinced. The boldness which ventures upon bearding the whole of our professional brethren, in their awful dereliction of physiological duties on the truths of Mesmerism, need not shrink at the simple facts connected with the classification of rheumatic and gouty diseases. When we say that the ancients were more philosophical in the alliance they established between diseases, we repeat only what Dr. Copland has said in his remarkable dictionary. We do not derogate from his merits, when we say, that notwithstanding his stupid and ill-mannerly personal conduct to me, he is still worthy to be considered as one of the most philosophical physicians of his age. He has a remarkable brain, and I do not despair of the grace of repentance reaching him. If he could reflect, as he should do, upon the discreditable exhibitions he has made at the Medical and Chirurgical Society, on Mesmerism, he would be led to effect a complete change in his views. He has too fine an organization, and, I trust, too large a mind to allow himself to be ranked, for the future, among the stolid stupidities of his age. I feel profoundly for any man, who, with a large capacity, can allow himself to be ranked in a grade far inferior to that which his talents entitle him.

We are now prepared to consider the subject of our inquiry in the statistics of catarrhal diseases. We are sure that no man who well reflects upon the necessity of atmospheric purity in relation to health, can hesitate about the numbers of those who annually fall victims to our climate. We think it a matter of ordinary occurrence when our Registrar-General publishes his periodical returns of the numbers who die from diseases acquired from habits of careless indifference to the wants of wholesome air. We seldom desire to look too minutely into a subject involving the loss of time, when the facts lead to no particular purpose. We little reflect that the purpose so nearly attaches to our own well-being. We are habitually given to think little of the atmospheric changes which are continually occurring. We are sure we cannot control them; and if we could, how should our knowledge enable us to take advantage of them. We are prepared to buffet them, if we do but guess that they are to assail us; but we are not often solicitous to use the best means of avoiding their noxious influences. Now, what is our conclusion from all these considerations? We fear only that we may be obliged to catch a cold, if we do not wrap ourselves up, with a due regard to dryness, and a sufficient degree of warmth. This is all we think of. We little think that, notwithstanding all our care, we may be overtaken by accidental changes which produce a catarrh. Now what is a catarrh? And how does it originate? As to what it is, the experience of most people will establish that it begins with a chilliness; a sensation of discomfort; a cold creeping down the back. These symptoms are followed by those we have before described, a stuffing sense about the head, pain at the roots of the nose; often across the forehead, a true rheumatic sensation, involving the fibrous tissues of the skull; here is enough to tell us that cold in the head is a gout, and not as some poor thinkers have said, a simple premonitory symptom of gout. Besides the symptoms we have enumerated, there are others deserving of attention. The disease is apt to spread, the dryness of the throat becomes a sore throat. The sore throat is sometimes complicated with inflammation of neighbouring parts. The tonsils may become the seats of disease. The uvula and the curtain of the palate, perhaps the whole palate itself may partake of the irritation. How are we to know where the disease will proceed? We try to reduce the inflammation, when it is present, by various means, but we must not conclude that every catarrh is attended by inflammation, any more than that every rheumatism is attended by an inflammation. Inflammation is a word little understood. Let us ask: What is inflammation? We are not bound to recur to the definitions given us by the most practical surgeons. We know, that if we were, we should soon be as bother-headed as they are. Some of them are liberal enough to have accepted Professor Macartney's doctrine. Others in a mulish spirit of silly vanity adhere to the illogical conclusions they had formed in opposition to all common sense. Inflammation is not a single phenomenon. It is the result of numerous concurring events. It can take place only in certain tissues. It can arrive only from certain sources. It is a congeries of phenomena. No single symptom of this congeries can constitute inflammation. We all know that pain accompanies gout; but pain, in gout, is owing to an affection of fibrous tissues. Fibrous tissues are never seats of inflammation any more than the woody fibre of trees is a seat of inflammation. Who ever heard of a tree dying of inflammatory disease? The structure of the compact part of an animal is just the counterpart of that of the compact tissue of an oak. Nobody thinks of bleeding an oak to cure it of pleurisy. We proceed to regard inflammation in another phase. We wish to exhaust our subject in order to shew how deplorably ignorant surgeons have been of their profession.

When the stag sheds his horns in the season of rutting, he furnishes an example of increased animal heat in the part which is to furnish the new horn. It is a fact and a truth. If we examine this subject, we shall soon perceive that we do not require much thinking to distinguish it from inflammation. There is much heat, and some redness, but no inflammation. The animal is more vigorous, more apt for all the healthy purposes of his existence, than at any time of his life. Inflammation does not appear to be attended by health. We are at a loss to account for our errors, when we examine logically the questions on which we have blundered. No subject in surgery has been so much blundered as this. Who would have thought of saying that a disease was a necessary part of a healthy process; and yet this is just what surgeons have said, when they have asserted that inflammation was necessary for the cure of wounds. No one could conceive why such an idea had entered into their heads. They must have sadly mistaken the phenomena they were studying; or they never could have arrived at this conclusion.

We are about to discuss this matter in another point of view. What could induce men to imagine that the phenomena of inflammation were to be accounted for from those of exhaustion. We know that when pus is present in a part, it is attended by a sudden prostration of strength. No one who ever saw a patient suffering from an abscess, could say that that person was necessarily getting stronger as the abscess increases in size. No one could imagine that inflammation ending in abscess could strengthen a man; yet, logically considered, this is the conclusion to which even the great John Hunter's doctrine tends. We are not now to derogate from the high reputation of that great man. Had he been now on earth, he would have, long ago, discarded most of the errors he fell into. Great minds are not mulish in their adherence to fallacies; and we have a right to pay them the compliment of supposing they would still have been in advance of their age.

Rheumatism is not a disease of inflammation. When inflammation is present, it is a phenomenon belonging to neighbouring structures. It is not a part of the disease. It is an adventitious circumstance, and were I called upon to define the true characters of rheumatism, it would be said rheumatism, gout, and all rheumatic diseases have, in common, the striking symptom of pain, seated in fibrous tissues.

1. This, though not without exception, may be regarded as their first characteristic.

2. A second is a diversity of forms.

3. A third is intermittent, or a tendency to recurrence, rather than to relapse; with disappearance and reappearance, at intervals, more or less distant, and the recurrence more or less sudden and frequent.

4. A fourth, a tendency to change the locality of the phenomena; a tendency known under the terms mobility, or metastasis; and this includes the tendency to excite abnormal growth, or to irritate parts to abnormal

secretion; or to weaken textures, so as to allow, passively, escape or effusion into organs or tissues, other than those in which the presence of pain was first declared; and thus, giving rise to changes of structure, constituting a vast variety of disease.

We are ready to awaken attention to the more immediate relations of our subject, considered in reference to Animal Magnetism. We have all along held that our views on inflammation were not those of the practising physicians and surgeons of England. We have had many opportunities of knowing how vague and illogical these are, but we desist from further exposure of them, having established the leading fact relating to their want of logic. We are now to inquire into the subject of pain. What is our first question. Can any one tell us what pain is? We may say we never met with any person, who could give us a definition of the word pain. Much has been written upon it, but no one has yet properly defined it. It may be said, that to do so requires a knowledge of Animal Magnetism. What is pain is a question that is more easily asked than answered, unless we have recourse to the phenomena of magnetism. We must then recur to the ideas we have left behind us in the introductory part of this treatise. We must reflect upon the phenomena of magneto-electric currents. These explain our subject completely. When an individual sleeps, he is under the influence of attractive forces; the particles of his brain and nerves are approximated to each other. When he is awake he is under the influence, to a limited extent, of repulsive forces. These operate in keeping asunder the particles of brain and of nerves. When the individual is in a healthy state, his tonic forces are engaged in making him quite unconscious of his own existence. He becomes conscious only when some impression is made upon his nerves. He is quite ready to allow that impressions may be pleasureable or disagreeable. In either case, pain is not necessarily present. When does an impression become painful? We are accustomed to reflect that sensation is not complete unless the brain is conscious of the impression which causes the sensation. The impression affects the functional extremity of the nerve, and that being influenced propagates the fact to the sensorium. This must arise from what we should be disposed, in the language of Magnetism, to call an inverse current. The direct current along a nerve conveys the influence of the will to the extremities. The inverse current conveys messages to the brain. We have regarded man as a magnet. Every particle of a magnet is magnetic. We are engaged in tracing the influence of impressions upwards; but knowing how far every joint of the body is an union of two magnetic poles, we know what must be the inevitable consequence of a disturbance of the polarities of parts. When polarities are quiescent, or to speak more properly, when the magneto-electric currents are normal, the individual is not conscious of any sensation. When new impressions arise, the polarities being reversed, the currents proceed upwards. A sensation is experienced. In order to constitute that sensation disagreeable, the degree of inverse force must be increased; but when it becomes increased to an extreme degree, there recurs the phenomenon of pain. How is this ascertained? We must again have recourse to the phenomena of sleep and wakefulness.

If we wish to witness complete anæsthesia, we reduce a magnetised patient to a state of rigid stiffness. Opisthotonos is such a state. The individual is quite unconscious, and Mr. Rutter has ascertained that there is a difference between a person in this condition, and one who is awake. It is this, that in the person awake, poles can be detected in his body, and he is subject to the laws of normal currents. When he is completely rigid, he is simply a conductor of electricity, and does not manifest poles. This curious fact is highly illustrative of a theory of pain. Men are little aware that when they suffer pain, they have the polarities of the part where the pain is situated, reversed to an extreme degree, for the unconscious sleeper has no poles, and cannot suffer pain, but upon being awakened, his poles are re-established, and if they be reversed to an extreme degree, he may suffer agony. We have a further illustration of this fact in a lacerated wound. The hand may be torn by a large nail, by the bite of an animal, or may be smashed by the crushing accident of a hammer, or other impinging weight falling on it. The pain we know is agonising. If in this state of things, we can realise the polarities so as to have their normal direct currents in operation, the pain ceases immediately. Professor Macartney lectured on this subject many years ago, when I was his pupil. It has required nearly half a century to give currency to his doctrines. He invented a very simple apparatus for the purpose of keeping up steam at a temperature of 96° or 98° Fahrenheit. This was applied around the lacerated part by means of an ample tube of serge, or other woollen material, to ensure an equable temperature, and the pain immediately ceased. I can testify to its marvellous effects upon my own hand and arm, when I was suffering almost unbearable excruciating agony, from having injured one of my fingers in dissecting a woman who had died from puerperal fever, and the pain was removed instantly. I little thought then that it would fall to my lot to immortalize my beloved master's fame by explaining the magnetic philosophy of his simple, but most important, invention. This led to his discovery of Water Dressing, which is, in fact, the same affair in principle. Water Dressing, properly applied to a surface of the body, which has suffered solution of continuity, immediately relieves pain; and its *modus operandi* is simply the application of an armature to the severed poles of the magnet. It acts quite in analogy with the armature by restoring the normal direct currents. This may have appeared digressive, but it is not so. It relates to the explanation of the phenomena of pain.

We proceed to examine the relation of pain to the phenomena of rheumatism, for by this time our readers must have perceived that Chomel's doctrine of the identity of *rheumatism* and *gout* is here fully admitted. Indeed to the philosophic mind that has had ample opportunities of investigation, his conclusions are undeniable. We are now inquiring how pain occurs in rheumatism in any of its forms? There cannot arise a doubt that there is an extreme reversal of polarity in the part afflicted with the pain; but the question is, how does the polarity become reversed? We are not tracing the pain beyond the painful part. We have now nothing to do with the nerves conveying the sensation of pain to the brain.

We must explain that the joints are furnished with poles. The upper condyle is, when in a state of health, free from pain, as well as the lower. Being opposed in the polarities, they attract each other, but disease is always a reversal of polarity, and is attended by increased fullness of the veins of a part. All disease is attended by an alteration in normal currents. In health Tone prevails, which means a direct course of the magneto-electric currents from the brain down the trunk and extremities. The inference is obvious that disease is attended by Clone on the removal of polarity, the currents proceeding inversely. These facts are hardly worth repeating, but for the explanation of the rigors that occur in the common cold in the head, and in fevers of every kind. We need not recur to the persistence of those inverse currents which I have named lethal, from their invariably ending in death in all those extreme cases where polarities cannot be made to become normal. In extreme cases of typhus, I have by steadily pursuing mesmeric passes, for several hours, restored patients to health; thus establishing a complete cure according to the scientific principle of Animal Magnetism.

Amidst the numerous temptations occurring to the

mind, on the illustrations elucidating the subject of effusions, for we have always to carry in our view that the magnetic forces governing the phenomena of rheumatism and gout encourage effusions of fluid into textures, perhaps remote from the seats of pain, and yet that all the diseases analogous to those we are considering are more or less attended by effusion; we cannot look at the common cold in the head, without being aware that copious effusions of fluid are taking place; we cannot think of bronchitis without the certainty of a quantity of fluid in the bronchial tubes; we cannot think of metastasis, of rheumatism to the heart without the probabilities of effusion of fluid; nor of analogous affections of the peritoneum, without analogous rheumatic effusions, which we know as the dropsy called ascites. Regarding the philosophy of this subject in this strain of generalization, we may be content to consider rheumatism as almost an universal disease.

We are arrived at a point of our inquiries which will not allow us to proceed without further investigation into the causes of rheumatism. How many theories have been advanced on this subject. We are at a loss to reckon them. We are sure that we are tired of reading the unsatisfactory solutions offered by pathologists of the difficulties of our subject. Our latest writers are quite as unsatisfactory as their predecessors. If we were to point out a man who had really done some service to medicine as a science, we should name with much respect Dr. Prout. He stands alone, as not only a pathologist, but as a deep-thinking, far-seeing physiologist. We are not sure that we are right when we single him out as the only man who has given us safe bases for chemical facts. We have heard much stress laid upon the reputation of German chemists. Prout thought them too eager to establish new views; and he had very cautiously gone over all their ground, not once, twice, or a dozen times, but many more; and he concluded that Liebig, especially, would be found to be a very unsafe man. My first acquaintance with Prout was when we were both members of a club composed of Licentiates of the College of Physicians. That, was many years ago, and I have since in the intercourse of medical life had occasion to meet him often in consultation ; I never felt that I was in the company of a man who gave me the idea of any thing, but of a character distinguished for deep sagacity and philosophical humility. The vanity which characterises many of those now in the first walks of our profession is wanting in the dignity which ought to distinguish the high calling of the physician. We are not to disparage individuals, but we may remark that the less men are distinguished for scientific thought, the more apt they are to give themselves the airs of professional pride. The reason is obvious. Self peeps out instead of that dignified regard to philosophic thought which should characterise the career of the medical student. We are not sure if Prout was not too much of a humble philosopher. He gave himself up to trains of thought that left him without refuge from new ideas. He could never listen patiently to the facts of Mesmerism, and this was quite unworthy of a man of his genius for one of the most remarkable admissions he makes in his great work, is on the vitalizing power of the vessels receiving the product of assimilation. We are not able to trace the point at which vitality reaches these products. The probability is, that light, in some form, exerting an electric agency, operates in changing the product of assimilation, several times, before the higher forces of vital influence are able to act upon them. They undergo, in fact, many chemical changes, for we are always to remember the law of series; a law regulating the upward progress of all organic changes.

To say that Prout was struck with the idea, that vitality was communicated to the organic molecules resulting from the progress of the products of assimilation, is not saying enough for Prout's sagacity. He affirms in as many words, that a force, or agent, analogous to electric force, may be supposed to influence those molecules from the sides of the vessels through which they travel. There is not much harm in supposing such a thing, but it was actually supposing that the molecule became mesmerised. The word was not used, but what is a word? Dr. Prout unwittingly meant that he saw further than many of his neighbours, who would be apt stupidly to deny that the fact is so. It matters little who denies the truth. Luckily, God's universe is based on a law of truth; and whether stupid owls, or learned societies, deny that fact, is of not the smallest consequence. We are proceeding in our inquiry into the causes of rheumatism. We must be allowed to plough up our ground, and to shew which of our labourers has done his work well. As to those who have ploughed badly, we must leave them to fight out the causes of their bad furrows among themselves. Prout's, then, is the only name for the prize.

We proceed to inquire whether he deserves our unqualified praise. We are far from going with him through all his speculations. His doctrine of intelligent agencies, however ingenious, is quite untenable. Much that he has written, both in his Bridgewater Treatise, and his book on the Stomach and Renal diseases, is quite out of our power of comprehension. Perhaps the fault may be our own, but not a little time, and a great deal of labour of pen and thought have been devoted to the study of his very meritorious and elaborate works, and the conclusion is, that, as a whole, there is no work of the age so sagacious and profound, but which may at the same time be said to contain matter so difficult to comprehend. We are not bound to praise that which we do not understand, but we should be sorry not to confess that Prout has been, in Physiology, the beacon light of his age.

There are few men who could be compared to our author as a man of deep philosophic thought, and we are about to shew the influence of his discoveries on the cause of Rheumatic Diseases. It is well known that he was the first who suggested the ingenious idea that all the processes of cookery were based on the principle, that the elementary aliments of which they were composed, must inevitably imitate the composition of milk. No one but a very deep philosopher, possessed of an original power of thought, could have suggested such an idea. It was so true in all its bearings. We are not now to inquire into the principles which he classed under the head of alimentary. To the whole of his facts and reasonings, in this department of his work, we must cordially assent. He is too captivating in the truth of his details; too lucid in his arrangement, not to require us to bow to his superior knowledge. All we have now to do with, is the fact that the debris of his secondary processes of assimilation are accumulated in the venous system. This striking fact brings us into contact with Animal Magnetism. The process of reasoning to explain what here appears as a non-sequitur, must first be entered into. That process is not very difficult.

We are aware that arteries are accompanied by nerves. We knew that, in health, the currents along those nerves are in the direct course. When the nerves are called upon to carry influences in a course contrary to those of their usual habitudes, they become the vehicles of lethal currents. We have learned that Matteucci was sorely puzzled to account for the current in the frog travelling in a sense contrary to the normal one. He did not reflect that lethal currents were those he was observing. If he had reflected upon Sir Benjamin Brodie's experiments, he would at once have perceived that in a dying animal the currents are always reversed. We are not anxious to trace the causes of the confusion that has perplexed this philosophical inquiry. We only know that the inverse current along nerves is attendant upon disease, and that when the veins receive their share of the influence they are apt to retain a much larger portion of the debris of the secondary processes of assimilation, than when their sides are stimulated by tonic currents. The magneto-electric or odic influences pervading the whole system, in health, are always tonic. When the veins become overloaded, they must necessarily become clonic. We are not prepared to occupy space by going too minutely into the proofs of this; but a long course of experience has established the proposition, that whether disease is productive of reversed currents, as in the catching cold suddenly, or in contracting a fever from contagion, the events we have insisted upon, are to be uniformly relied on.

We are now able to perceive why in rheumatism, authors have insisted upon premonitory symptoms. Many have written on these, who have no idea of their real meaning. A premonitory sign may be quite unperceived by the patient. A man about to be attacked by Acute Rheumatism may not be aware that his friends have observed him, unconsciously to himself, applying his hand to parts which subsequently become the seats of severe pain. The patient did not think that fully twenty-four hours before his attack he had suffered the warning which Professor Macartney aptly enough, called organic or instinctive sense of injury. The fact is, the impression, which a reversal of polarity had established, was only incipient. The inverse current required time to produce the *pain*. Here was really a premonition of a reversed current, unknown to the individual, but known to the instinctive sensibility of a part. This fact, so pregnant with instruction, on the phenomena of vitality, has escaped the notice of most writers, and when they write of premonitory symptoms, they blunder awfully in designating an actual cognate disease as a symptom of another disease.

It is not our object to display the confusion of wretched thinkers. We desire only in pointing to the fact to clear the subject from extraneous considerations; a chilliness, a rigor, a sense of sickness, may each and all be premonitory symptoms, but they belong to a number of other diseases as well as rheumatism, and hardly deserve to be put down, except for the purpose of defining our meaning. When the disease arrives in full force, we know that the premonitory signs have preceded it.

Continuing our observations, we must ask, how does it happen that we find rheumatism spoken of as a disease mostly attacking youth, and hardly ever affecting persons of more advanced periods of life. There is a distinction between youth and age, and the distinction is important. We come to consider a very curious subject. Dr. Bright, in his quarto book of Reports, has been at much pains to point out a connexion between a degenerated secretion of the kidneys, and inflammation of serous membranes. His words are, "When the secretion of the kidneys is greatly deranged, the serous membranes seem always ready to become the seat of inflammatory action." (Pleuritis, Peritonitis, Capsular Inflammation, &c.) page 10. We might quote from page 8, from page 59, from page 66, to shew that Dr. Bright thought much of the important influence of deranged kidney in the production of *serous inflammation;* but the truth is, there was no inflammation in any of the cases. We are about to shew that many years ago, after no small amount of deep thought on the numerous cases we had observed in a very large field of practice, we had had occasion to note the same coincidences without believing in the necessity of inflammation, for the production of lymph on a serous surface. We must particularise. A boy, fourteen years of age, the son of a celebrated medical practitioner, an intimate friend of mine, had a play-fellow staying in his father's house. The two boys, full of mischief, chose to amuse themselves in warm weather by splashing each other with water in a large shower-bath they had used. The pastime lasted for more than an hour. The one boy suffered nothing from his idle imprudence. The son of my friend, a boy of fair complexion, light hair and strumous habits, was dull, and in the evening had rigors. The pulse became frequent, the bowels costive; the mouth clammy; the tongue loaded. There was no pain to be detected on the closest examination in any part of his body. The urine was brownish red, depositing a pinkish brick dust sediment. The father regarded the case as one of fever, and treated it with aperients, sudorifics, and copious diluents. A sense of malaise was the principal symptom, and though no pain was observable, there was the peculiar anxiety of countenance which belongs to intestinal mischief or to incarcerated hernia. This condition remaining stationary for ten days, a very eminent physician again examined the whole body without detecting the source of disease. The next day, some uneasiness and pain were felt all over the abdomen, and the day after the boy died. On a post mortem examination, it was discovered that the whole peritoneal surface of the intestines was glued together, except where a little serous fluid, not amounting to half a pint, had been effused. The kidneys were degenerated, the liver was not unhealthy, the lungs were studded with abundance of small tubercles. If this intestinal lesion is to be called peritonitis: where redness, pain, tumour, and all other signs of inflammation, except the one of abnormal growth, were absent, how is inflammation to be distinguished? Like those cases observed by Dr. Graves, of Dublin, of so-called rheumatic pericarditis, without previous rheumatic symptoms, this was peritonitis, if it please pathologists not to distinguish the process of lymph deposition, without a symptom of inflammation, with one to which the nosological termination of *itis* might be applicable. To account for *growth*, *nutrition*, or *secretion*, abnormal in serous surfaces, and on mucous surfaces accompanying rheumatic disease, they are as already observed, only a *characteristic* of these diseases. It is a tendency of Rheumatism to form deposit or secretion in structures, other than those in which pain had declared itself. How was the above case to be classed with Rheumatism, seeing no pain and that a mere tenderness, thirty-six hours before death, had been observed? The causes of Rheumatism were abundantly in operation. The premonitory signs of the malady lasted ten days. The reversals of polarity were there. The signs of secondary mal-assimilation were there. The kidneys told of venous fullness from the debris of the secondary assimilation.

Reflection on the philosophy of growth and nutrition will satisfy the inquirer that it is quite unnecessary to seek for inflammation, in every case of disease, in which either deposited, or abnormal serous effusion is present. The influence of cold is, in many cases of unsuspected degeneracy of kidney, sufficient to stimulate organs to rheumatic trains of disorder, without looking for inflammation, which is, in fact, a word accounting for nothing. If habit must require the resilience of the mind to be subservient to the idea of the word inflammation, then let the votary of habit detach portions of the phenomena from other comprising parts of the aggregate necessary for the favourite idea, and he will find that we may have portions of a morbid condition, called inflammation, and not the whole. Normal growth or healthy deposition may proceed well, under the laws regulating healthy or normal magnetic polarities of growth. Deposition of lymph, in excess, may take place in the delicate structure of the kidney, or on the surface of serous membranes without any of the other phenomena of inflammation. It does not follow that depositions of lymph and all instances of hypertrophy result from inflammation. Hypertrophy of the Heart, so commonly coincident with some severe cases of rheumatism of the joints, does in all probability result, in the majority of cases, from no inflammatory process, but simply from the individual being, for a time, placed in those magnetic relations, in which the currents are directed in abnormal trains. It is no where contended that muscular growths result from abnormal stimuli, unless it be in such cases as those in which the heart becomes subjected to the influence of abnormal magneto-electric currents. The nutrition of the heart has some peculiar conditions which render that organ liable to abnormal growth under rheumatic trains, or what is analogous to them, abnormal trains stimulated by the loaded Vena Cava and right auricle. These effects are caused by the local arrangements for the supply of arterial blood to the muscular substance of the heart. Other muscular structures are liable to rheumatic inconveniences which are not attended by hypernutrition, but on the contrary, sometimes with a condition of atrophy, arising from the diminished supply of arterial blood, consequent upon loaded veins.

To clear the subject of inflammation from much confusion, it may be stated that Professor Macartney never allowed the word congestion to be applicable to arteries; veins may be congested, arteries never. A full state of arteries favours nutrition, and up to the publication of his remarkable book on inflammation, this subject was too frequently confounded.

All growth, in every step of the process of nutrition in any part of the body, must result from a stimulus of evolution, which is necessarily caused by a magnetoid current, or evolution of polarity, producing the deposition of a particle, according to a law, analogous to the law of deposition in Electro Metallurgy. When, under normal trains, this law obtains its full development in living organisms, there must exist polar arrangements constantly operating to produce a balance of absorption.

Small as are the aspirations of the authors of books on Rheumatism, they ought to perceive what a vast field is opened for them by the labours of such men as Mr. Rutter, and of the Baron von Reichenbach, whose last great work is a monument of careful research and transcendent genius. We, of this country, can little appreciate the discoveries made by this noble benefactor to his race. We are apt to devote ourselves to hero worship, and we lose sight of the careful, deep-thinking men, who work with slow industry in unravelling problems we would be glad to understand. Volumes of verbiage are poured out on the medical world by fashionable writers, who are not aware of the discoveries alluded to. They would stare if they were told that much as has been done at our Royal Institution, much more profound and important discoveries have been made on the relations of Electricity to Animal Light by the Baron von Reichenbach. They, ignorant of these discoveries, go on pottering at subjects which can receive no illumination but from the very facts we are now discussing.

Light is of all subjects, at present, the most interesting to the medical philosopher. We cannot go astray in believing in its vast importance. We are obliged to contract our observations into a limited space. We declare ourselves unable to ratify all our facts, simply because some of them have reached us at second hand. These facts are, however, so explicit, that we do not hesitate to adopt them, because they are quite consonant with the observations we have for years been in the habit of making. So that no ratification would be needed, but that we are willing to yield the palm of discovery to one who has shown his superior logical power in dealing with every subject he has examined. It is not to be understood, that having heard of the Baron von Reichenbach's discoveries, I can vouch for all the facts he has published; but that making allowances for the superior nature of his mind, I am to be considered as a labourer in the same vineyard. There is not much to be said on the subject of the philosophy of light, on which he has forestalled all his cotemporaries. We may state that he has laboured not in vain when he tried to establish the separate qualities of the magnetic force to which he gave the name of OD. He then produced another work, in which he proved that he had deeply reflected on more forces than Mr. Faraday has ever dreamed of. They are not made up of facts limited to the one idea of Magnetism and Diamagnetism. This is a hobby which will soon be ridden to death, for it depends upon arrangements capable of being illustrated only by certain tricks, if the expression be allowable, of electric batteries in relation to the fields of magnetic force. There is no doubt that these fields exist in Nature. I have shown that Mr. Faraday was anticipated in the whole of his Diamagnetic facts by M. de Thouvenel so long ago as 1784. He may be assured that the matter will not remain a secret in my note published in the English Translation of the Baron von Reichenbach's Researches. It will form the subject of more extended inquiry, and the facts of the existence of certain metals placing themselves, or being placed by superior agencies in certain directions in the bowels of the Magnetic Earth will not militate against my arrangement of forces, conceived in a spirit of philosophical investigation long before Mr. Faraday thought of the conservation of Force, which is only in fact another expression for the same idea; but which has been put forth with a pretension of originality to which he has no claim. In a paper which appeared in the Zoist, in April, 1846, I wrote these words. "Some of the most interesting as well as the most important among the facts which have come under the observation of the cultivators of mesmeric science, relate to the analogies between the mesmeric and magnetic fluids. That these exist, is an assumption based upon the direction taken by certain currents of forces, and by the light emitted under certain circumstances, when particular arrangements of matter are attended by its evolution. Abstractedly considered, there is no proof that magnetic matter exists, and its entity is granted, only to account for numerous phenomena which require explanation and which can be accounted for only by a hypothetical reality. . . . . . Assuming that electric and magnetic currents exist, we may infer that a fluid in many particulars analogous to magnetism may be proved. We may be led to infer from numerous observations, made at different times by numerous persons, that certain effects are produced by the operation of an influence from one living being upon another. An endeavour will be made to show that this influence, indebted for its existence perhaps to pyschological causes, operates by the agency of a supposed fluid, which producing physiological phenomena, similar to those produced by the magnetic fluid, may be inferred to be analogous to it. Striking facts may be adduced, too, which tend to the conclusion that the exercise of the faculties of the human mind, and particularly that of the will, is attended by the emanation of a fluid from the brain, from the fingers, seats of the functional extremities of nerves, or from some parts of the person who may be exercising the mental faculties. I propose to show that the same series of events may be produced in individuals of a certain nervous diathesis by the impingement of a fluid evolved by the will of another, or by manipulations attended by the emanation of the same fluid; or by certain emanations of magnets, or from some metallic wires through which currents of electricity are passed; or from the direct application of certain metals. I do not attempt to establish the identity of these fluids, for the facts daily developing themselves tend to show that the distinctive properties of these fluids are as various as the substances from which they emanate, and it may be that the great power antecedent to all consequents may ordain the simplicity and unity of one electric, gravitating with centrifugal force, and evolving an infinite complication and variety of magnetic cohesive and repulsive agencies; the entire system emerging from the *Volonté directing*, 'La Grande Formule.'"

If we were to attempt to trace the ideas which pervade Mr. Faraday's confused abstract of his Lecture on the Conservation of Force delivered at the Royal Institution on

Friday, 27th February, 1857, we should find that all his views were anticipated eleven years before in this paper, published in the Zoist, only that Mr. Faraday had never conceived the idea of so regular and complete a gradation of forces as those pointed out in some late essays I published in the Spiritual Telegraph, which contain ideas entertained many years before, while I laboured at Chemistry, at the Middlesex Hospital. People think these ideas are new. They occur to various minds, and are so lightly estimated that they are hardly thought worthy of publication until one finds that they are claimed as important discoveries. We are far from being disposed to resist the claims of discoverers, and should be sorry to be misunderstood. Accorded to each claimant be the meed due to his own labours. We are too well acquainted with the subject of forces to yield the palm to either Mr. Faraday, or the author of a book on correlative forces, in which propositions are stated that would be considered too strongly tinctured with materialism, were they to be stated in their palpable nakedness. We do not, however, pretend to cross-question a lawyer. He is the best judge of what he means, though that meaning may not be the most philosophical. We do not pretend to criticize our neighbours, unless we have found them sneering at our own labours. They have a perfect right to their own opinions. Time will establish the truth.

We are now bound to proceed with our inquiries into the influence of light in the production of the phenomena regulating the polarities of the human body, for this is closely tied up with the Philosophy of Rheumatism. We have seen that when an individual sleeps, he is under the influence of attractive forces. When he awakes, he is influenced by repulsive agencies. This is a fact of vast importance in considering the phenomena of the influence of light on the system. If we were expected to account for sleep without considering that there is a great distinction between day and night, we should be told that we only half understood our subject; and yet we should find few medical practitioners who could inform us why this difference, in our diurnal periods, should influence our propensity to sleep. It is evidently not necessary to go to sleep when night supervenes on day, or otherwise our fashionable world would be baulked of half their amusements. We may say that night disposes us to sleep, because we have been subjected all day to the influence of light. Nothing is so fatiguing as daylight. Is that a fact? Undoubtedly, for day is lighted by the sun. The sun is the source of all the magnetism which the earth receives. Light is the great magnetic agent. The light we see around us magnetises the air we breathe, and that air contains more or less of Ozone. Every one who knows anything of Ozone, knows how exhausting it is to inhale it. It eats metals with astonishing avidity. It prompts us to feel weary and ready to lie down and sleep. Why is this? Because it contains a force which may be said to be the actinism of light, a magnetic agency, highly attractive, when it influences the nervous system. We are at no loss to account for the manner in which Ozone acts in inducing sleep. We recur to the subject of Animal Magnetism. Very sensitive persons, who are easily mesmerised, cannot inhale a single whiff of Ozone without falling into a profound sleep, most frequently so deep as to be attended with rigid spasm. This is a sufficient proof of its attractive agency. Phosphorus in a bottle, containing oil, may be applied to the nostrils of the same class of subjects, and the same effects will be produced.

Our knowledge of Ozone may be said to be yet in its infancy. Very few experiments of any importance have been made on the inhalation of this wonderful substance. The curious electrical agency it exerts in sensitive cases, would seem to warrant us in the expectation that it shall become a powerful health bestowing means of inducing anæsthesia. Our ethers and chloroforms may be objected to on the grounds which have been previously pointed out of producing peripheral forces within the bony vault of the living skull. If Ozone shall be found to induce healthy tonic direct currents in the system, it will possess a great advantage over the poisonous agents we have been accustomed to employ. Thus the advocates of Animal Magnetism will have another advantage over their unreasoning and obstructive opponents. One of the neatest summaries of our knowledge on Ozone is from the pen of the gentleman to whom this book is dedicated. I found it in the newspaper entitled *The Brighton Herald*, dated Saturday, 24th December, 1864.

"A correspondent in the *Brighton Herald* of the 17th instant, who designates himself "Inquirer," wishes for some information about Ozone. I believe it impossible to give correct answers to all his questions. The little I know shall be communicated; and I hope other observers may be induced to do likewise.

"The discoverer of Ozone is Professor Schönbein, of Bâle. Everyone who has used an electrical machine must have noticed the peculiar odour which is diffused as soon as sparks are emitted through the air. This is due to the presence of Ozone. Many different opinions had been held, and efforts made, in respect to this odour —the constant attendant upon electrical experiments. It was reserved for Schönbein to find the *true* cause, and to give the right explanation.

"The name '*Ozone*,' is derived from a Greek term which signifies, '*I smell*.' The exact nature of this wonder-working substance (?) is unknown. It is easier to say what it is not, than what it is; and not easy to give an intelligible account of it either way. The best we can do is to try and make ourselves familiar with some of its effects; those which are of most frequent occurrence; and in this way we may learn what Ozone *will do*, although we remain ignorant of *what it is*.

"By watching some of the effects of Ozone, and recording the dependent conditions, we discover that it is endowed with remarkable properties. It is fair to conclude that it performs important duties in natural processes; that it is the constant attendant upon man; and connected with, and contributes to, the preservation of health and the enjoyment of life.

"The best definition that occurs to me, or that I have ever seen, is to call Ozone electricized oxygen. What does that mean? Oxygen (the life-sustaining element of air) in a different condition to what it is ordinarily. All the best authorities agree that Ozone *is* oxygen; but how its habits and properties are so greatly and suddenly changed, excepting by electrical agencies, is inexplicable. But, when we talk about Ozone as electricized oxygen, we have need to do it almost in a whisper, lest, after all, we should find that we had better have said nothing about it. Still, with these acknowledged difficulties, there is no necessity for feeling discouraged. The greater the difficulties the greater the pleasure, and the greater the merit, in getting over them. We may go on thinking, and guessing, and speculating; and, what is still better, observing, experimenting, and recording; and, if we don't get all the light we want ourselves, we may add a little to that possessed by others.

"What will Ozone do?

"Some of its properties are known, and can be illustrated by direct experiment. Ozone possesses bleaching, oxidating, and deodorising qualities. In former days we used to hear of "bleaching grounds," and Irish linens were frequently stamped with the words, "Grass bleached."

"Before the discovery of chlorine, woven fabrics of linen and cotton were whitened (bleached) by exposure to light and air and moisture. No doubt the active agent in this process was Ozone.

"Ozone exists in the atmosphere, but in variable quantities. The quantity, at any given period, depends chiefly on the direction of the wind, and the relative proportions of moisture. It is influenced by temperature and by local conditions; as, for example, thickly populated districts, in contrast with those at a distance from human habitations,—the latter being most favourable. Judging by my own observations, I believe the only way by which the habits (so to speak) of Ozone can be understood, is to look at them locally, instead of generally: because the sudden changes which occur, as to the quantity present, are explainable only by reference to local circumstances, limited to a comparatively small area.

"Let me try to make this intelligible. We will suppose the wind to be blowing from the south-west, and the air highly charged with Ozone. A change takes place; the wind veers gradually to west, and then north-west, and the presence of Ozone, at Black Rock (eastward of Brighton,) can no longer be recognized; but on the westward of Brighton, the Ozone test will still be affected. Another example:—During several successive days the wind has been from the north-east, and during the whole of that period not the slightest trace of Ozone can be detected. The wind changes from north-east to south-west, and in ten minutes Ozone re-appears.

"Referring these examples to the theory that Ozone is electricized oxygen, we are not much wiser than we should be without any theory at all. What electrical condition is oxygen likely to assume when it possesses the properties of what we here call Ozone? Is it overcharged, or undercharged, with electricity? Is the electricity what, for want of better terms, is called positive or negative? Is it not deserving of special notice that the presence of Ozone should so much depend upon moisture? In our ordinary experiments with common (frictional) electricity, a moist atmosphere is exactly that which causes difficulty and disappointment, and very often total failure. When we know so little, it is best to confess it. Perhaps Ozone may consist of more than one element. Why think only of oxygen? There are other materials in the atmosphere as susceptible of electrical influences, and electrical changes, and whose activities are just as likely to be aroused. Nor are we obliged to refer the phenomena exhibited by Ozone to atmospherical electricity, properly so called. Other kinds (forms) of electricity are as likely to be in operation in the air, as we know them to be in the earth and the sea.

"After an attentive observer has had some experience of the manifestations of Ozone, he will be able to detect it by his sensations. When present in large proportions, it produces what we call a balmy state of the atmosphere. Perhaps it is wrong to suppose that, when we are unable to detect it by any, at present, known test, it is not there. I believe it impossible to conceive of the comparatively trifling conditions which affect Ozone, causing it to pass from the active to the passive state as suddenly as a flash of lightning.

"Other circumstances being the same, there is more Ozone in the air in summer and autumn than in winter and spring. It is also abundant in that state of the atmosphere which we feel to be favourable to the diffusion of odours (good and bad,) and which is, probably, especially active in causing the fermentation of vegetable, and the putrefaction of animal substances. In this newly-discovered deodorizer may we not recognize a special provision for neutralizing miasmata when most numerous and most noxious?

"At Brighton there is more Ozone in the air when the wind blows gently (not violently,) from some southerly point, between south-east and west, than at other times. Of the health-inspiring influences of the Downs at such times, too much can never be said; and even when the wind is from the north or north-east, the air there is so pure and refreshing that for those who can bear it I can think of no better description than to call it food, medicine, and exercise, beautifully combined.

"Now I must conclude—not because I have said all there is to say on this interesting subject—but because I wish not to be wearisome.

"The test for Ozone consists of pure iodide of potassium one part, starch ten parts, distilled water 200 parts, to be boiled together for a few moments. If slips of white (unsized) blotting-paper be dipped in the preparation, and, when dry, kept from light and air, they are ready for use. In exposing them to the air the test-papers must be sheltered from sun-light and rain. According to the quantity of Ozone present, the paper, in a given period, will change from white to violet, fainter or deeper, as the case may be. To bring out the colour fully, the tests are usually dipped in water at the end of the experiment, and compared with a coloured scale graduated from No. 1 to 10.

"The best tests I have used are prepared by Mr. Lowe, Observatory, Beeston, near Nottingham. They are sold, with proper directions, by Messrs Negretti and Zambra, Opticians, 1, Hatton-garden."

J. O. N. R.*

Black Rock, Brighton,
December 20th, 1864.

The doctrine of the production of disease in the human body, by a reversal of the normal, healthy, or direct nervous currents of the system is one of sufficient importance to be illustrated by events that are of no uncommon occurrence. Long ago I had made up my

* The philosophic caution which characterizes Mr. Rutter's communication should induce one to be very chary of emitting an opinion on this subject; it may, however, be simply suggested that those electricians, who like the late Mr. Crosse examined the conditions of atmospheric electricity have found that a negative state of electricity is present when hail-storms occur that are attended by a damp state of atmosphere, or what some people call a catch cold state of weather. Some philosophic electricians have gone so far as to conclude that colds are caught most frequently when there is an abundance of Ozone in a damp atmosphere, notwithstanding, however, the possibility of denying this fact, it may be averred that in north-east winds catarrhs and rheumatism are undoubtedly prevalent, and this may have some relation to the inverse currents connected with what some are pleased to call a negative state of Ozone.

I can have little doubt from my own observations of the truth of the general facts stated by Mr. Rutter which are strongly corroborated by the influence of Ozone on sensitive mesmeric subjects.

mind that certain poisons applied to the external surface of the lungs by inhalation, or even by application to the surface of the mouth were sufficient to produce peripheral forces that had the effect of producing mischief in the magnetic brain. Facts have been discussed in the chapters of Sleep and Wakefulness which explain the course of those forces that operate centrifugally instead of attractively. Impulsive or repulsive agents operating within the sphere of the brain must be productive of reversals of normal polarity. To illustrate this point, a case of the inhalation of chloroform is now presented to my readers. The patient was a highly sensitive individual, whose constitution renders the illustration of my doctrine more forcible. In December last, the operation of extraction of the cataract from his right eye was determined upon, nothing could be more skilfully performed, in less time, or with more surpassing dexterity, than this surgical feat. He was placed under the influence of chloroform by the most experienced manipulator of the time; but such was the influence of the poison upon the patient's highly sensitive nervous system, that an instantaneous reversal of polarity after the anæsthesia was the result. No illustration of the rigors attendent upon a fever or a catarrh could be more complete, a sickness of the stomach produced vomiting, the hands and feet remained cold, and an intense headache supervened. The traumatic pain in the eye became extremely urgent, and continued almost without intermission for a couple of days; the nerves of the fifth pair in the forehead, cheeks, and side of the head became irritated to such an extent as to cause no small agony, meanwhile the smarting and aching in the wound continued unabated.

Sir C. Bell has pointed out to us the close connexion between the nerves of expression and those of respiration; the polarities of all these were disturbed, and in the course of the night the sufferer gasped for breath like one afflicted with asthma. No man need desire a more complete illustration of the reversal of normal polarities in the nervous system. A gouty man seized with a catarrh which has taken possession of all the compact textures about his forehead, suffering agonies in the supra and infra orbital nerves, could not be a more apt illustration of the fact of polar reversal, had he, at the same time, a congested condition of the chest; his respiratory system of nerves could not have responded more completely to the exigencies of his position. Then let us ask, what analogy can there be between gout and the administration of chloroform? The gout arises from a full condition of the veins of the body, this is the preliminary stage of the pathologic condition, the analogy in the presence of chloroform is the creation of a condition of heart which fills the veins on the periphery of the brain. Now this is a condition not to be coveted by any man in his senses. During three weeks this patient declared that he repeatedly feared the loss of his senses; that he not only suffered agonising pain, but that he had traumatic iritis and inflamed conjunctiva. May we not inquire whether some other mode of inducing anæsthesia would not be acceptable to the surgeons of England. Suppose, for instance, that upon inquiries being instituted, it were found that by a combination of the processes of Animal Magnetism and the inhalation of ozone, we could produce a state of perfect anæsthesia! We cannot doubt of the advantageous position we should be in, for in addition to the known fact that our processes are attended by direct nervous currents, normal conditions of tone of health; we should have no imitation of the dire diseases to which humanity is liable. I call upon my friends to form a committee of inquiry, in order to ascertain if their scientific labours cannot result in some great improvement in the production of anæsthesia, not one of disease, but of one which would ensure a healthy recovery.

We are not now to mention the light which phosphorus yields to oil. Every body knows that it has been used to smear the face, in order to produce a luminous effect in the dark; but the singular part of the question on the solution of phosphorus in oil relates to the disappearance of the heat which is well known to be evolved when phosphorus unites with oxygen. Why should it not evolve heat in oil? Because it becomes decomposed, and thus gives us reason to conclude that it is a compound substance. I believe it to be a compound of the matter of light combined with the matter of heat. What becomes of your doctrine of forces? *Non constat,* that both light and heat being material, should not be associated with forces. On the contrary, it may be a powerful argument to prove that forces are independent of matter, since the attractive power of light, and the repulsive power of heat being dissociated, each pursues its own proper function when liberated from its material thraldom. When phosphorus acts as a medicinal agent, we are obliged to deprive it of its heat, or otherwise it would act as a poisonous escharotic. We deprive it of heat by dissolving it in oil; and it has been my lot to produce some very remarkable cures of the results of chronic gout, as well as of scrofula, by adopting the practice of downward frictions along the spine, trunk and extremities of a weak solution of phosphorus in cod's liver, or preferably of neat's foot, oil.

The marvellous cases of cure in spinal disease, whether occurring in the incipient lateral curvature of the spine, or in the scrofulous hump-back productive of such serious mischief as to lead to loss of life, has been often successfully treated by gentle downward frictions of the fingers smeared with this oily phosphoric solution. The proportions which I generally recommend, and which have been attended with great success, in the light and gentle downward frictions from the top of the spine to the crupper bone are one grain and a half of phosphorus dissolved in one ounce of cod's liver, or neat's foot oil. Medical practitioners may be assured that they do not require much deep thought to cure such cases empirically.

It becomes a scientific application of magnetic agency in the cure of disease; for the light introduced into the system is the source of an attractive agency, which results in the accumulation of tone; a process superseding the clone which had previous possession of the system. These views will be startling to medical practitioners, who have hitherto studied very little on the *modus operandi* of frictions, and consequently adopt without much examination the empirical ideas of those who advertise their practice. We are not pretending to condemn these men. We are anxious to urge medical men to indulge in habits of philosophic thought. So many have been our reflections on the various relations of light in reference to the treatment of disease, that we are at a loss how to class them, so as to produce the most striking effect.

We are arrived at a point, whence we must necessarily diverge a little from the train of our argument. Engaged in instructing, if we may be allowed to use the expression, a body of men who have long refused to be taught, it is requisite to shew that the subjects we have studied have not been slurred over, but have been laboured at most assiduously. The proof of this assertion is in the application of the numerous facts, which have been adduced for the sole purpose of illustrating our theme. It would have been easy to overlay these pages with an abundance of quotations from authorities. Many interesting facts, culled from nosological and pathological works, are stored up in notes, which, if printed, would swell out the work to sixteen hundred pages. The object is to condense, not to expand, and still reiteration and passing reflection must be resorted to, or the due impression of the trains of argument would not be made. We are conducting our readers to understand the principles which have so forcibly struck us, and this is to be done only by avoiding too great a minuteness of detail. Light is an agent we have to study under various aspects, and the influence of colours in modifying the habits of animal life. We very little think of the influence exerted on our own minds by the colours which surround us. If we were to sit in a room, the walls of which were painted of a decided yellow, we should soon find ourselves uncomfortable, as bees do when inhabiting a glass beehive, and a yellow silk cover is thrown over their habitation. They cease to build their cells, evidently forgetting the duties they have to perform in storing up a provision for the future support of their community. It is not necessary to go over numerous facts to exhibit the influence of colours on the living economy. We are aware that colour is only a part of the spectrum into which all light is divisible. Our inquiry relates to light as it comes from the sun, and that contains not only colours, but heating and cooling rays. Rays which affect metals, and which change previously unmagnetised metal into magnets. We are not discussing the course of treatment by which most usually pieces of steel are made to become magnets, or the foolish proposition that a force must be derived from some matter, which having that force inherent in itself can communicate it to another matter. Here is a force, in light, travelling at the rate of nearly 200,000 miles in a second, impinging on a metal, and rendering it magnetic. If we were called upon to explain the fact, we should say that magnetism comes from the sun—that light is the vehicle which conveys the force, and that the sun is the source of that force. But at the same time, we do not aver that the sun may not be material. We have only contended that we are not obliged to regard magnetism as being inherent in any matter on earth. That all of us are apt to be influenced by the light which emanates from the sun, is the point on which we are now insisting; and it is an interesting fact that the sun's rays are so important to our well being. We may inquire why, if the sun's rays are necessary to our existence, do we not live in perpetual sunshine. All nature is positive or negative. Nothing about us is stationary, and it is easy to perceive that if we get too much of any thing, we retire from the world disgusted and surfeited. There is no form of our existence that does not exemplify this fact. The sun's rays, in tropical climates, are a great source of the diseases affecting the natives of the more temperate regions of the earth, who happen to be temporary residents of the tropics.

We are not now obliged to make ourselves acquainted with the causes of Asiatic or other tropical fevers, or of the many forms which these assume. We must be content at present with our own subject, remarking only, that it would be well if the medical investigators in those climates would study Animal Magnetism. Considering it philosophically, they might discover that in Magnetism, they would find the causes of all the reversals of polarity promotive of the diseases which they cannot unravel, and the cure of many which now terminate hopelessly.

Many times we have wondered, in India, over the influence of light on our nervous functions. At the time of our sojourn in Bengal, we knew nothing of Animal Magnetism, and in writing to Dr. Macartney on the causes of Cholera, we spoke of Shakespeare's idea of airs from hell, little thinking that we were so near the mark in our guesses. Many a time reflecting on the phenomena of that curious disease, I felt certain that Telluric influences produced the phenomena. We do not reflect on this subject enough. If we thought of the effect of the sun's rays on the earth, we should place the subject in all its various points of view. The sun being the source of all magnetism, affects the earth in one of two ways. It emits rays which attract, and the repulsive rays, or forces, of the inverse current which repel. We have only to regard the repulsive agencies of these rays to understand how disturbances in the earth's crust may yield morbific poisons. The airs from hell is only an expression of an idea, typifying the result of what may be supposed to arise from a source under our feet. We point significantly, with our forefinger downwards to denote the abode of the spirits of evil, or to that of the gentleman in black, and we are not far wrong.

Many are the mistakes in which we indulge, and the silly prejudices against the facts of Animal Magnetism are really a disgrace to the advanced civilization of which we so much boast. We are sorry thus to disturb the equanimity of our learned colleges. They have so behaved in this matter, that they must be content to be regarded by the public, as they assuredly will be, when this book appears. They cannot hold up their heads and plead their contempt for the author. They may be told that the author has coursed all the subjects which they propose to have exclusively studied. He may not be so very profound a scholar in the learning of Greece and Rome, as some few of them may be, but he has not been idle in his time, and he has devoted more and deeper thought to anatomy, physiology, and physics, than several sections of their professors put together. Which of them has studied Animal Magnetism in relation to these philosophical sciences? They need not sneer at the acquirements, or at the morals of those who have upheld the Truth. They may know that the pursuit of Truth is the most intellectual, as well as the most ennobling exercise of the human faculties. They will not much longer persuade the public that Truth is to be scoffed at, vilified, and sent to the limbo of the ignorant and the prejudiced.

Sometimes the hope is indulged in, that a few noble spirits will arise, and advocate a full examination into the various and extensive relations of our subject; but as time advances, we see no signs of such exalted feeling. The world is occupied with thoughts of war, and the men of science with the claims of their own distinction.

Little is thought of those higher considerations which lead men to the more important thoughts connected with the only source of permanent fame. The views of men are devoted to the gross rewards they can obtain on this earth. Fame, the idol of the giddy, is not that which rewards the real votary of science. He must have a far higher reward, which reaches not only to the approbation of men, but to that of higher spheres of intelligence. Magnetism is not confined to the laboratory, or the lecture-room of the Royal Institution. It pervades Nature, and "looks through Nature up to Nature's God."

We are engaged in a task by no means irksome. We look not for our reward to the baubles men can give us. We look to higher objects. He who studies light, should not be disappointed if the votaries of darkness do not appreciate his labours. He must be content to be led on by the attractive agencies of light. These are very powerful. We have seen how they induce sleep, but we could illustrate their attractive power in various more curious and certainly more important relations, but we should shock immeasureably the prejudices of the ignorant. We are not called upon to make greater sacrifices than we have already made to the gross ignorance of our race.

If men are desirous of being instructed, they will take up our subject at the elements of our knowledge. Electricity and its ramifications, chemistry and its ever changing facts will fully reward the industry of the candid inquirer, and if his aspirations soar to the higher subjects of human inquiry, he will find ample scope for his investigations without looking into the absurdities that have been written on Mental Philosophy, and the bothering pursuits of Metaphysics. Reasoning power devoted to objects of real science will amply reward the reasoner, and he must indeed be a poor philosopher, who is deterred by fear from pursuing his inquiries into the ultimate depths of nature's domains.

Our senses are invigorated when we have indulged in a correct train of reasoning. We are more rational beings. We are not bound by the trammels of sects, or by the prejudices of the ignorant. Notwithstanding our position in society, we feel that we are to be made answerable for acts, and that to support a high position, we are bound to yield to no one in the desire to advance the cause of Truth.

It is quite clear that the progress of our inquiries has convinced us that the views of the late Professor Macartney, on the subject of inflammation and its consequences, has opened up many new ideas on the nature of disease; we may perceive that insanity is in most cases no way associated with inflammation. The old nosologists would have a hankering to attach the termination "itis" to all aberration from the healthy state of brain, and the wishy-washy ideas of those profound philosophers, the Psychologists, would lean upon some substantial blunder of inflammatory condition, or meningitis of the cerebrum, to account for every aberration from the healthy standard of the mental faculties; in fact, so interwoven are material ideas with the whole fabric of medical thought, that we must rely upon some Lawrence Brodie union of Psychology and Materialism to account for all the phenomena of insanity, or we must cease to be considered philosophic thinkers of the present age.

Gout and rheumatism are often attended by distressing symptoms of mental irritation; gout and rheumatism have in fact, in the majority of cases, nothing to do with inflammation; we are anxious to eliminate all ideas of inflammation from ordinary cases of insanity, in order to separate that which actually exists, as the truth, from that which is uncertain, and from much of that which is untrue. How can we speak of insanity as dependent upon a condition of brain, which is not present or existent; our anxiety must be to make the whole subject of the aberration, from a healthy standard of intellect, clear. We desire to know why the healthy body should not always be accompanied by the sane mind; as far as we know at present, the healthy body is always accompanied by the sane mind, and it is only in those cases in which there is some aberration from a healthy state of body, that we find anything like an aberration from a healthy state of mind; those who can reflect no further or not more deeply than this point, immediately conclude that they now have us on the hip as confirmed materialists.

All profound knowledge must for ever belie those who hold such a doctrine, for the established fact of the inertness of matter points out the solid objection that for ever exists to their bold assumption. Mind is superior to the structure of the body, and mind would never give way, if there always existed a habit of regulating and controlling its operations, so as to control the passions of the soul which must always regulate the appetites of the body; we little think, as we ought to do daily, of our moral duties—our moral duties are in fact our religious duties, and these in truth lead us on to that conservation of force, which obliges us to regulate all the appetites of our bodies, and all the aspirations of our souls. This then is true religion; this then is prayer; for it inevitably leads us on to cultivate those graces, which make us acceptable to our fellow-beings, as well as to our Creator. It is idle to talk of curing rheumatism, gout, scrofula, or any analogous disease, unless there is not only a personal, but a family disposition to yield to the principle of self-control.

It will now be perceived that our rules for the cure of disease are far more severe than could have been suspected by those who have taken up our work as an object of curious inquiry. We profess to treat of Animal Magnetism, and our subject is interwoven with clairvoyance, table-turning, magic, intercommunication with the spirits of the dead, spiritual manifestations, and God knows how many other vague notions that are floating among the vulgarities of the world. We do not shrink from an inquiry into any one of these subjects, for we boldly assert our right to pursue our inquiries into all those matters which shall enable us to improve the condition of human knowledge, into all those subjects connected with the amelioration of our race on the science of mind, on the art of medicine, and on our relations with a future state of existence.

Those who are well acquainted with the deep philosophic thoughts which pervade one of the most admirable works in modern surgery, the Treatise on Inflammation by Professor Macartney, will be able to estimate the veneration which has at all times been entertained and expressed by the author of this work for that exalted character. Rheumatism and cognate diseases, regarded as they have been and will continue to be in the present work, owe much of their interest to the points of view in which Macartney placed inflammation and congestion; the simple fact that arteries are accompanied in their course by nerves, and that veins have no such companions, may lead to the reflection that nervous currents must be very important to the functions performed by this portion of the circulating system. The whole subject of the reversals of nervous polarity becomes the more important, when we consider that congestion is so common a cause of the origin of disease. A case of that kind of rheumatism which is said to have its seat in the sclerotic coat of the eye, which has been strongly illustrative of the facts relating to that extreme reversal of polarity which constitutes pain, occurred in a person who suffered agonies while electrical currents passed from the caudal extremities to the head, but was immediately relieved by either mesmeric passes made downwards or by electric currents passed from the head to the caudal extremities, *i.e.*, the pain caused by the inverse or upward current, was relieved by the establishment of the normal or downward current. It must be noted that this case would be characterized by Von Reichenbach as one of a highly sensitive temperament. The whole subject of premonitory signs in rheumatism may be regarded as a portion of the philosophy of the influence exerted upon the human system by direct and inverse currents.

Acute observers have in their treatises noticed those premonitory symptoms which constitute the instinctive sense of injury.

Chomel* says that "articular rheumatism seldom attacks without these signs, such as uneasiness, general torpor, or even irritation, stiffness and especially in the joints that are to suffer, commonly on waking or getting out of bed; or having walked and felt a weary desire of rest; on wishing to walk again, this stiffness becomes more remarkable. The patients themselves do not seem to be aware of the events. Their hands are applied almost instinctively to the parts suffering the discomfort, and the bystanders remark upon them, and point out to them the existence of malaise."

This instinctive sense of injury, as Macartney calls it, originates in a reversal of polarity, which has not yet gone to the extreme local effect of pain. Let us suppose the fact to be regarded in another point of view. Instead of the occurrence of reversed polarity in a joint, it may be exemplified by the influence of cold applied to the feet. An individual susceptible of the reversing influences of cold applied to the lower extremities, feels the effect on the surface of the skin on which the polarities have suffered a reversal. In this manner a catarrh is established. "Persons who are in the habit of going barefooted never catch cold from wet applied to the feet, while others who are accustomed to wear shoes are extremely liable to catch cold from this circumstance." In such persons the revulsion takes place immediately; the individual has caught cold. A person being placed in a situation where some part of the body has been exposed to "a stream of cold or damp air, passing through a crevice or a window, not properly closed, is first sensible of that part being chilled, and soon after

* In his *Idées*, as published by Requin, page 159.

feels as if the temperature was departing from the rest of the body, especially along the region of the spine, although the body at the time may be as warm as usual. The immediate effect of the insidious abstraction of heat is the collapse of the arteries in the part; this extends to the rest of the vascular system, especially of the skin. The surface of the body becomes pale and shrunken, . . . the muscles begin to tremble involuntarily, particularly those of the jaws and of respiration . . . producing that agitated respiratory movement, and chattering of the teeth which is called rigor, a clonic spasm. The patient shivers, sneezes, or makes an ineffectual attempt to sneeze, an effort accompanied by pain in the eyes, eyelids, forehead, nose and eyebrows. The fibrous tissues here are more or less affected, and as in rheumatism, so in all other parts, abnormal secretions are established in the vicinity of the pain; from the mucous membrane of the nostrils a large quantity of a fluid unusual to them, a hot saline liquor often burning the upper lip, and inner membrane of the nostrils, from its acrid qualities. The eyes water, and when the rheumatic paroxysm of pain is at its height, the orbicular muscles of the eyelids are forcibly closed from the stimulus of pain."

In some cases, the catarrh or cold begins with an almost unconscious sense of chilliness, observed by others more than by the patient. The clone is there. The organic instincts have suffered the sense of injury. The sneezing, the dry state of the fauces, sometimes amounting to a feeling as if a sore throat were approaching, sometimes the same kind of feeling that people experience from grief. Some cases begin with severe hemicrania, attended by a pain in the nostrils of the affected side, extending to the palate and the fauces, dryness, soreness, redness of these parts, said by some to be symptoms *premonitory* of gout, and truly so, for these preliminary symptoms arise from one cause, and the case may either go on to catarrh or gout. Both are rheumatic diseases. A gouty belongs often to a catarrhal diathesis. Many a one has caught gout in the foot or in the hand from exposure to precisely the same causes as those we are considering, and many a cold has ended in a fit of the gout.

"The sudden change from a heated room to a cold or damp atmosphere is another way in which a cold is caught; and breathing a chill air at night while the rest of the body is heated will also give cold to many persons; but most frequently people catch cold by standing or sitting in what is called a draught of cold air when the body has been previously heated, in which case, the impression has been made for some time before the person is conscious of it." The skin has suffered in its organic instincts, and the shiver is the first sign of the nerves being involved in the phenomena.

"Of all parts of the body, the back is most susceptible of cold thus insidiously applied, and the chilliness is sometimes even first felt in the back, although it may not have been immediately exposed." (Macartney, Inflammation, page 88.)*

In considering a catarrh thus as a purely rheumatic affection, it becomes necessary to trace not only the apparent causes of its phenomena, but to try how far its other analogies hold good with the phenomena of rheumatism in other parts of the body. We have seen that the mucous tissues in the immediate vicinity of the textures are prone in this complaint to an abnormal excess in out-pouring. This is one result of reversed normal polarity. In a condition of Tone, they ought to produce only a normal quantum of mucus.

Dr. Prout's secondary formative processes of malassimilation are here in great activity if much debility can deserve to be so characterized. In the snivelling of old people, of gouty or rheumatic diathesis, whose distilling rheums are so troublesome, there is an unconscious state

* Who that has suffered from Hay fever in summer, will deny that in this rheumatic or gouty complaint the symptoms above described have not been fully experienced!

of venous congestion, one principal source of the disorder called a cold, this congestion is always present. It is succeeded by a cutaneous reversal of polarity or rigor; inordinate secretion, oppressed nervous state, and oppression about the head.

When rheumatic reversal of polarity affects a non-secreting tissue, the result is increased nutrition, growth, an abnormal formation, and deposition, not compensated by the due or perfect performances of the secondary destructive processes of assimilation. This constitutes hypertrophy, and pain may not be necessarily an accompaniment of this state, even when as in the hypertrophy of the heart, a serous membrane partaking of the tendency to deposition is found covered by an adventitious membrane or coagulated lymph; and among the intestines, where rheumatism may be rapidly generated from cold feet in travelling, extensive depositions of coagulated lymph have glued the serous surfaces of the intestines together.

In many cases of catarrh, primary and secondary assimilation being at fault, fever supervenes. Pain, the characteristic of rheumatism attends. The urine is loaded with lithates. Some cases have proceeded to violent spasms of muscles at the back of the neck, the intercostal muscles and the muscular apparatus of the abdominal parietes with bearing down, the antagonistic peritonœal forces yielding from an irritation established by the presence of loaded veins in the rectal mucous membrane, and if the veins of the portal system and of the intestines have been equally abnormal in their contents, equally full of the debris of the secondary processes of assimilation, most distressing spasms of the pelvo-femoral muscles have been present. Severe cramps in the calves of the legs, and in the feet have been symptoms in a case which has alarmed all the friends of the patient, who suffered extreme agony, which was described as a sensation from a sharp knife passing up the rectum. All this pain was relieved by continued mesmeric passes and the strong will of my friend Mr. Henry Thompson.

Action and reaction are terms so frequently made use of, that it would be very desirable to understand the meaning of them. We are often told that colds are caught in a large proportion of cases rather by going from a cold to a heated atmosphere than from a warm to a cold temperature of the air, and it is said that such a result must stand to reason, for the reaction of the system is greater in the one case than in the other. This is not exaggerated, but an ordinary expression made use of with the intention to explain very clearly an unusually vague idea.

All catarrhs as well as rheumatisms depend primarily for the presence of their phenomena upon an enfeebled condition of the system. That condition is due to the tendency to load the veins beyond their healthy quantity of blood by the debris of the body, thrown into them during the secondary processes of assimilation. All debilitated states of the frame favor the injurious influences of cold. There are circumstances in which the influences of cold, artificially employed under proper regulations, may be rendered very salutary; but among the unreasoning and the falsely reasoning portions of the large public, it might not be difficult for bold and daring assertors of dogmatic propositions, to make the conviction spread very widely that cold was healthy and comfortable, while heat was the great source of unhealthy enervation and the origin of all diseases. There would seem to be no proposition too absurd, where the object is to prop up for a time any system *ad captandum vulgus*, any views to gull the *gobe-mouches;* thus they would say, It stands to reason that actions must exist in the system, for animal heat results from chemical action, a fact abundantly proved by Liebig and his followers. If chemical action be the source of animal heat, chemical reaction should be the source of animal cold. But perhaps it is not fair to assume propositions for those who delight in hobby-horse theories. The meaning of action and reaction must be sought for not in chemical, but in vital phenomena. If this be so, the active agency of a force called vitality must be granted, and vitality must be investigated either instead of, or as well as electricity and chemistry; but men fear to investigate vitality. The bigotry of false views of religion, that of a false philosophy, and last, but not the least important in its present results, the bigotry of real atheism forbids men to make themselves masters of facts which are evolved in a legitimate logical series from those actually in the possession of mankind. The errors in reasoning on the subject of a common sneezing cold in the head are thus more or less dependent upon the faulty education of mankind, which encourages the fear of carrying propositions on all subjects to their legitimate conclusions.

Sir Benjamin Brodie, a prince in the school of bigots, though deemed in some respects an estimable man, discovered, unknown to himself, that Mesmerism was the source of Animal Heat. He exploded some captivating hypotheses relating to the source of heat in animals. He devised some beautiful experiments to show that where respiration was artificially kept up after the decapitation of animals, their heat gradually diminished. The *cold* in fact, which ensued, was not long in reducing the subject of his experiment to the condition of dead animal matter. Dead animal matter be it remembered, has the astonishing power of quickly bringing the pendulum of Mr. Rutter's Magnetoscope to a dead stop.

The living brain, the magnetic source of the warming electric currents of the body—the direct currents, the tonic currents—was removed in Sir Benjamin Brodie's experiments. The cooling currents were established in their stead, the inverse currents taking their courses upwards; and the rate of cooling determined the time when the vital phenomena were to cease by the repulsive agency of the lethal currents.

It would be well if the words action and reaction

could be replaced by the words attraction and repulsion; but the time for the change is not yet arrived.

How is the development of catarrhal symptoms on entering a warm room from a cold external atmosphere to be attributed to reaction. The cold must have *acted* upon a system deficient in vital vigour, in healthy tone. Cold must have been productive of a certain tendency to debility. A vigorous individual in good health would not catch a cold in coming out of the open air on a winter's day into a warm house. Many an old person with loaded veins has a catarrhal rheum with distilling nostrils on going from a warm room, when no such symptoms were present, into the cold air. It is where Clone prevails, where the veins are more full than they ought to be; where debility instead of tone is present, that what is called reaction takes place. If people would comprise a sufficient scope in reasoning on the phenomena before them, they would find that whether a subject of catarrh be transferred from a warm to a cold, or from a cold to a warm atmosphere, the catching of cold is in fact not so paradoxical in its causes. The development of the phenomena may take place in a warm or in a cold situation. The change of atmospheric temperature is but the exciting cause. There can be little doubt that when the cold is once established, the patient will be wise to cultivate moderate warmth, and all that tends to produce healthy tone, the origin of the moderate warmth of the body, than to expose himself with congested veins to a cold atmosphere. In most cases of catarrh, it is exposure to cold which generally aggravates the symptoms of the complaint, not having a care to remain in a comfortable warm temperature while the increased susceptibility is present.

It cannot be denied that this form of rheumatism, which we call catarrh, originates in persons, more or less prone to the malady. Individuals in vigorous health may render themselves liable to the causes which favor the catching of cold in others, and yet not incur the penalty. In most cases, there is, though the patient may not be aware of it, a congested state of liver, that is to say, of *veins*. All congestion of veins is symptomatic of a reversal of normal polarity; and is commonly, when to an extent more remarkable than usual, accompanied by rigor, a stage further of the evidence of polar reversal.

All gouty and rheumatic diseases agree, in this respect, with catarrh. They must be preceded by an accumulation of material in the venous system.

Dr. Golding Bird in his most valuable work on Urinary Deposits, has this paragraph, (4th Edition, page 27, paragraph 31. "It is well known that our bodies are always in a kind of transition state, that during each moment of our existence, every atom of the frame is undergoing some change or other; the old matter is absorbed and thrown off at one or other of the excreting outlets of the body, and new matter is deposited from the blood to supply its place. The old and effete atoms of the animal structure are not excreted in the form of dead tissue, but becoming liquefied, they re-enter the circulation, their elements being re-arranged. One series of combinations thus produced, rich in nitrogen, is excreted by the kidnies, whilst those products which contain a preponderance of the inflammable elements, carbon, hydrogen, and sulphur, are called upon to perform, chiefly through the medium of the liver, an important office, previous to their final elimination from the system. Thus the blood is not only the source of the elements derived from the food which serve for the nutrition of the body, but it also serves, like a sewer, to receive the matter arising from the waste and liquefaction of the old and exhausted tissues."

If the veins form in fact a portion of the drainage of the body, it is not difficult to understand how the blood of the veins may become the source of the various predispositions of the system to form disease. An abnormal excess of material in the sewer is sufficient to influence various parts of the frame which the blood of the loaded veins traverses, and the expression, abnormal polarity, is intended to convey the idea that an altered state of venous system accompanies an altered condition of vitality.

All physiological changes in the human body are referable to the influence of the two opposite poles. To those who have closely studied Mesmerism, there can be little doubt that like the influence of the two opposite poles of crystals, which are magnets, there are two opposite poles regulating the influence of man upon his neighbour. The one is tonic—the other is clonic. The one leads to quietude, the other to agitation. The one is soothing, the other irritating. The one is healthful, the other is lethal. The one gives increased strength and health, the other induces weakness and disease. Is it necessary to ask to what good account can this knowledge be turned? If only to the explanation of the phenomena of health and disease, that would be one important good. If the statements now made be truths, the trains by which medicinal agencies produce their effects will be explained, and examined by the light of physiological facts, will result in these conclusions.

1. Agencies stimulating or exciting to what is called high action, affect the brain and nerves by sthenic influences. If fever result, it is a sthenic or synochal fever. If delirium be induced, it is active, with development of great force. The arterial system becomes full. The capillary arteries of the brain carrying much pure blood, the cerebral organs become excited; most frequently the animal preponderates, and rage and ferocity characterize the actions.

Some poisons appear to influence the higher organs, and death overtakes the patient in an ecstacy of moral and intellectual manifestation.

2. Agencies depressing the powers of the brain and nerves. If fever result, it is asthenic, adynamic or typhus. The arteries diminish in size, carry less purified blood, the veins are enlarged, and full of impure blood. To how

many degrees between the disintegration of the blood corpuscles by lightning, and the changes of polarity present during the passive clonicity of typhus fever, may the corpuscles of venous blood be subjected? Each degree must be productive of a difference in the magneto-nervous condition of the individual. Each degree must involve a vital change. Look at the course of a lethal typhoid fever. Each change is one leading to more depression, as long as the veins continue to become full of the debris of the body. The brain becomes more and more affected by clone. Its force diminishes; sleep results only from the unhealthy pressure of the full veins upon the substance of the brain. Healthy sleep implies a normal magnetic polar force of attraction, which enables the brain to regulate the contents of the blood vessels. In healthy sleep, the arteries are not too full: the veins obey a law of balance. It is in a condition of disease that the system gets into wrong trains. When in typhus, the clonic condition of brain produces wild, low, muttering delirium, force gradually abandons the patient; the muscular system demonstrates the presence of *subsultus tendinum:* great weakness is a marked symptom, and that kind of sleeplessness comes on in which the individual expresses anxiety for repose; and night after night, day after day, the wild, staring, anxious eye asks for the soothing aid of that tonic sleep which Mesmerism alone can give. The solemn but ignorant physician knows not that he still has the power to cure, and to restore a valuable member of society to his family; he mocks the agony of surrounding grief by his pretence of sympathy, and allows his patient, after intervals of venous coma, alternating with that ghastly stare, to sink exhausted, and killed by serous effusion from those overloaded veins.

In all very large towns the inhabitants are more or less subjected to the asthenizing influence of impure air. They more or less become affected by those events which determine, according to different degrees of susceptibility, a more full condition of veins than is compatible with perfect health. This condition is favourable to the reception of impressions which determine the fibrous tissue of some part to an extreme reversal of polarity. The individual may be attacked with catarrh, with bronchitis, with chronic rheumatism, with scrofulous disease, with gout, or rheumatic gout, with sub-acute, or with acute rheumatism. In any one of these states he is suffering from a modification of the rheumatic disease.

When the veins from whatever cause become abnormally full, whether from the gormandizing habits of the individual, or from his habit of stimulating his nervous system with any of the alcoholized drinks, from tobacco, or other narcotics, from the stimulus of tea or coffee, taken in immoderate quantities, the digestive power must suffer, for the nervous influence necessary for the maintenance of due force is not possible with overloaded veins. The progress of error may not be so rapid as to attract notice. It may often be very slow, but in the majority of cases it is sure. The regulation of appetite is a moral duty, and the violation of the moral laws is as sure of legitimate consequences as the violation of physical laws; so intimately blended in the organization of man are the moral, intellectual, and physical necessities of life.

The train of events constituting a cold may lead to sore-throat, redness of fauces, from gouty tendencies of the digestive organs, accompanied by an abnormal state of the secretion from the liver.

Some persons have, every winter, sore throat, accompanying a cold, pain in some of the fibrous parts about the throat with redness, extending to the fauces, tonsils, and one or both eustachian tubes, ear-ache, acute fibrous pain in the forehead and temples; or in the root of the nose and brows.

But there is, most often, no inflammation in these cases. The throat presents to the view a diffused deep redness, more livid or purply than florid, sometimes crimson. The tonsils are not enlarged, yet the difficulty of swallowing is great, and the pain is more continued and acute than when inflammation is present; and to the patient seems as if it extended further down the throat.

The pulse is seldom much affected, often it is full and soft, and has its characteristic number of beats in a minute. Characteristic of what? Of a disturbed state of alimentary canal. This pulse seldom beats less than 78 in a minute, more often 84—sometimes 88 or 90. When it reaches 96, there is generally, not always, more complication than trifling venous congestion and disorder of the abdominal viscera. Simple disorder is yielding the *pas* to a state of disease.

The bowels are most frequently costive. The excretions being of one of two states, indicate hepatic disorder by being too light in colour, or deficient in the quantity and character of the bile, or otherwise by a vitiated muddy darkness. If they be open, the headache and tendency to sneeze are relieved.

The skin is, most frequently, not very dry, but at intervals dryish, and is said to be like goose skin. Chills often coming on. Perspirations are almost instinctively encouraged.

In most of these cases the urine is scanty, red, and abounding in lithates; the mouth is clammy, the thirst not very great, but the patient drinks often, and thus increases the oppression. When the head is much stuffed, the breathing is oppressive in paroxysms.

It is a mistake to suppose that this form of rheumatism is not subject to metastasis. Its tendency to shift its locality is remarkable. How often the cold in the head ceases, and cough comes on. In cases where the lungs are prepared by tubercles, an inflammation or series of inflammations succeed to an apparently trifling bronchitis or rheumatic cough, or gouty cough.

Metastasis from catarrh may attack the heart, and in young subjects, unsuspected. This result is more frequent than the practitioner not vigilant in his calling would allow. Who would have thought it. He was not alive to observe symptoms. So the child must have had a *natural* tendency to enlargement of the heart before he could possibly have suspected it.

Very common metastasis is the disease misnamed bronchitis.

We have to consider the philosophy of a very common complaint of the lungs that has terminated the lives of a great number of individuals. We have already said that it is misnamed bronchitis, it has no relation to inflammation, and is really a disease of congestion, a rheumatism, or a gout, and has, therefore, no right to any name terminating with "*itis.*" Whatever may be the complications which involve suffering from disordered conditions of the æsophagus, duodenum, or liver, we may be assured that these established only its relationship to gout. In all the diseases we have been considering under the head of rheumatic or gouty, there is the tendency to effusion or to abnormal deposition; in fact, typhoid symptoms accruing from a congestive condition of veins, produces the typhoid form of influenza or low catarrhal fever, in most cases of bronchitis.

We may illustrate some of these observations by the following case:

A few years ago, my friend, Dr. Neil Arnott, being absent in Scotland, one of his patients, Mr. C. I., of D., in Aberdeenshire, a most learned and interesting man, was attacked with bronchitis; his malady assumed, what I should call, a very clonic or asthenic form.* He coughed incessantly, had a feeble and frequent pulse, the mucus excretion from his lungs was frothy, and as he complained, very salt, and contained no mucus; he would give the world to cough up some good solid phlegm, but he

* Asthenia, in typhoid fevers, always bearing in mind Macartney's distinction between congestion and inflammation, attends upon congestion either of the whole system or of any organ. Those currents that give tone or strength to the frame, the direct currents of magnetoid force, must be inverted in all cases of congestion. Venous fullness in the brain forbids that organ to send down the normal or direct current in sufficient intensity to give sthenic power to the frame. Asphyxia, dead

said his weakness was so great that he could not realize the happy condition. Though he sent for me, he would not obey a single injunction I laid down for him; being a dear friend, I visited him several times a day, and I found that, notwithstanding all my protestations, he would insist upon drinking three quarts of brown stout daily. Feeling convinced that his bronchitis was a true gout, I administered rhubarb and magnesia with large quantities of potash-water frequently in the day; I gave him colchicum with bicarbonate of potass as frequently as he could be induced to take it. I told him that if he insisted upon porter, he must put large quantities of bicarbonate of soda into it, but all my efforts to reason on the saturation of his system with alkalies were in vain; he died on the fourth day, suffocated by extensive effusion on the lungs. Here was a man imbued with the high taste of a thorough scholar, who had often argued with me on the superiority of the apothegms of Horace, and on the wit of that wondrous poet, above all that Shakespeare ever wrote, and who had such an admiration of Juvenal, and of the higher order of Latin classics, that his original thoughts gave a deep charm to his conversation, and this man was destined to die from the irritable obstinacy of that aberration of mind which has been alluded to as a characteristic of the gouty condition of frame.

No man who has bronchitis, or any other form of rheumatism, or gout, need flatter himself that he is free from insanity, for if the abnormal inverse currents—which favour effusions—affect his brain, he may be assured that he is very nearly becoming a lunatic. Bronchitis is merely a catarrh of the lungs, and the pain attendent

drunkenness, anæsthesia from chloroform or ether, are only degrees of extreme congestion; with the exception that the chloroform has the most marvellous variety of magnetic properties of any substance yet examined; although there is no doubt that when it produces death, its operation is to asphyxiate, and in many constitutions in which it does not kill, it is apt to leave a loaded state of veins that becomes the antecedent to serious disease.

upon it is a pain affecting the compact texture of those organs. The congestion accompanying this disorder is a venous congestion attendent upon inverse currents of animal magneto-electricity. We are not now to arrest the progress of our inquiries by asking who may approve of the views we put forth, we are quite sure that the analogies we have traced in considering gouty and rheumatic diseases are correct, and lead us naturally to highly philosophical views. The termination of such cases is most frequently fatal, unless very powerful and very energetic applications of Animal Magnetism be had recourse to; these are the cases in the presence of which the most experienced physicians are apt, in their ignorance of the value of God's holy gift, to sneer at Mesmerism; when the measures that afford the only chance of recovery are adopted.

We have considered, but not too minutely, the natural course of rheumatism when it appears in those forms we designate as catarrh, cynanche, or sore throat, coryza, cough, or bronchitis. Maladies each and all attended with pain, venous congestion, absence of inflammation, clonic course in relation to vital force—in other words, want of tone. We now come to consider this disease in the form in which it attacks fibrous textures of other parts, and this gives us a view of rheumatism in four principal forms, which, for convenience of description, will be arranged as—1. Acute; 2. Subacute; 3. Gouty; and 4. Chronic.

Rheumatism is not to be designated as a disease, until pain is present. Then, indeed, it is a fearful disease, unsettling every function of the body, every faculty of the mind. "Pain," says Shakespeare, "is perfect misery, the worst of evils, and excessive, overturns all patience." We hope to show that we are right in our assertion that Shakespeare defined the pain of rheumatism, when he gave so accurate an idea of this evil.

Many tell us that rheumatism comes without any previous imprudence, that any man may have an attack of rheumatism; that it is an adventitious, or accidental disease.

They may be right in individual cases, in which a previous generation has produced the proclivity to the complaint, but as a general rule, the condition of the venous and absorbent systems form the remote, or more correctly, the undermining cause of the explosion. A train is laid sometime before the spark ignites it. The patient is not aware of it. The friends even know little of it, unless it be from some trifling accidental observation. When the spark is applied, however, fearful are the results. We are now telling of acute rheumatism, or as some like to call it, rheumatic fever.

The symptoms are unmistakeable, shall we allow the word premonitions to escape us. Acute Rheumatism commences with premonitions of an unconscious or unperceived sense of uneasiness and chilliness; soon followed by a feverish, animated flushed countenance, and often a head-ache, dry tongue and thirst. Red urine, a stiffness in the joints that are to suffer, occurring on waking or getting out of bed. Rigors follow, sometimes so severe as to resemble the shiverings of ague. All this takes place before the great agony of the limbs. Not always, however, for sometimes these phenomena are simultaneous; and then we may say, that great agony is present. The patient suffers agonising pain in every joint. The joints are usually the first parts to be affected with pain, which extends to the aponeurotic expansions of the thigh or arm; or may be, in some cases, confined at first to one lower limb, and then is found affecting the upper limb of the opposite side of the body, Sometimes, three limbs are at once affected, and the fourth follows soon after. Unless persons in attendance observe these things, the patient is too much preoccupied to give a correct account of the sequences.

The affected joints become hot, swollen, round, and after a while flushed outwardly. There is a sense of internal as well as external heat.

The pain is intensely severe. By some it is described as lancinating, burning, tearing. Others say, it is stretching, plunging, gnawing, aching, as if a blunt instrument were screwing into the part. The suffering is so acute as to deprive the patient of all power of movement.

There are varieties in these symptoms, according to the idiosyncrasies of constitution. A gentleman of nervous temperament was affected with pain to such an extent that he fainted repeatedly when the nurse shook the bedstead, by accidentally touching it while engaged in administering to his thirst, or if some heavy step crossed the room producing a vibration. A lady, with the same sensitive system, screamed, and expressed, by her countenance, the most excruciating suffering when the selfsame events happened to jar the bedstead on which she lay. A young man thirty years of age, who was a servant in a gentleman's family, had *every joint* of the body attacked with pain, which soon to a certain extent subsided leaving a great swelling and heat. He was timidly treated. Metastasis to the heart took place. When I saw him, within thirty-six hours of his attack, he told me he had suffered some pain, but was now only oppressed and stiff, unable to move himself in bed, but otherwise in no suffering. These are examples of the varieties in the same disease.

In some cases, the pain attacks only a few joints, and others are gradually added. In others, the intensity of pain is confined to a few joints, for six or seven hours, and then shifts quite suddenly to others. The varieties in this respect are numerous; acute suffering in one wrist and the knee on the opposite side of the body. Some suffer acutely in both knees, and afterwards in the ankles; then the shoulders, elbows and wrists come in for their share of agony. In aggravated cases, the helpless being lies on his back suffering in almost every joint, ankles, knees, hips, fingers, knuckles, wrists, elbows, and shoulders. In some cases, the jaws are affected, in others, different parts of the trunk. The muscles of the thighs, and those of the arms are very tender. The weight of the bed-clothes aggravates the pain, and it is unnecessary to repeat the facts, which have occurred within my own knowledge, of the acute suffering from vibration of the floor and bedstead; but it may be stated that Chomel had witnessed in the exquisite suffering of acute rheumatism, tetanic spasms seizing the patient from the fear of a person approaching too closely to the bed.

There is a remarkable alliance between the character of the fever in acute rheumatism, and the quantity and degree of pain suffered by the patient. In some cases the premonitory fever will announce, by its severity, that a number of joints are to be attended by pain, and the fever; though apparently asthenic, is not so in reality. Its relation is to the number of parts, that have their polarities reversed, and its approach to an inflammatory character, is in the ratio of the severity of the aggregate amount of pain in these parts. All the conditions of inflammation not being present, but on the contrary, a tendency to *mobility* in the pain, so little characteristic of a true inflammatory state; when the reversed polarity is at its extreme, when the pain is most severe, then the character of the fever and the tendency of the pain is to be *remitted*. The veins are always full, and the stimulus afforded to the brain by the excitement of agony in the fibrous membrane, is directive of a tendency to promote perspiration. The pulse is indicative of that tendency. It is mostly bounding, and excited; full and compressible like the pulse of a person preparing to perspire freely, from the use of diluents, or of antimonials, or of Dover's powder. In frequency, during the nocturnal increase of pain and fever, it ranges from 96 to 108, occasionally to 116; commonly in the day not much higher than that of a person who has a certain amount of chylopoetic disturbance, *i.e.* from 84 to 96, except at the periods of exacerbation in the day, then it mounts up perhaps 8 or 10 beats more.

In the course of an acute attack of rheumatism, the pain may cease, and the fever continue. In such instances, the probabilities are in favour of a metastasis to some internal organ, occasionally effusion into a serous cavity, this is rare, but if this do not take place, the pain invariably returns. In all such cases, the chest should be frequently examined to detect any change in the action of the heart, or in the condition of the pleural or pericardial membrane. The advocates for an inflammatory state may point to the condition of the blood abstracted from a vein during acute rheumatism. The coagulum is firm and thick, and presents a buffy coat, which, contracting at the edges, forms the surface of the crassamentum into a cup, exactly as in pleurisy. (Sydenham, Cap: Rheum: "Utpote qui pleuriticorum sanguini tam est similis quam ovum ovo.") An accumulation of fibrine in the venous blood may show an abnormal condition in the nutrient pabulum of the body, but it does not establish the presence of inflammation. It establishes, however, the fact that the secretions and excretions must be in an unhealthy state. In acute rheumatism they are all impaired. There is much thirst, especially at night. The tongue is dry, furred, and loaded with a clammy coat. The skin at first hot and dry, appears to loosen under the severe pain, and an unctuous perspiration is established very profuse, of a peculiar odour, acid, and giving no relief to the sufferer; but, on the contrary, becoming an additional source of discomfort.

The urine is scanty, of a deep red colour, and without sediment during the continuance of the high febrile symptoms. As these subside, it exhibits a copious brick red deposit.

The bowels are sluggish and even costive, and the evidence they afford of the condition of the biliary secretion is always unfavourable as to the state of the liver.

There is no appetite; at night there is always an exacerbation of fever. The symptoms increase in urgency, the pain is more intense, and the patient can get no sleep.

Most often there is no disposition to sleep; the patient fearing to aggravate his pain by the occasional startings that overtake him in his drowsiness. In some cases severe cramps attend the sleep, and the torture becomes exquisite.

There is some compensation for all this suffering from the fact that the termination of the disease, unless it be interrupted by a casualty arising from some complication, is always favourable. If no metastasis take place to the heart, or other internal organ, the duration of the symptoms appears to be about fourteen days; but by judicious treatment, in many cases, the period can be shortened to one half. Some writers have given from twenty to thirty days as the average period of duration for an attack of acute rheumatism. Even after the severity of the symptoms has passed away, the patient is left with uneasy sensations in the limbs, which when these are moved at night, are converted into actual pain, though not of so severe a kind as before. There remains, even after these sensations have disappeared, an extreme degree of weakness. While the symptoms are all becoming less marked, the perspiration remains offensive, and the urine deposits its sediment more copiously. In this state, the judicious physician continues to watch over the progress and state of the secretions; and if the biliary indications afforded by the evacuations are pronounced healthy, the patient is saved from the risk of the acute attack subsiding into the sub-acute, or the chronic form of the disease; an event very likely to occur unless the health be well regulated and established. It has been asked whether acute rheumatism has ever terminated in suppuration? It can easily be answered, that true fibrous rheumatism has never terminated in suppuration. But we have the authority of Dr. Macleod and Dr. Copland for the fact that capsular rheumatism has terminated in suppuration. When rheumatism commences, a change occurs in fibrous tissue. This may not be pain at first, but may be the stimulus to secretion in a mucous, or in a synovial structure. Pain may succeed, and the character of rheumatism be present, but not that of the simple disease known as acute rheumatism. Synovial membrane stimulated to secrete, has other phenomena developed, and inflammation may possibly result. What is called synovial rheumatism becomes arthritis. Let us ask what are the characters of this inflammation? and wherein do they differ from those of acute rheumatism? When capsular rheumatism attacks an individual, the symptoms are eventually localized. There is no shifting from joint to joint. The synovial membrane becomes actually *inflamed*, constituting a case of arthritis, and goes through its several stages. The fever is more intense, and more severe than in rheumatism. If rheumatism be confined to one or two joints, and travels no further, the fever is considerably less; often there is little or no fever. When capsular inflammation of a joint terminates in the formation of pus, the pain becomes fixed, and the fever runs very high. It has already been stated that acute rheumatism always terminates favourably. In a case without complication, there is no danger. In inflamed capsule, or arthritis with suppuration, the end is fatal. Some cases have occurred in which pus has been found in the capsules of joints after death, in which the fact has not been suspected. The articular abscess has been attended with no pain. These cases may possibly have been examples of rheumatism, but mostly, have owed the presence of pus to phlebitis. Some writers conclude that because in the course of acute rheumatism, the parts surrounding joints become swollen, round, hot, and occasionally the skin becomes covered with slight erythematous redness, this must necessarily be inflammation. The swollen condition is caused by effusion into cellular membrane, lying contiguous to fibrous tissue. The heat belongs to the phenomena attendent on the reversed polarities of the fibrous membranes. The redness is a blush. No one tissue is inflamed, and, consequently, not one tissue will suffer the consequences of inflammation.

Was there ever a person attacked with acute gout, who could not fairly be said to have had an attack of Sub-acute Rheumatism? Sub-acute Rheumatism is a conventional expression, marking an intermediate degree of severity between the form of disease which we have been considering, and that known as the chronic form. We do not divide the form of rheumatism, known as catarrh into acute and sub-acute, and chronic catarrh, because we have not occasion to visit patients, who would call upon us to distinguish between their degrees of suffering in that often most distressing malady; but when the fibrous structures about joints, and aponeurotic expansions of muscles are affected, we are called upon to distinguish, in order to assure anxious inquirers as to the probability of one form of the disease assuming the severity of the other.

That which is called Sub-acute Rheumatism, generally has frequently its origin in a previous attack of the acute form. When the disease has been treated by a physician, who has some pet series of forms whereby the cure is to be unfailing in a prescribed number of days, colchicum, calomel and opium being the principal ingredients in the nostrum for cure, the probabilities are that the conditions of the secretions have been of secondary importance. Some of the symptoms are wonderfully relieved; but the bile, the perspiration and urine remain unhealthy. In such a case, the patient is cured of the acute attack, to remain an indefinite time, as a sufferer from sub-acute, or perhaps from Chronic Rheumatism.

Let not our readers imagine that in this enlightened age such cases as now described can never occur. They do occur, papers of importance containing admirable views being read before learned societies, and being published in their Transactions notwithstanding. There are many intermediate grades of severity, between the extremely agonising and wide seizing acute disease, and the very bearable chronic; all more or less dependent upon the disordered functions of the chylopoetic system. The name of sub-acute rheumatism belongs only to those cases strictly when severe pain attacks two or three joints.

When this form of rheumatism is not the immediate successor to the acute form, there are great varieties in its modes of attack. Primary sub-acute rheumatism may occur with or without premonitory rigors. But there is a sense of stiffness in some joints, or wandering pains in the limbs. The loins may suffer for a few hours, and pains accrue around the abdominal muscles, and at last a pain in one knee and ankle; or the pain may remain fixed in the trunk. This is said to be rare, but any man of tolerably large experience must have met with it, more than once, in persons of a gouty diathesis. Sometimes we find the pain severe in two or three joints of the upper extremities.

In the evening, there is perhaps increased pain and fever, with thirst and heat of skin. The pulse rises from 84 to 96 or 100. A bad night is passed. The patient has a clammy mouth, an unctuous acid sweat, with unhealthy odour. The urine is scanty and of a brown brick red colour, depositing a sediment. Sleep comes on, and the pains are mitigated, but not gone. The bowels are generally costive, and the evacuations exhibit a most erroneous condition of biliary secretion.

As far as we have gone, the experienced practitioner would find himself puzzled to say whether we were describing gout, or splitting the symptoms of rheumatic disease into some fanciful form, with a new name. Still, it is what may be called a primary sub-acute rheumatism; such a disease as may occur in a patient who has never suffered from gout in his hands or feet. Persons who have recently been subjected to the vicissitudes of tropical climates, or of malarious influences, are very apt to suffer in the way we have attempted to describe.

Where ague has, for a while, taken its seat, and departed, for a time, its first cousin, sub-acute rheumatism, may pay the next visit; and in the such cases, notwithstanding good and careful treatment, it may degenerate into long enduring chronic rheumatism.

We are no advocates for the doctrine that gout can show itself only in small joints, in the feet or in the hands, nor on the other hand would we be understood to express a very deep respect for any medical inquirer who could be satisfied that a man had no pretension to a gouty diathesis, who had at no time redness or swelling about his knuckles or toes. The tendency to reversal of normal polarities; the tendency to fill the venous system with the debris of the secondary processes of assimilation, so apt to occur in those individuals who have in the East or West Indies, been accustomed to take the stimulants of beer, wine, or spirits, or otherwise to have impaired the tone of their digestive organs, amid the vicissitudes of malarious and telluric nocent influences, are just as productive of sub-acute rheumatism or gout, as the tendencies to the same events arising from habitual potations of Port or Burgundy.

The distinctions which have been laid down in books between gout and rheumatism are most unphilosophical and unsatisfactory. They prove convenient to nosologists, but in determining the nature of a disease, they are very insufficient. The distinction of the phenomena as occurring in a large or a small joint is not worthy of a minute's consideration. It is quite untenable, for in the greater number of persons affected with gout, both large and small joints are apt to be involved. It is not uncommon to find a patient attacked in both knees; and where too, one of the alleged characteristics of gout, the redness, is wanting. Gradations of redness, which are certainly most remarkable in small joints, is not distinction enough to establish a complete difference of disease.

The structures affected in gout are fibrous tissues, and they are the same in every form of rheumatism. These structures being the seats of the disease, the analogies between the twin brothers should rather be regarded than their trifling differences, if we wish to arrive at a correct view of their relationship. Surely gout is more like articular rheumatism, than this latter can be said to resemble muscular rheumatism.

If there be two maladies, they can be made out to be two only by the trifling distinctions afforded by the results of chemical refinement. Dr. Prout saw, in this point of view, one of the most important distinctions offered to the consideration of those lovers of hair splitting diagnosis, who believe that the great art of the physician lies in exquisitely minute distinction. He thought that gout was more or less connected with errors in the primary assimilative processes, in which hydrochloric acid was developed, and that rheumatism was allied to the disordered states of the secondary assimilative processes in which lactic acid is developed.

It is much insisted on that gout, in the great majority of cases, is preceded and accompanied by a deranged state of stomach, with want of appetite, nausea, sour eructations and flatulence. Rheumatism has exposure to cold for its alleged immediate cause, and is said to be not accompanied with any remarkable disorder of stomach. This is an assertion rather too eager, for the bowels are mostly very torpid, and the secretions always unhealthy.

The duration of an attack of the primary sub-acute rheumatism must depend partly on the previous history and the actual condition of the patient. If the liver be obstinately deranged, the meaning of which vague medical language is, that if the source of the principal magneto-electrical currents, the brain, be wanting in efficient tone, the case may last several weeks; but if Mesmerism can restore that organ to its normal state, the cure may take place in eighteen days.

These facts may be illustrated by cases, perhaps better than by a continuation of details of symptoms.

A gentleman of fortune, 52 years of age, is subject to attacks of gout; he is of an anxious disposition, and is engaged in pursuits that take him much into the open air; he lives well, and is a wilful advocate for a moderate quantity of wine every day. People are little aware of the influence upon the nervous system, and upon the vital electrical currents, of a few glasses of generous wine, daily swallowed at a late meal. The tendency, after some years, is to affect the primary processes of assimilation, and to produce an evolution of muriatic acid in the first passages. A necessary consequence is, that the debris of the body loads unduly the veins in the secondary assimilative processes. The bile yielded by the liver is deficient in quantity, and abnormal in quality. This gentleman exhibited the symptoms of such a state, and being very liable to be affected with catarrh, he was not unfrequently seized with it. Lately, going out in a north-east wind, he came into his house with a chilled right hand, which soon developed an attack of gout, affecting the thecæ of the extensor tendons, the knuckles, and the palmar fascia. The redness at the back of the hand was highly characteristic of gout. In essentials, the malady was rheumatic. The venous accumulation, the condition of the bowels, the exciting cause of the attack, the tissues affected, the agonizing extreme reversal of polarity, all belong to one disease. The age of the patient and his wine bibing disorder of stomach claim the case for gout. When books of reference are consulted, additional arguments are afforded to establish a marked distinction, and to establish two diseases; for the nervous temperament, the irritable constitution, and the state of opulence are considered as predisposing circumstances in the scale of differences.

In this case, cold has been the exciting cause to the development of perfect gout; whereas it is laid down as a direct distinguishing cause of rheumatism. Really, all the distinguishing facts rest upon the two processes of assimilation, and whatever may be the arbitrary distinctions between the form of malady of the fibrous tissues called rheumatism, and that named gout, it is very certain that neither age, nor wealth, nor exciting causes are of any value in a philosophical estimate of diagnosis.

The following case, called one of sub-acute rheumatism, has been kindly given me by a medical man, who was himself the subject of it.

"You are aware that I have always been partial to good wine. For a good many years of my life I accustomed myself daily to one or two glasses of sherry at dinner, and four or five glasses of port, afterwards. You attributed my late attacks of illness to this habit. Be that as it may, when I was about forty years of age, I was one morning, in washing my face as usual with cold water, seized with such a severe attack of hemicrania on the right side of my head, accompanied by agonizing pain in the right eye and nostril, that I felt astonished at the suddenness and severity of the new visitation, which lasted for several hours, only subsiding very gradually. I left off wine at once, according to your urgent advice, and moderated the amount of my food. My health, digestive power, and alacrity were undoubtedly much improved. A year afterwards I began to drink wine again, taking a couple of glasses of sound light hock daily, and sometimes adding one glass of old sherry. This did not appear to disagree, except that an occasional visit from a very teazing, irritating complaint to which I have been been more or less subject from boyhood, whenever I ate acid vegetables or fruits, occasionally attacked me. This was urticaria-febrilis, sometimes intermixed with purpureal spots. As a child, this complaint used to break out in wheals all over my body, and even then it was observed to be always connected with acid ingesta, apples, pears, peaches, plums, oranges, cake made with butter, or light French wines would bring it out on my skin. I was very subject to headaches, and these were always relieved by the presence of the eruption. When the purpura accompanied the urticaria, hæmorrhoidal losses occurred, apparently always connected with a congested state of veins, by relieving which, a buoyant feeling of health and spirits resulted; whereas, before the relief, there existed a feeling of melancholy, and a want of mental energy.

"Every winter I have been a martyr to catarrh, terminating in teazing bronchitis. Ten years ago I was seized with very severe and painful gout in the right hand; which, with the aid of colchicum, colocynth and blue pill, with due quantities of magnesia, potass and rhubarb, yielded in two days. Again I left off wine, and became a water drinker; but according to your view of gouty and rheumatic diseases, I must have had something very like gout again, although it was said to be primary sub-acute rheumatism. On the 1st September, (1847,) I walked along the east cliff at Brighton, meeting the wind from the north-east. After awhile, I felt a pain in the front of each shoulder, and as it was becoming very unbearable, I turned about and walked westward. In this direction I felt no pain in the shoulders. The next day the weather was hot, and I walked a good deal, not caring for the north-east wind. I perspired, and walking on the down against the wind, the wet perspiration felt cool, not remarkably cold. I thought no more of it. In the evening, the window of my lodging looking upon the sea, was open. I sat there for some time, and fell asleep; waking in about a quarter of an hour rather chilled. An hour afterwards there was a miniature likeness of a rigor. At night, when I got to bed, I had a real shivering, and found myself with a sore throat. The uvula and the velum were very red. I felt much oppressed, as if catarrh were approaching. I hoped, by rest and quietude, to be much better on the next day, but on the 4th, I had stuffed nostrils, without increased secretion. The bowels were confined, the shoulders felt very stiff, and there was a sense of weariness down the back. After some hours, a severe pain seized my right shoulder in front. I could not bear to have it touched even very slightly. I took a compound colocynth pill; but not in sufficient dose to afford relief. A sense of fullness and tightness came over the chest, and round the region of the liver. At night I swallowed a pill composed of oxide of mercury and tartarised antimony, and two hours afterwards compound colocynth pills. Fever at night, with thirst, clammy mouth, and loaded tongue, were very troublesome. Nausea, headache, rigor, great increase of pain in the right shoulder, extending along the deltoid muscle, and engaging the biceps and its expansion on the inner part of the elbow. The elbow joint teazed me, but not nearly to so severe an extent as the shoulder. On the 5th, the medicine, though in full doses, failed to relieve the bowels satisfactorily. The urine was of the colour of dark brown sherry, and deposited abundance of lithates. Weary muscular pains affected the right hypochondrium. Husky sore throat, and tonsils externally swollen. A wretched restless night. The legs as cold as ice; on the morning of the 6th I took compound infusion of senna, decoction of aloes, and sulphate of magnesia in a full dose, which effectually relieved the bowels of a load of watery black mud. In the evening, some relief, although I felt very faint. Severe muscular pains about the right acromion, down the humerus, besides severe pains along the sterno cleido mastoid, and the levator scapulæ; distinct torticolis on the right side. Pain insufferable upon being touched. Took more aperient pills with some calomel at night; mouth very clammy; drank a good deal of water, feeling excessive thirst; free perspiration, offensive, and very uncomfortable. On the morning of the 7th, infusion of senna, decoction of aloes, potassio tartrate of soda. For three hours abundant catarrhal flux from the nostrils. Much teazing dryness of throat; cough in fits; great restlessness, complete inability to apply the mind to any intellectual exertions. Relief from pain while lying perfectly still. Secretions much the same. Aperients repeated at night, and in the morning of 8th, too feeble to get up, though better. Pains at times not perceived; at others very teazing. Perspirations very profuse towards morning; no sensible relief from them. Strong acid and peculiar odour of perspired fluid, very disagreeable. At times thirst has been troublesome; agreeably relieved by potass water. Evacuations to-day continued dark, watery, with shreds of mucus and bile. Urine scanty, deep in colour, red, lateritious deposit; at night a repetition of aperients. Morning of 9th, senna and potassio tartrate of soda; results beginning to improve; nostrils stuffed; cough in fits; dry sore throat; much debilitated from exertion of sitting up to write letters. At night medicines repeated. Little or no sleep from agony in the right shoulder; a wretched night.

"Tenth repetition of draught; pain in the right shoulder very severe, extending along the muscles of the humerus to the elbow; expansion of biceps very tender, olecranon tender; pain extending along the ulna to the carpus, to the metacarpus, and to the palmar fascia, and to the outside of the right hand, where certainly some redness was present. Mesmerism was recommended. A strong healthy looking woman made the passes from the shoulder downwards. The relief from pain, at first, was slow for about an hour, but after that, very decided. Potassio tartrate of soda, in small doses, every two hours, quantities of concentrated bile were passed. In the course of the day flying pains outside the left knee and metatarsus, and little toe of the same extremity, lasting rather severely for about two hours, and then subsiding. In the evening, catarrh and hoarseness. No sleep was obtained by the general passes, which were made by a vain conceited person, who laboured under the mistake that her experience, as a rubber, must enable her to become a perfect mesmeriser, but her local passes succeeded in allaying the pains considerably, and at night I slept soundly. On the next morning the catarrh was more marked. There was less pain in the shoulder, but great debility in the right arm; coracoid process tender, ulna and hand free, biceps a little stiff and painful when moved. I went on with the potassio tartrate of soda three, sometimes four times a day. I drank freely of aerated potass water, and was locally mesmerised. Decided relief was obtained, and the secretions became much improved. In six days more I felt quite well, and discontinued all treatment."

In such a case the trains were cheek by jowl, and eager, indeed, must have been the lover of hair-splitting, who could forget that this case of sub-acute rheumatism occurred in a man of a gouty diathesis, who actually had gout in his right hand while he laboured under an attack of rheumatism. All gout is rheumatism, and most of the cases of rheumatism seen by medical practitioners, whether cold has had any share in their production or not, owe their origin to the same causes remotely which favour the access of gout. The various shades of these are so numerous, that men of confused minds have a difficulty in distinguishing the name they ought to give to a pain they are called upon to prescribe for, and hence, for several years after the introduction of the word neuralgia, it was amusing to watch the gullibility of the patient, and the bold impudence of the practitioner on occasions when this name was given from sheer inaptitude to distinguish between a rheumatism and a nerve pain. The inaptitude was natural, for the mind of the practitioner had never reflected that nerve matter *per se* suffers no pain. The pain is in the theca or compact tissue, which embraces the organism called nerve. This is not now the place to enter into the physiological question, but it may be simply stated that neuralgia is always a rheumatic or gouty pain, having its seat in the nerve theca, and like pain anywhere, is due to an *extreme* reversal of normal polarity.

Is there such a thing as gouty neuralgia? Is there not? Sciatica, in a vast majority of cases, is a gouty neuralgia. That it is a true case of extreme reversal of polarity has been often proved by the quick removal of the agonizing pain under the hands of Mr. Capern, who appears, in these cases, to have a remarkable power of curing this description of pain. I have seen enough of this extraordinary man's power to be fully convinced that he does not exaggerate the truth in his statements. He has done an enormous deal of good among the poor, in various parts of the country, as well as in London. People may talk of his failures in speedily curing cases here and there. They would do better to enlighten their own minds. It is ignorance of the fact that every case is not equally susceptible of a speedy cure, which hastens them so speedily into false assertions. I was requested by a physician, whom I have known for a number of years, to visit his wife, who sitting one Christmas day with a family party at dinner, was seized with agonizing pains over the abdomen, after drinking a glass of champagne. Various means had been resorted to in order to quell this pain, which I was told had been so urgent, as to resemble peritoneal inflammation; for the finger applied upon any part of the pained suface could not be borne. Physicians and surgeons of eminence had been called in, and declared their inability to produce the desired relief. The pain travelled downwards, along the course of the sciatic nerves. Spasms of a very distressing nature affected this severely afflicted lady. I heard that once she had been curved backwards, the spine being bent, as we see it, in lock-jaw. Were I to mention the name of her last medical attendant, it would be a sufficient guarantee for the fact that no appliance, within the resources of our art, had been wanting to effect a cure, *except that one* which he has not yet learned to recognize as a blessing from God to man. I was called to this lady fully ten days after she had been subject to these pains.

I requested to have the assistance of Mr. Capern, and he came, but without any hope of curing her in ten minutes, a period of time in which he had reversed abnormal polarity in many a case of sciatica. This lady's was a genuine case of gouty rheumatism. I myself mesmerised her a considerable time, and having succeeded in producing drowsiness, I knew that perseverance only was required in order to affect a perfect cure. Great and immediate relief accrued, and in little more than a week

I had the gratification of seeing my patient surrounded by her children in the drawing-room. Did she take any medicine? I am sorry to say, that my good friend is like a retired coachman, he loves the crack of the whip. He felt obliged to interfere, and to give a dose, perhaps two. He is a wonderful stickler for professional etiquette, and would, I believe, be really unhappy if he had not had the opportunity of being able to say that some doubts might be entertained whether the cure might not be due to his remedies. If my friend would allow me to banter him, I would say, of course, he had a right as a husband, to give his wife either a dose of physic or not, as he pleased; but as a physician, observing his favourite principle of professional or medical ethics, he had no right to give the drugs without my permission. Turning upon me he might say, "I was the physician, I considered you as the *mesmeriser* simply, or as my wife called you the magician!" Humility is the duty of him who takes up my mission. I am content. The patient knows that God sent me as an instrument for his own purposes; and if I have failed in making her husband reflect that the higher interests of his profession demand of him the duty of studying the physiological and pathological bearings of Animal Magnetism, it has not been from an effort to draw his attention to this holy subject being wanting on my part.

Bothering names lead to no good end. Neuritis has been proposed for neuralgia. Coincident with an attack of most acute pain along the course of the left sciatic nerve, a gentleman 35 years of age, of dark hair and eyes, of sensitive nervous temperament, had an obstinately costive state of bowels for three days and nights. Aperient medicines failed in their usual results. At last calomel and colocynth pills in bold doses, succeeded in six hours by compound infusion of senna and salines, acted on the bowels. But not the slightest relief was obtained. The slightest pressure of the forefinger in the course of the nerve and its ramifications occasioned agony. It must be *severe neuritis*, and cupping was directed. At first, thirty ounces of blood were in this way abstracted. Next day, no relief; eighteen ounces more, by cupping, were taken; no relief. Opium was objected to because it had once produced delirium. The objection was overruled. I did not then know of the efficacy of mesmerism, or the risk would have been avoided—a drachm of the London Tincture was given; *a deep sleep* of eight hours, which toned the exhausted patient, and enabled a reversal of the abnormal polarity to allow the normal currents to resume their course, and the patient was cured. My colleagues in the treatment of this case were the late eminent Sir Charles Bell and Dr. Southey.

It may be said we have wandered from Sub-acute into Chronic Rheumatism. We are more intent upon pathology than upon names. Chronic is a word implying a relation to time. We have been considering the severer forms of the disease, and they are as to duration far more limited than that which we know as Chronic Rheumatism.

Acute rheumatism is decidedly too severe to be confounded with sub-acute rheumatism, nevertheless it is the most severe form of gout. We hear of acute gout where one foot is very painful and inflamed. It is the degree of pain which warrants the use of the word acute? Acute rheumatism should be confined to the disease when a large number of joints is affected, sub-acute when fewer joints are involved, and chronic rheumatism may be the term applied to those forms of gout or of rheumatism less severe in symptoms, occupying fewer joints, but often wandering over divers various parts of the muscular system. The conventional term sub-acute would not exist, if the vitiated condition of the secretions did not warrant a distinction of gradation in the urgency of the symptoms of the one common disease, rheumatism or gout.

Chronic rheumatism is either primary or secondary. It is said to be secondary when the more severe forms subside into the less severe. But between the primary and secondary there is no difference of symptoms, except perhaps that the secondary may now and then be said to be characterized by reminiscences of some of the signs distinguishing the severest cases. To a great extent, the fever may be regarded as in some degree the measure of the severity of pain. This is a general rule: consequently, before sub-acute rheumatism becomes chronic, the fever must entirely have subsided. No case of chronic rheumatism is accompanied by fever, and the degree of pain must have become much diminished.

The secondary chronic rheumatism following upon a more severe form, has for a time the dry and loaded tongue, the clammy mouth, especially in the morning. The case is but a continuation of the uncured rheumatic fever. The bowels are wrong. The biliary secretions are vitiated, and as this sign of health or disorder is unerring, until it can be brought to assume its normal condition, the limb or joint is left with pain, differently described by different individuals. Some say it is tearing, or wearing, or gnawing.

Those more severely afflicted, complain of its boring nature, aggravated at night with reminiscent restlessness and want of sleep. Some less severely pained, complain of aching sensations relieved when they become warm in bed.

The pains are more fixed, and they have a continued character, not always, however; for in some cases they remit in the morning, ceasing the whole day, to return when night sets in.

The cases of chronic rheumatism in which this remittance is present, occur for the most part in those constitutions which have suffered from the thousand causes that entail cachexia. The difficulty of establishing a diagnosis in such instances is great, except to the pragmatical practitioner.

The gouty or scrofulous; the scrofulous gouty; the victim of ague and gout, or of gouty ague; the poverty stricken labourer or artizan, whose energies are depressed by loaded veins resulting from damp habitation, insufficient warm clothing, or telluric or malarious influences, all those who from asthenia form extreme gout imperfectly, and yet have disordered chylopoetic organs attuned to disorder of the secondary processes of assimilation, for there is harmony in disease, as there is normal order in health—all such individuals are prone to have chronic rheumatism, attended by the modifications we are considering; and among the poor especially, the complications produce very obstinate persistence of symptoms. In various parts of England, these chronic cases have reference not only to a lack of sufficiently warm clothing, but to causes that produce obstinate forms of rheumatism affecting the muscles of the loins, the back and thighs, in some cases affecting the theca of the spinal cord, and entailing distressing and slowly advancing paraplegia. In districts where much cyder is swallowed, and in others where hard beer is much drunk, and where exposure to damp atmosphere and telluric mal-influences favour the progress of this disease, a persistent form is common, which no one thinks of distinguishing by the name of gout, although it is the result of primary mal-assimilation, as well as of disorder of the secondary processes.

Sometimes the phenomena of chronic rheumatism observe a marked periodicity, especially in those persons who have suffered from ague; but even without any previous attack, they sometimes assume the true intermittent type of character, referable in a greater or less degree to the digestive functions of the individual; and having some relation to the nocturnal disturbances allied to restlessness and want of sleep, cold lower extremities, rigor without fever, slight head-ache, and uneasy desire to shift the position in bed. Chronic rheumatism is most often a remnant of gout. It is infinitely more common in persons who have passed the middle term of life, and whose primary as well as secondary processes of assimilation are in an abnormal condition. But the redness has completely left the affected parts, not always however; the swelling which though less than it had been, is sometimes more diffused, and occasionally œdematous.

It is often difficult to determine whether a case of chronic rheumatism be primary, or secondary, for it is occasionally secondary to catarrh, and often secondary to gout; that is to say that although either of these maladies may have subsided with apparent indisposition to return, and neither of them may at once have glided into chronic rheumatism, yet the processes of primary and secondary assimilation have not been restored to their healthy condition of tone; and, consequently, the chronic rheumatism is but a renewal of the phenomena of disease belonging to the unfinished, or incomplete, train of events. It may exhibit itself in a muscular part; or in the joint of the shoulder, or of the elbow, or of the wrist, in the knee, or in the ankle, in the loins, or higher up in the back. A lady, whose digestive functions occasionally suffer from errors of assimilation, has chronic rheumatism referred to the ischial tuberosities. It begins with gnawing pain, and becomes more severe if medicine be neglected. When the biliary secretions have become improved, the pain ceases; to return when the same set of functions are again in disorder. This illustrates the case of primary chronic rheumatism in a person who has for years had recurrences of the same symptoms, and whose occasional bilious disorders have been attended with cramps, cold lower extremities, and muscular pains in various parts of the body. In early life, she experienced a very severe and painful attack of rheumatic opthalmia, and for years afterwards had no recurrences of rheumatic disease. Rheumatism now visits her by attacks in the fibrous parts about the ischium; in various muscles, and occasionally in the subcutaneous bursa near the joint of the great toe. It is a matter of little moment, what name it may please certain individuals in the medical profession, arbitrarily to allot to such a case. Premonitory symptoms of gout, one would say; another would be satisfied with the fact that the bursa was painful to constitute a case of gout. The question is not what shall satisfy a bad reasoner, however exalted may be his professional position, but whether, in fact, the indications of gout and of rheumatism, in a vast number of cases, are not such as to warrant the conclusion, that they are most frequently identical, and that they vary only from modifying circumstances dependent upon age, habits of life, peculiarity of constitution, and the numerous trifling causes that impel the nutrient forces of the body to march in various lines aberrant from their normal routes.

Can an individual be subject to gout who never developed a true fit of the malady? The following case illustrates the dogma of the great demarcations between gout and rheumatism. The Marquis of ——, a man of very active habits, a keen sportsman, fond of athletic pursuits; much on horseback, a keen huntsman, an early riser when in pursuit of wild ducks; no drunkard, yet fond of jovial company, in which he would occasionally drink more wine than was good for him; ate moderately in regard to quantity, yet was careless of the hours of meals; would sometimes drink a cup of *café au lait* in the morning, and eat nothing until one o'clock, when he breakfasted on well peppered pigeon pie, washed down with brandy and water. His appetite for dinner was never very robust, and he considered himself a very moderate feeder. At the age of 35, he consulted me in 1844, for pains in the muscles of his limbs, which were never very serious, but sometimes rather inconvenient. They were not of long duration, and the pain which fixed itself for the longest period, and that was a few days, was in the muscles and fascia of his right foot. The pulse was always good. He declared that he was, and had been a number of years, in excellent health, and was desirous of knowing what his pains could mean. When it was hinted to his lordship that he had premonitory symptoms of gout, he laughed at the idea. There was no swelling, no redness. His boots were admirably made, yet there seemed to be pressure from them because of the tenderness on some muscles. It was a fact that he never had a return of the symptoms, and his medical friend in the country could not be persuaded that he had had any signs of gout. A year afterwards, he had oppressed breathing, with valvular disease of the heart, and died after a three months' illness. I was in Germany, and the eminent physician who attended him during my absence, was of opinion that the disease was metatasis of gout to the heart.

In what respect could this case be considered as different from a very slight attack of chronic rheumatism? And if the heart were attacked, whence did the metastasis result? The tendency to deposit lymph on serous surfaces is part and parcel of gouty disease. It is very probable that in children, when rheumatic hypertrophy occurs, it may be due to severe accession of that form of rheumatism known as catarrh.

There is another point of view in which these cases of metastasis may be regarded, we know that in abnormal dentition children suffer much from chondroma of the gums, and from various rheumatic pains in different parts of the body; nay, children are not the only victims to severe rheumatic pains under these circumstances. A lady who had been eight or nine months confined to her bed in Liverpool, was brought up to London that I might be consulted upon an abnormal growth of a wise tooth; she had suffered almost unbearable pain in her shoulder, elbow, wrist, hip, knee, and ankle of the left side; I had the assistance of the late Mr. Cartright, who, after twenty minutes of exertion, extracted a wise tooth that was growing aberrantly. At the end of the operation all her pains ceased, and she was restored to perfect health. It is a question with me whether, in some analogous cases, metastasis has not taken place to other organs of the body. I have witnessed instances in which abnormal cysts have been found on the surface of the pleura, where the wise teeth have been abnormally developed. And one extracted by a very expert dentist, Mr. Hayward; relieving at once a young lady who had had her head, shoulders, and ribs on the right side pulled down to the hip by ungovernable spasms.

Having taken a view of the origin and relations of rheumatism different from the accepted idea on the subject, it will not be regarded as more singular that, under the head of complications, only those results of rheumatic irritation which have been before alluded to will be noticed as belonging naturally to our subject.

By far the most frequent of these is catarrh and catarrhal fever, or influenza. Chomel, quoting Stoerck, goes extensively into the question of the epidemic character of rheumatism, under certain conditions of atmosphere, &c. If influenza, which is really a rheumatic febrile and epidemic catarrh, had been regarded in its natural point of view, the inferences on the cases related by the German physician would not have been so difficult to comprehend. If the peculiar constitution of the atmosphere retains at times unexpected subtle nocent forces, it is easy to conceive that numbers of individuals may labour under a load in the veins which shall engender a proclivity to catarrhal or rheumatismal affections; and as such conditions of atmosphere are known to be often limited to localities, there is every reason to conclude on the endemic characters of a disease afflicting that locality. Ague is thus often a complication with rheumatism, and in some cases this complication renders the cure of both diseases very tedious, and Dr. Todd has lately considered a complication of this disease with the disordered state of the catamenial discharge. An idea relating to this peculiarity of the female constitution is seized upon by practitioners as well as patients, and the fictions of the imagination are pressed into the service of the favourite dogma by ladies as well as their apothecaries. There is no doubt that disordered states of the veins of the womb, when coincident with an asthenic condition of the liver, force upon the uterine system functions that are abnormal to it. A woman may often be carrying on the depuration of her blood by a passive menorrhagia, or by an abundant leucorrhæa, and when a certain point of debility is arrived at, the fibrous structure of the womb may call into sympathy the fibrous parts of the cloacal system of the pelvis, and rheumatic pains of joints and aponeuroses may result, but by far the most common results of abnormal depuration by the uterine system consists in those asthenic and clonic states of system which induce trains of disease, commencing in catarrhal affections, proceeding to strumous rheumatic symptoms, and terminating in those dire forms of disease induced by long cachetic suffering.

The most distressing form of uterine rheumatism is that which occurs in labour. It is, fortunately, very rare. A lady in Regent Street, subject to frequent bilious attacks and accumulations about the hepatic system, which were attended, when active, purgative, and depurative treatment became necessary, by very copious evacuations as dark as pitch, requiring sometimes from twelve to eighteen days to produce a healthier character, was taken in labour of her fifth child. She had neglected the precautions repeatedly insisted upon regarding the use of aperient medicines. Rheumatism of the uterus attended the labour; no case of severe peritoneal inflammation could have been more exquisitely painful to the touch. The pulse was soft, and in frequency not more than 84. Each pain was so agonising that the patient's screams were terrific. Very bold doses of opium and calomel did not in the least assuage the suffering, though I believe they imparted healthy tone, and enabled the patient to conclude her labour in four hours. The pain upon pressure remained, and did not vanish for several days, when in result of active depurative practice they quitted their seat.

Diaphragmatic rheumatism may, in similar circumstances of hepatic congestion, be present during labour, and it forms a very painful complication. Though I have seen numerous cases of all the varieties of puerperal peritonitis, I cannot recal to my mind a single case in which it originated in a rheumatic affection. The more ordinary cause of the disease is phlebitis, but there is no reason why a serous membrane of the abdomen should not be inflamed as well as one lining a joint, in result of rheumatic irritation.

Where the pleura becomes affected with inflammation in result of neighbouring irritation, it is seldom a rheumatic irritation that induces the inflammatory state. Pleurodynia is not an uncommon complaint, and may easily be mistaken, from the severity of the stitch, for pleurisy. It catches the breath very suddenly, and the pain is agonising. It is most frequently dependent on a disordered condition of the bowels. If properly treated by purgatives, pleurodynia seldom lasts more than twenty-four hours, leaving some tenderness in the pained part. The countenance does not denote an anxious state, and the pulse is seldom more than 84, soft and full.

The most serious of the complications of rheumatism are those which involve the inner and outer surfaces of the heart. The endocardium, in some few cases, is the seat of changes of structure apparently synchronous with acute rheumatism, and in others the affection of the heart involves simultaneously the pericardium and endocardium.

It has been a question with some pathologists whether the affection of the heart may not have been the primary disease, and rheumatism the secondary. Dr. Copland says, "that many cases of this *complication* present the heart as the primary seat of the disease is by no means improbable," and Dr. Todd's view is similar, for in repudiating the idea of a metastasis, he says, "the truth is that the cardiac inflammation may be primary, it frequently exists at the same time with the articular affection, and dates its origin from the same period, as it derives it from the same cause." Dr. Graves has gone so far as to say that rheumatic disease may exist without its external manifestations, and then the natural conclusion is, that arthritic rheumatism is not necessary for the existence of that state of heart which has been said by many writers to be a metastasis from that rheumatism. Palpitations of the heart are put down by some authors as premonitory of gout; and some cases have terminated fatally without forming full endocarditis or pericarditis, merely a bare attempt at effusion of lymph on the cardiac surfaces. These have been set down as suppressed or undeveloped gout. Without contending for theoretical ideas on the facts of depositions of lymph on the surface of the endocardium or of the pericardium as connected with articular rheumatism, without embracing the idea of metastasis, or of a materies morbi; it appears rational to be content with the fact that in some cases of severe acute rheumatism there exists a tendency to throw out lymph on the endocardium or pericardium, or both,—the remark being allowed that the surfaces like those of the interior of articular capsules and bursæ are those of serous membranes; and a further illustration is offered of the doctrine that in rheumatism there exists a tendency to irritate tissues, to deposit or to secrete, they being other than those in which the pain first declares itself.

It is unnecessary to moot the question, whether rheumatism may exist without pain, or whether the play of Hamlet is complete without the character of Hamlet. This would lead us into a discussion on the necessity of a reversal of polarity being a necessary condition of the existence of disease. Pain being an extreme reversal of polarity, it is not absolutely necessary that it should be present in every case of reversal, and yet all the signs of rheumatism, save that one of pain, may be present in a case that shall end in dropsy. It is unnecessary then to be so severe in our definitions as to appeal to the fact of an analogy between a complete rheumatism, and an incomplete Hamlet.

No illustration of the identity of the phenomena, and of the causes of gout and rheumatism, is more striking than that offered by lumbago.

An individual of a gouty diathesis, apparently in good health, but who has been drinking a full allowance of Port wine, or of excellent Burgundy, fully confident that nothing disagrees with his stomach, is suddenly seized, on endeavouring to rise from his chair, with a severe pain in his loins. He is fain to resume his seat, when he is quite free from pain, but on the slightest effort to raise himself the parts affected with the pain in the loins remind him at once that his health is not so robust as it was. Most people under such circumstances who have had two causes operating to derange the economy of their chylopoietic system, namely, a costive state of bowels, with a disordered biliary secretion have experienced some degree of lumbago ; for we need not specify whether the bile has contained an excess of taurine, or whether some other cause of mischief, as an exposure of the loins and back to the influence of a draught of cold air, or having been seated in an open carriage with the back to the east, or north-east wind, has given rise to it. Many persons, said to be of the gouty diathesis, who have eaten more than they can easily digest either at breakfast or dinner, or who have been subjected to the irritation of the mucous surfaces of the intestines, consequent upon fulness of the venous system, loaded with the debris of the secondary processes of assimilation, which not unfrequently occurs in cases of hæmorrhoids, are apt to have what the Irish would call a touch or a taste of lumbago; this condition of things which gives the idea of a slight taste of a pain is sometimes not very severe, and calls upon the patient for a fit of laughter; but it may increase to a very agonising extent and is then no joking or laughing matter. The patient is obliged to go to his bed where he may get some solace from this pain, but where very often he endures all the distressing symptoms of a true and severe gout. There can be no question that now he may have all the symptoms of a true gonorrhœal rheumatism. We do not allude more explicitly to the sympathetic pains which occur from certain morbid conditions of mucous membranes irritated in gouty or rheumatic subjects, because the pathology of all these cases is so thoroughly interwoven with inverse currents of animal electricity, dependent upon the laws of Animal Magnetism. If medical men would but study their profession philosophically, they would find the views here exhibited to be perfectly correct, and would be enabled to treat their cases with great advantage. One might be asked at once how such cases should be treated; the reply would be simple; induce always as speedily as possible a condition of Tone, by passes, by the force of the human will, or by the well known appliances of Animal Magnetism. Men of experience on this subject can easily and speedily cure such cases.

Pain in the loins when once established in a severe form, like a bad attack of catarrh, is apt to recur in milder form upon the slightest exposure to cold, or to the irritations of drastic purgatives that do not satisfactorily cleanse the veins. Perhaps no form of rheumatism depends more completely upon errors of the lower bowels, the due performance of the functions of which are so intimately allied to the secondary processes of assimilation. The cloacal system, by which is meant the series of organs connected with the kidneys, the bladder, the larger intestines, and the lower abdominal and pelvic veins is often stimulated to undue or over-action, by gamboge, aloetic and other drastic aperients habitually swallowed by those who find a certain amount of relief from the consequences of heavy or hearty meals not completely digested, and in many of these cases the back is the part mostly debilitated and rendered liable to suffer from attacks of cold. Women suffer from pains in the loins from irritation to the pelvic nerves, and men are apt to have the same sets of nerves made to respond to irritations applied to organs whose veins are loaded. Lumbago is strictly rheumatism of the fibrous structures of the lumbar and dorsal vertebræ, called into action originally by exposure to cold, and often renewed by irritation to the membranes of the cloacal system.

Lumbago is thus both a primary and a secondary ailment, varying in its degrees of severity, and becoming sub-acute or chronic according to various circumstances.

A recent attack may be very sudden. It may or may not be severe. The patient finds himself one minute quite easy and free from pain. In the next minute he has stooped and cannot place himself erect without agonising pain in the back. He sits or lies down and is easy again. The moment he tries to move his body on the pelvis or the upper part of the back, the pain is severe. When warm in bed, or quiescent, the individual is generally pretty easy, but in severe cases and in those at all complicated with rheumatism of the sciatic nerve, every movement even in bed is agonising.

Rheumatic pains may affect a person in any part of the back, and are not confined to the lumbar regions.

With those liable to gout, the aponeurosis of the dorsal and posterior cervical muscles are very apt to be affected. Wherever fibrous textures are present, there rheumatism may find its way. Cervical rheumatism, or as the French call it Torticolis—is that familiarly known in England as stiff neck or crick in the neck. It results mostly from exposure to currents of cold or damp air. The pain sometimes effects the sterno cleido mastoideus muscle, and is then called wry neck, and sometimes the aponeurotic expansion of neighbouring muscles in the back. Like all other forms of rheumatism it varies in degree, sometimes it is severe with a very dry and loaded tongue, clammy mouth, and accompanied by a more frequent pulse than is usual. Mostly it is chronic, still varying so much in degree as to cause on some occasions great pain; while on others, the pain amounts to little more than an unpleasant sensation.

The application of cold to the mouth, cheeks, eyes, teeth or temples, either from draughts of cold air, or from plunging the head into cold water, will, in certain constitutions, be followed by rheumatic pains in some part about the head.

The tendencies of the system, when the state of loaded veins, and other circumstances, previously alluded to, as remote causes, are present, operate to produce severe and troublesome rheumatism in unexpected localities.

A lady, 29 years of age, anxious about her husband, taken suddenly ill in the night, went hurriedly and impulsively to the bottom of the house, into the kitchen, for hot water. She forgot her slippers, and her feet were bare on the cold stones. The next day rheumatism attacked the sclerotic coat of the right eye. The agony was almost insupportable. Opium, colchicum, calomel, alkaline remedies freely administered, prevented the disease from proceeding to any other tissue, and the cure was complete in five weeks. Rheumatism of the eye, like rheumatism in all other parts, by reiterated visits to an organ, weakens the texture of that part, and may, as we see in chronic rheumatism of the joints, alter the train of nutrition, so as in continual abnormal currents to become a serious disease. But when it attacks the sclerotic tunic of the eye, it seldom causes immediate change of structure in that compact tissue. Its recoupe is upon the vitreous humour, or upon the crystalline lens, or upon the iris. Deposit of lymph takes place in some of these parts, and we have not seldom in persons of gouty diathesis, glaucoma, cataract, or the dire consequences of what is misnamed Iritis. In fact we have depositions in structures, other than those in which the pain first declared itself.

The sympathies arising from irritations (reversals of

polarity) in the fibrous sacs of the teeth, and in the alveolar and endodental membranes are so curious and diversified, that I am hardly able to trust myself again with the subject, so many painful events connected with the treatment I have experienced in endeavouring to enlighten the crass ignorance of so-called eminent medical men, starting up to the memory. I wrote on dentition many years ago, at the urgent recommendation of the late Dr. Gooch, and to this day, the thoughts in my book are not understood. The subject has certainly not met with that deep attention which its importance merits.

The periosteum is a strictly fibrous membrane; very liable to be affected by rheumatism, when the loaded venous system of the individual has induced a condition of the solids and fluids which is known by the term, bad habit of body, Cachexia. There are many causes that are apt to impair the tone of the system, and to call forth specifically the pains which constitute rheumatism of the periosteum. The application of cold to this membrane, in any situation not protected by thick coatings of muscle, under the cachectic conditions alluded to, is an exciting cause of the pains, and it is easy to conceive that important as are the functions of the periosteum, it would sooner be influenced by reversal of polarity to become altered in its texture, than most other fibrous tissues. Accordingly we find that it becomes thickened, and that it communicates irritation to the bone under it; that if that bone is influenced by the irritation of some morbific poison, a destructive, though often slow inflammation is set up in it, and that not only thickened periosteum results, but nodes grow up protruding the periosteum, and forming very painful swellings. It is the fashion to make minute distinctions in the varieties of thickenings and nodes. Each owe their origin to the causes of rheumatism. Their progress, and the history of their origin may be modified by various causes, but they begin in rheumatism with pain, the extreme reversal of polarity, with disordered venous blood, and with veins loaded with the debris of the secondary processes of assimilation. Some of the varieties run a course more rapid than others. When syphilis has a share in their production, they proceed more speedily to inflammation, tending to ulceration. It is pretended that the pains arising from syphilitic rheumatism of the periosteum are more severe when the patient is warm in bed at night, than the pains of ordinary rheumatism of this membrane. Periosteal rheumatism is very apt to assume a sub-acute form, and then the nocturnal exacerbation is very severe, and the thirst, clammy mouth, furred tongue, restless state of limbs, slight degree of fever, disordered state of elimentary secretion, dark coloured urine, vitiated alvine excretions, are very similar in character to the syphilitic form. The fact is, that setting aside for the time the poisonous origin of the syphilitic form which allows of as much variety in severity of symptoms as any other form, all descriptions of periosteal rheumatism require very judicious management. A mercurial form of this disease is described, arising from injudicious and excessive use of mercury, producing a great number of nodes in the bones most exposed.

All forms of rheumatism commencing in the incipient stages of cachexia; it is not to be wondered at, that every variety of periosteal rheumatism should be attended with severe pain, and be tedious to manage. It is very wearing in the tibia. In the ulna I have seldom seen it produce such severe suffering. Pericranial rheumatism is attended with very disheartening pain. It sometimes attacks the bridge of the nose, and in very bad cases of scrofulous cachexia it becomes a most serious and appalling malady, lasting for several years, with very slow progress.

The pain in periosteal rheumatism is at first dull, and is referred by the patient to the depth of the bone. There is no redness, nor is there perceptible tumor in most cases. When the hand is passed over the skin of the ulna, if it be firmly pressed on the part, pain is complained of. Even when nodes arise, which are not always of syphilitic or mercurial origin, there is no redness, or discolouration. These are occasionally exquisitely tender.

When the nodes arise in consequence of changes induced by syphilitic poison, or from mercurial cachexia, redness often supervenes, and inflammation attended by troublesome ulcers results.

The case then is taken out of the category of rheumatic disease.

Had the range of our inquiries permitted, numerous cases, illustrative not only of sympathetic pains and formal disease, dependent on irritations of fibrous membranes could be adduced, but evidence of some of the trains of events whereby the progress of scrofulous disease and sacculated formations on the pleural membranes appeared to be dependent for their existence upon abnormal development of the four last permanent molar teeth. When Dr. Bright throws forth the hint that inflammation of serous tissues may often be coincident with renal disease; his speculation is not regarded as wild or chimerical. Pathologists may be assured that the field of inquiry embracing the sympathies of fibrous membranes is quite as philosophical an object of research, and perhaps far more extensive in its scope. The taste of the age has been for conclusions dictated by experimenters on the nervous system, and the confusion that has enveloped the subject of the sensations of pleasure and of pain, and their pathological results have considerably retarded the progress of the medical mind. The groove in which investigations relating to the nerves should be made is that indicated by Gall, which, as time wears on, necessarily allies itself to Animal Magnetism, the subject of such vast influence on all future facts in medical science.

It has not been forgotten that there is recorded a facial and a temporo-facial rheumatism. It need not be pointed out to the experienced medical practitioner, that in very many cases this has an intimate connection with the irritation of dental fibrous membrane, dependent on some unhealthy condition of a tooth, or of teeth. In such a case, the cause of agony being ascertained, if the application of leeches or of creosote, or of narcotics, do not remove the evil, the cause had proceeded to an extent that constitutes incurable inflammation of the endodental structure, and the removal of the peccant organ is the only measure for the removal of the pain. But in many cases, the rheumatism of the fibrous membranes of the teeth may be secondary to the rheumatism of the aponeurotic textures about the face and the temples. For the most part, this latter case can be distinguished from the former by the immediate cause of the symptoms. In both conditions, the remote fact of the loaded venous system is the same, but in the primary affection of the aponeuroses, there has generally been a direct application of cold, in some form, either of draught of air, long exposure to north-east wind, or the injudicious use of cold water.

Neuralgia or Tic Douloureux is distinguished from the pain we have been considering by the paroxysms of its visitations, its temporary acuteness is hardly greater than the permanent severity of rheumatism, but appears to the sufferer to be worse from the contrast between the intervals of ease and the plunges of the pain. It is often a result of dental mischief. The fibrous tissue of one part affecting the fibrous theca of nerve; often rendered permanent by disease of the bone through which the nerve passes.

The observations applicable to facial rheumatism are equally so to epicranial rheumatism. It may prove to be primary or secondary; sub-acute or chronic; mostly the result of a disorder of the primary and secondary processes of assimilation, loading abnormally the venous system. In such cases it is apt to be sub-acute, and often to be complicated with a diseased state of the teeth and their sockets. A remarkable case occurred to the author in the year 1814. A currier in Whitecross Street, Barbican, had been for years a martyr to cephalalgia rheumatica, associated with severe asthma. He had tried the suggestions of various medical advisers to no purpose. I perceived that he had a very sad state of mouth from numerous decayed teeth and stumps. These were all removed, and for several years afterwards he enjoyed undisturbed health; his asthma and cephalalgia having completely vanished.

It is not uncommon to find rheumatic cephalalgia affecting only one side of the head. Hemicrania is the title it then receives, and it is set down in some books as a premonitory symptom of gout. Rheumatism and gout being the same disease, with simply the difference of age and further disorder of the digestive functions, and the consequences of that disorder to constitute rheumatism or gouty disease, the question will not here be further discussed. Hemicrania is occasionally an alarming symptom of gout, for it has been attended with an amount of cerebral congestion threatening paralysis.

The seat of hemicrania and of rheumatic cephalalgia is always in the aponeurotic expansions about the head. The pericranium is never, as far as I have learned, affected with rheumatism, except in scrofulous or syphilitic cases, and in these, it is generally attended by exfoliation.

The changes in the structure of joints resulting from the trains of events, consequent upon extreme reversal of polarity in the affected parts, have been noted with great care by eminent pathologists; but their labours have not traced the mischiefs, which have commenced in rheumatic phenomena, to the relations of a wide family circle. Dr. Prout's acute conjecture of the intimate alliance of gout with scrofula, and Chomel's philosophical view of the identity of the general pathology of gout and rheumatism, have met with but small approbation among the eager diagnosticating physicians of our day. The taste is for concentration and minute division rather than for scope; and most valuable as are all contributions of minute facts to the general storehouse of medical science, yet too exclusive an importance given to them, precludes the power and the habit of the mind for generalization.

Dr. Todd, Mr. Adams, Dr. Colles and Cruvelhier are among those who have distinguished themselves by very clever and clear descriptions of the changes that take place in articulations from the long persistence of chronic rheumatism of the joints; but there are cases of disease due to this variety of rheumatic maladies, which from influences having reference to constitutional peculiarities of frame, proceed to a certain length on the same road, and then turn off to travel in a different direction.

Morbus coxæ, or hip disease, white swelling of the knee; the vertebral disease, or tubercular abscesses of the spine, or psoas abscess are recognised as scrofulous diseases as much as mesenteric atrophy, or tubercular consumption, or submaxillary, and cervical glandular swellings, and collections of pus. Each and all of them commence from the same causes that give rise to catarrhal and rheumatic diseases. Why they do not follow the same course is a pathological question. There are strong analogies in some of the changes that occur in white swelling, and those which are observed in chronic rheumatism of the knee joint. Dr. Todd wisely declines to consider this malady as an inflammation, and if Mr. Adams's name of chronic rheumatic arthritis is to be applied as a synonyme in any case, it should be transferred to the scrofulous inflammation of joints, which is first pain and tumour without inflammation, but proceeding, in some constitutions, to develop inflammatory phenomena as disease, when its destructive tendencies march forward.

Inflammation is a series of phenomena, taking their turn, primarily, in tissues other than those which have been first affected by the causes productive of catarrhal, or rheumatic, reversals of polarity. Pain and absorption may occur without inflammation, as we may perceive in cases of chronic rheumatic gout, one of the most common varieties of rheumatic disease, as well as one of the most intractable and obstinate, continuing for weeks, and even for months, without alleviation; and if improperly treated, becoming daily more serious and formidable. Here the reversal of polarity may start from the aponeurotic expansions where the pain is first felt; but the periosteum and articular cartilages soon share in the rheumatic troubles, and an absorptive process begins in the synovial membrane and cartilages of the interior of the joints. In the course of the disease, these structures disappear, and in place of the elastic cartilage, which with the synovial membrane has been changed, there is deposited a porcelain, or ivory-like, thin coating on the articular ends of the bones.

This destructive process is a chronic affair. It does not take place at once. The articular tissues undergo some remarkable changes which have been carefully described by Dr. Todd. He remarks that "suppuration or ulceration seldom occur, and where they do, it may be fairly conjectured that they proceed from a venous inflammation coming on in the course of a paroxysm;" but he describes changes taking place, in some cases, "in the ligaments of the joints, in the periosteum of the articular ends of the bones, and in the tendons of the muscles inserted into them, and sometimes the fibrous fasciæ which invest certain joints." Although these structures are thickened by an "abnormal deposit," and suffer in consequence of it, it is not pretended that the change results from inflammation. "The synovial membranes appear thickened, probably, by a deposit in the sub-synovial areolar tissue, which as being the nidus of the blood-vessels, whence these membranes are nourished, is very likely to become the seat of such effusion." Up to this period, the malady is chronic rheumatism; but whether it shall continue to be subject to the laws regulating rheumatism, or become white swelling, or strumous arthritis of the joint affected, must depend upon the animal force, or tone of the individual. Inflammation of all kinds is destructive disease. The complaints we are considering may terminate in a destructive process locally, unattended by the phenomena of inflammation, but formidable as are the consequences of rheumatism, they are as nothing compared to the progress and effects of scrofulous inflammation. To a certain extent, however, in the incipient stages, there is a strong family resemblance in this group of diseases. The causes which excite to a reversal of polarity; the errors of the primary and secondary processes of assimilation; the loaded venous system, and its inevitable influence in promoting asthenia; the sense of injury, first in fibrous textures; the thickening which takes place in them, the loss of their natural flexibility, and as Dr. Copland remarks, "the impaired state of their *tonicity* and *cohesion*," connect the lethal influences among their molecules; the consequent effusion of lymph into the sub-synovial tissue, abnormally thickening the synovial membrane; the effusion of fluid into the synovial cavities; all these events are those of a train belonging to the scrofulous inflammation of one joint as well as to chronic rheumatism of joints, as well as to gouty disease. If it be shewn that a certain number of diseases commence their career in the same phenomena, that not only in the remote, but proximate causes, they resemble each other; that in their early stages the tissues affected are the same; that modifications arise in the severity of their symptoms from causes which we have not yet sufficiently studied, that if in their course of progress, strong analogies can be traced in the chemical substances developed by their inflammatory, or congestive changes; we have reason to group them into one family. When the individual maladies of this family emerge from their parent roof, and catarrhal affections continue an intimate acquaintance and alliance; while catarrh, bronchitis, rheumatic opthalmia, acute and sub-acute, and chronic rheumatism in all their various degrees, chronic rheumatism of joints, and gout, march on without inflammation of neighbouring tissues, they are but the incipient types of more serious diseases. Where the presence of any of these stimulates neighbouring tissues to inflame, we have new phenomena, and the members of our family become associated with new classes of disease. Scrofula, so intimately allied to gout, proceeds, if not checked by tonic currents, to awful consequences. Local lethal currents, prolonged in time, produce not only inflammation of proximate local tissues, but a condition of veins that abnormalises the functions of the stomach, duodenum and liver, or in other words, irregulates the primary processes of assimilation; all destructive processes, in the course of time, so load the common sewer, as Dr. Golding Bird called the venous system, that the secondary processes of assimilation cannot escape damage, and in the language of practitioners, constitutional disturbance is set up.

Every practitioner in recurring to the events of his past life, must remember many cases that began with simple catarrhal or rheumatic symptoms, and that after a chronic course, proceeded to the formation of capsular disease, unchecked by the best treatment of the most eminent surgeons—leeches, cupping, blisters, poultices, or water dressings, frictions or setons. Some of the cases extending to venous inflammation, scrofulous abscesses under fasciæ, and in the cavities of joints, destruction of bone, and development of tubercular inflammation of peritoneum. These trains of disease are trains of events beyond the facts of rheumatism. It is attempted to be shown that while rheumatism persists in its course, the train of events is not one of inflammation; that when inflammation sets in, the errors already existing in the primary processes of assimilation induce a new series of events, and when these arise the most ordinary phenomena are those resulting from loaded veins and errors in the albumenous depositing structures.

The chronic rheumatism of the joint affects the knee more frequently than other joints. It is common to find both knees affected with pain at once—but not to the same extent, one being less severely pained than the other. The next phenomenon consists in an effusion of fluid into their synovial cavities, which if not excessive in quantity would appear to alleviate the pain; of course, in such a state of things, the pain is increased if the limb be moved, or if the part be pressed upon.

It is only upon repeated attacks of chronic rheumatism that these phenomena supervene. A long continuance of the diathesis would appear to be necessary for effecting the morbid changes of structure, and Dr. Todd makes the remark that in these respects the analogy with gout is obvious, and that rheumatism cannot severely or frequently attack the joints "without impairing to a material extent the nutrition of their textures."

Pain, which is extreme reversal of polarity, cannot repeatedly occur in the same part without impairing the nutrition of the textures of that part. The normal, or vital, or nourishing currents is the direct one, the abnormal or lethal current is the inverse one, under the influence of which serious damage is done to the periosteum, to the synovial membranes of the cartilages, and to the articular extremities of the bones.

Dr. Todd's description of the absorptive changes in the articular cartilages is deeply interesting. "The cartilage divides into fibres vertical to the surface of the bone, and the change resembles that which maceration is apt to produce in articular cartilage. Grooves or depressions may be seen upon it, which gradually enlarge, unite, and leave the bone uncovered."

As the changes proceed, the articular surface of the bone is deprived of the whole of its cartilage, and at points where constant pressure and attrition have exerted an influence, it becomes polished like ivory, or porcelain; and when that influence has not operated, "the fine cancelli of the articular extremity of the bone are exposed."

The change that is effected in the nutrition of the serous texture is an enlargement and alteration of shape, owing to excessive and irregular bone deposits around the articular apophyses, and not only is a deformity produced by them, but they impede the movements of the joint.

The periosteum having been damaged by the progress of the disease, and the articular end of the bone having been altered in shape, a curious change has been described by Mr. Adams as belonging to the synovial membrane which besides being thickened, assumed the character of villous processes "soft . . . and entirely of a red colour," dipping into and completely occupying depressions around the neck of the bone. Dr. Colles's remarks are alluded to by Dr. Todd, that when the ivory formation on the heads of the bones does not take place, there is a deposit of ligamentous substance, consisting of productions, of most probably, altered synovial membrane, dipping down into small cavities in the head of the bone, as if to absorb the bony particles. This altered synovial membrane is precisely the *vascular investment* described by Dr. Macartney in page 73 of his work on Inflammation, and in page 68 of my essay on Dentition. I have pointed out that an analogous account of an absorptive organ in the removal of deciduous teeth was given by Bourdet under his description of the fleshy tubercle, and by Laforgue as the fleshy button.

The small cartilaginous bodies in the cavities of joints, are but separations of the cartilaginous structure resulting from absorptive activity, and are certainly not to be attributed to a chronic inflammation unless all the processes connected with nutrition, the deposition of new bony matter and the absortion of what may be considered sequestra are regarded as specimens of inflammation.

Wherever rheumatism occurs for a longer or a shorter period, this formidable phase of it may take possession of the individual, sparing none of the joints which may be liable to be affected by rheumatism.

The analogies between the phenomena of this form of rheumatism and those which attend scrofulous disease of the articulations have not escaped the attention of observers. All the large joints of the extremities, the smaller ones of the hands and feet, the temporo-maxillary, and the vertebral articulations are liable to be affected by each of the two forms of disease, and the question as to the progress or course of the trains of symptoms will, in some future time, be settled by the early application of proper magnetic treatment.

Asthenia is very common in early life. Coincident with it, is more or less of venous congestion, the principal remote cause of rheumatic, gouty, and scrofulous diseases. We have the authority of Dr. Todd for the assertion that chronic rheumatism of the joints is by no means uncommon from the age of thirty-five onwards, chiefly among the labouring poor, among whom being exposed more than others to the proximate causes of the malady, it is very rife, but the earliest age at which he had known it was about twenty, and the subjects of the disease were women. In the earlier part of my medical career, I was for several years physician to the Parochial Infirmary of St. Pancras, where I witnessed many cases of the disease. If I were to state my impressions, I should say that the greater number of these cases occurred subsequent to the middle period of life. I saw much of chronic rheumatism among subjects of both sexes, for my attention was attracted to it specially from the accidental possession of a very large electrical machine, with which, I am sorry to confess, I made out no new fact, and had very little success in the cure of cases. About that period, I was consulted by the friends of a young lady, aged eighteen, who was a subject of chronic rheumatism in the joints of her right and left hand and in the claviculo-sternal articulation.

I lately had a case, L. R— a young lady, thirteen years of age, who had suffered for years from a succession of strumous affections, of which the neck and left side of her face exhibited evidence. Marasmus with epigastric pain, aggravated by pressure on the part; rheumatic pains, first in one ankle, then in the other, aggravated by the same description of pain in both wrists, and in several spots in the course of the spine, threatened to wear out the poor sufferer with want of sleep; urgent hectic fever with rapid pulse, distressing thirst and perpetual desire for relief, kept the attendant on the alert six or seven times in the night. The remedies usual in such cases had all been had recourse to, in vain. Months rolled on, and the disease marched without a prospect of alleviation. Additional advice was obtained, but no bright hopes could be indulged in. If the arthritic symptoms were those of strumous inflammation, why did not a serious disorganization take place? There was abundance of pain in fibrous tissues with slight increase of heat, and so much tenderness that the joints could not be handled. Month after month, pain, the extreme reversal of polarity, gave the local symptom of premonition; the premonitory advanced guard of the enemy—inflammation; the rest of the troop of symptoms had not yet joined their comrades, pain and slightly increased heat. Then it was the presence of scrofulous rheumatism in several joints; but rheumatism, with a certain amount of error in the primary processes of assimilation, is gout; and though in the scrofulous variety, there is not, after months of suffering, a deposit among the fibrous textures, or in their immediate vicinity, of lithate of soda, or carbonate of lime, there is a slow albuminous accretion; a thickening of membrane, the first mischief; not yet an inflammation, but a growth; an abnormal nutrition, where none ought to be, and always, according to the law of Geoffroy St. Hilaire, at the expense of some other part, local abnormal hypertrophy with marked atrophy.

In endeavouring to seize the salient power of various diseases to analogise them in order to deduce laws from the effects of groups of sequences, it is to be expected that much opposition will be offered to the liberty taken in placing some of the symptoms of scrofula in a point of view in which they must be regarded as symptoms of gout. If we could proceed to identify the diseases more closely, we should simply be affording a testimony and a homage to the genius of two of the largest, or telescopic, minds of the last fifty years among the physicians and physiologists of these islands — Macartney and Prout.

That pain is dependent on a reversal of polarity, that the pain in analogous conditions of gout and scrofula is subdued by measures which restore the healthy polarities of organs is obvious, from the fact that it is necessary only to mesmerise the pained parts for a sufficient time downwards, in order to overcome the suffering. The process of cure consists in establishing the healthy direct current in place of the inverse, or lethal, current. The young lady alluded to was submitted to downward passes for an hour at a time, twice a day during three months, and she was cured; restored by God's blessing of Mesmerism to such health as to be able to enjoy life with all the exhilarating influences and impulses of youth. She was a pattern of good health and spirits.

The tendency that exists in all rheumatic diseases to separate from the blood either an abnormal material of growth, or an abnormal fluid secretion, which in joints, according to Dr. Copland, must necessarily be of an irritating nature as derived from tissues in a morbid condition, leads us to a closer view of the nature of gout than we have hitherto enjoyed. Dr. Hawkins, Dr. Macleod, and some other writers on rheumatism have distinguished between acute, and what they have been pleased to call capsular, rheumatism; between, in fact, what the French more accurately distinguish as articular rheumatism and arthritis. None of them venture to carry out the analogy to gout, or they would be landed in Chomel's conclusion of the identity of the phenomena in gout and rheumatism. Dr. Prout has shewn that the effusion in which gout terminates, is one that has a strong analogy to albuminous deposit. The desire to distinguish capsular from ordinary fibrous rheumatism has been thought highly sagacious, and it is for the amiable Dr. Chambers that the intellectual credit for the distinction is claimed. The fact is, that capsular effusion is an adventitious circumstance in the history of rheumatism. It belongs to the malady from the tendency where fibrous membranes are irritated as a consequence upon reversal of polarity, to pour out fluid somewhere in their neighbourhood. In the nostrils, a very abundant quantity of salt and irritating fluid is poured from mucous tissues, when in catarrh, rheumatised fibrous textures are in their immediate vicinity. In acute rheumatic affection of the knee, it is, for the most part, the cellular membrane, rarely the capsule of the joint into which the lax condition of the blood-vessels allow their pores to be permeated by serous effusion. When the capsule is the recipient of the abnormal fluid, the malady is a subacute form of rheumatism. The number of joints affected is fewer than in the acute form. It is highly probable that, while the system is marked for its cachetic condition, the fluid effused is morbid in its nature and very irritating.

Dr. Macleod is eager to tell us of Dr. Chambers having observed, (Macleod on Rheumatism, page 96,) "in several cases of old synovial rheumatism a white deposit of carbonate of lime," and he states that in other cases he had witnessed collections of pus in cases of capsular rheumatism. Dr. Copland is clear upon the occasional occurrence of purulent collections in the knee joint in cases of acute rheumatism. He himself having witnessed two cases. There is a striking difference between the symptoms of gout affecting primarily the fibrous textures around a joint, and that affecting secondarily the interior of a capsular synovial structure. Gout terminating in an effusion of pus into a synovial cavity is said to be rare. It is perhaps not so uncommon as is supposed. The conditions required relate principally to the degree of asthenia in the constitution of the subject. Gout being in fact essentially rheumatism occurring at a period of life generally beyond the middle term, arises from the causes of rheumatism, those causes having a relation in the first instance to loaded veins, to pain, the extreme reversal of polarity, and in addition to these a debilitated condition of the digestive function which favours the production and accumulation of lithic acid. It is one of the characteristics of rheumatism that it may cause a growth, a deposition, a secretion in structures other than those in which the pain first declared itself, other than the fibrous tissues. Rheumatism sometimes favours an enormous effusion of fluid, abnormal to mucous membranes, from the nostrils, and from the bronchial tubes, sometimes favours accretions about the heart, sometimes so relaxes the vessels in the neighbourhood of joints that they shed fluid into cellular structure, sometimes causes irritation on the surface of synovial membrane, which sometimes terminates in deposit of carbonate of soda, or lithate of soda. Beware of lithates, for there it must be admitted that rheumatism and gout are identical. She was better for a time, but died abroad two years after of scrofulous disease.

This form of rheumatism is most apt to occur in persons who have had repeated attacks of acute, or even of sub-acute rheumatism, and in these cases the progress of mischief in the joints is very slow. It has been before remarked that the fact that sub-acute and chronic rheumatism, in many instances, are consequent upon really uncured acute rheumatism, and the valuable remarks of Dr. Copland on this subject, cannot be too frequently urged on the attention of medical practitioners. Albeit there exists great cause for pity, that a man with so much philosophical capacity should not have seen his way to the most powerful means of removing the abnormal venous condition which is always at the bottom of the rheumatic diathesis.

It is clear that in the course of these observations no undue value has been attributed to the chemical changes observed in the blood, and in the secretions of the subjects of the diseases under consideration. It may be of considerable value to be assured that if we were to examine chemically the urine of such or such a patient, we should find the presence of a large proportion of uric acid. We know that under certain other conditions, a paralyzed state of the kidneys might prevent the formation of that uric acid, and that the urea might act as a poison in the system, or that other abnormal events might occur if the blood were not properly depurated by the kidneys. There is no end to the value of minute diagnostic facts. All that is here contended for is, that *pari passu*, the law obviously regulating or modifying the depuration of the blood, the laws of molecular forces should obtain a due share of the attention of the pathological physician.

Our purpose in the considerations we have indulged in, has been to connect the philosophy of Animal Magnetism with that of the diseases of which we have treated. It is hardly necessary to point out that Animal Magnetism gives us the best therapeutic agency to our hand in all these diseases. The pharmacopœa may assist us with a few alkalies, with lemon juice occasionally, with colchicum, and a few other medicinal remedies, but for the most part, these are unnecessary, if we can render a patient sufficiently sensitive by the energetic mesmeric passes of a good mesmeriser, who has sufficient power of will. In the application of Animal Magnetism as a means of cure, the great difficulty has always been to obtain sufficient energy and power of will in the operator.* All the disappointments that have taken place from the obstinate opposition of ignorant and determined enemies of our

* The well known verse in the book of Ecclesiastes, Chapter 9, verse 10, runs thus, "Whatsoever thy hand findeth to do, do it with thy might."

cause, have accrued from the crushing resistance that has been almost organized by the low feeling of a desire to rule, even when the gratification of self-will has cost the life of a dear child. No power of reasoning, in the case of an ignorant and wilful mother, with a preconceived idea in her head, has the slightest influence.

Medical life, with its multitudinous experience, must have convinced many an observant practitioner of the truth of this fact.

Rheumatism is a malady of fibrous textures, arising from reversals of polarity, depending upon an accumulation in the venous system of the debris of the body, poured in by unusual activity of the secondary processes of assimilation. The reversals of polarity may have taken place from the submission of the patient to cold, or to draughts of malarious air. The stomach is said to be not so soon affected, nor so much affected as in gout; it is a mistake to suppose that it is not at all affected. No individual can have constipated bowels without some error in the first process of digestion, and healthy digestion implies the existence of normal currents downwards, and consequent tone. At the ages at which acute rheumatism most usually attacks individuals, between 15 and 30, or 35, the errors existing in the secondary processes of assimilation do not very summarily affect the stomach, but at a later period of life they entail more quick, as well as more serious consequences. Every one is aware that in general, as age advances, people find their powers of digestion become more easily impaired. In very vigorous youth, wine which the stomach can with facility perfectly digest, would, even in many cases of reputed vigour in age, become acid in the stomach, and prevent the healthy process of first assimilation. If, for a course of years, wine be taken into a stomach unable to digest it, and consequently it is rendered unable to digest completely other material that has been swallowed,—we perceive a very good reason why the gouty subject labours under a disadvantage compared to the subject of simple rheumatism, the one has, with less vigour of nervous power, tried his constitution more than the other. He has superadded to the venous accumulation from one source, the debris of the secondary process of assimilation, another accumulation, added from the primary processes of assimilation. The first blood is impure, the result of the process of secondary assimilation is also impure. The gouty subject may well look more livid or purple in the cheeks, more suffused in the veins of the whites of his eyes than the simple rheumatic. His good eating and drinking have accumulated mischief, that has loaded his veins with dark blood. One of the curious phenomena attendant upon this part of the gouty diathesis is, that the fact of venous fulness does not more speedily reduce the tonic force of the system. There is a reason for it, and the reason is found in the two different states of polar condition of the brain and nerves in gout and in typhoid fever. The blood is more poisoned in the low fever, that is, has received the depolarizing lethal current to a more decided extent. The facts on the striking differences of existing amount of tone in the two diseases will not be disputed. People are little aware of the extent to which an unhealthy proportion of venous blood in the circulation determines the irregularities of deposition and absorption.

Dr. Prout was of opinion that gout and scrofula originated in the causes dependent on a depraved condition of the processes of assimilation. (Renal Diseases, 5th Edition, p, 495). He says, "Where too much food is taken, relatively to the constitution of the individual, either the primary or the secondary assimilative processes, or both, may, more especially suffer. In some instances, the primary assimilative processes are so weak and so easily deranged, that individuals are constrained to be careful, both with respect to the quantity and quality of their food, and such individuals often escape the more serious and deeper seated diseases of a secondary kind, arising from excess. On the other hand, there are individuals whose primary organs will permit them to take enormous quantities of all sorts of matters. In some of these instances, such matters pass off by the bowels, very little assimilated. In others, a large portion of them undergo more or less perfectly the primary assimilating processes, and are carried into the mass of the blood; and individuals in whom this takes place, suffer more especially from derangements of the secondary assimilative processes, as from hepatic congestion, gout, &c., particularly about the middle periods of life, when the consequences of excesses of all kinds begin to be manifested." It is at these periods of life, when, if the veins have acquired a tendency to disproportionate fullness, all the inconveniences of hepatic disorder are experienced, gall stones in some, hæmorrhoids in others, before the habit of indigestion and its more serious consequences are known. Gout is one of these consequences, and how often is it spoken of as a safety valve. There are two modes of regarding such sources of safety. One is, that most local pains and irritations arising from the consequences of disordered stomach divert mischief from the brain; another, that the discipline necessary to abolish the evils of local pain place the system in a better state than it enjoyed before that pain arrived.

The connexion between gout and hæmorrhoids has been extensively acknowledged. Quickly suppressed hæmorrhoidal flux has often produced a fit of the gout, and no gouty individual, who has suffered from hæmorrhoids, will deny that a free discharge from the hæmorrhoidal veins has often relieved him, for a time, from severe head-ache, and the miserable sensations that occasionally precede a fit of the gout. But hæmorrhoids are signs of venous congestion, and always that amount of it that belongs to the two processes of depraved assimilation. When this state of accumulation is present in the venous system, the arteries carrying less pure and healthy blood, there exists necessarily a tendency to slow deposition. There is more fibrine in gouty blood than is wholesome, and Dr. Prout has the remark, that mal-assimilation of the albuminous principle is the cause of the scrofulous, of the lithic acid, and of the gouty diathesis. These, according to that high authority, often gradually run into each other. "Thus gout and struma are frequently, if not always, associated, and the gouty chalk stones of old age may be considered as little else than modifications of the scrofulous tubercle of youth, both being formed from the mal-assimilation of the albuminous principle," (Fifth Edition, p. 492).

Every one is aware that in scrofula, the processes want tone, the absorptions and depositions proceed slowly from want of vital force, and wherever the formation of strumous tubercle occurs, it is a deposit of albumen producing a slow change in the part, and leading to a degeneration of structure. If we contemplate the change that takes place in fibrous textures, where chalk, or lithate of soda deposits take place, as in the knuckle, or theca of tendons, or even in the capsule of the knee in cases of gout, or capsular rheumatism, we must acknowledge that the phenomena are manifested in seats that form a favourite resort for the visits of scrofula. Those familiar with the slow progress of mischief in the painful affections of scrofulous joints, will acknowledge that before inflammation is fairly set up, there is often a continued suffering for months with no further organic mischief than is observed in gout, or its twin brother, rheumatism.

A case of gout was lately brought under the notice of the author, terminating in a collection of pus in synovial subcutaneous bursa of the great toe joint, which was attended with great pain for a time, but which subsided after being opened under the judicious treatment of the medical attendant. Except this painful local affection of gout, the gentleman, a vigorous man of about sixty-eight, never complained of illness as long as he could remember. Head-ache or any other sign of indigestion, he has never known, consequently in the case of this vigorous individual, an attack of rheumatism, called gout, because inflammation in a bursa of a small joint is present, terminates by exciting an inflammation ending in the formation of pus. In a patient more asthenic the analogy to scrofula would be more complete, and the lithate of soda would have been deposited instead of pus. Is gout quite identical with rheumatism then? or are they as different in their characters as to warrant nosologists in placing them in as distinct categories as they would place scarlatina and variola. The Natural History of gout and rheumatism would make them identical. The different periods of life at which they most frequently occur, the differences that most ordinarily characterize the respective states of digestive function between youth and age, most frequently the difference in the size of the joint affected are the only warrants for insisting on the importance of the chemical bases of diagnosis. The sister of rheumatism in asthenic youth, is scrofula. The brother of rheumatism in asthenic age, is gout. These are so closely and intimately allied that it appears marvellous the family relationship should have been for so many years ignored.

Suppuration would be a very heterodox termination for a fit of gout. Dr. Macleod (on Rheumatism, page 97,) had seen several cases in which capsular rheumatism had terminated in suppuration, but never saw any other form of rheumatism in which this had occurred. The fact is, that capsular effusion is an adventitious circumstance in the history of rheumatism, as it is in the history of gout. It belongs to the malady in either form from the tendency which exists to pour out fluid somewhere, irritated by a reversal of polarity.

Fewer joints are affected than in the very acute form. It is highly probable that when the system is marked for its cachectic condition, that the fluid effused is morbid in its nature, and highly irritating; and when the sub-acute slides into the chronic form, gives origin to the series of changes that follow each other in such vexatious, and frequently uncontrollable sequences, so ably described by Dr. Todd and by Mr. Adams. Gout itself may be seen in some cases to exhibit all the changes described by these pathologists. The term rheumatic gout, applicable to such cases, is a pleonasm.

The numerous ideas which pervade the chapter we have here laid before the public are sufficient to prove that our life has not been an idle one, but we have to apologize to our readers for leading them to wander in a somewhat discursive manner, through so great a range of subject. Our work is now concluded, and though it has been a laborious one, has, at the same time, been one that has afforded us the excitement of many agreeable hours, and of much pleasing reflection.

FINIS.

Printed in Dunstable, United Kingdom

91220124R00268